Alexander Gegov

Fuzzy Networks for Complex Systems

Studies in Fuzziness and Soft Computing, Volume 259

Editor-in-Chief

Prof. Janusz Kacprzyk
Systems Research Institute
Polish Academy of Sciences
ul. Newelska 6
01-447 Warsaw
Poland
E-mail: kacprzyk@ibspan.waw.pl

Further volumes of this series can be found on our homepage: springer.com

Vol. 243. Rudolf Seising (Ed.)
Views on Fuzzy Sets and Systems from Different Perspectives, 2009
ISBN 978-3-540-93801-9

Vol. 244. Xiaodong Liu and Witold Pedrycz
Axiomatic Fuzzy Set Theory and Its Applications, 2009
ISBN 978-3-642-00401-8

Vol. 245. Xuzhu Wang, Da Ruan, Etienne E. Kerre
Mathematics of Fuzziness – Basic Issues, 2009
ISBN 978-3-540-78310-7

Vol. 246. Piedad Brox, Iluminada Castillo, Santiago Sánchez Solano
Fuzzy Logic-Based Algorithms for Video De-Interlacing, 2010
ISBN 978-3-642-10694-1

Vol. 247. Michael Glykas
Fuzzy Cognitive Maps, 2010
ISBN 978-3-642-03219-6

Vol. 248. Bing-Yuan Cao
Optimal Models and Methods with Fuzzy Quantities, 2010
ISBN 978-3-642-10710-8

Vol. 249. Bernadette Bouchon-Meunier, Luis Magdalena, Manuel Ojeda-Aciego, José-Luis Verdegay, Ronald R. Yager (Eds.)
Foundations of Reasoning under Uncertainty, 2010
ISBN 978-3-642-10726-9

Vol. 250. Xiaoxia Huang
Portfolio Analysis, 2010
ISBN 978-3-642-11213-3

Vol. 251. George A. Anastassiou
Fuzzy Mathematics: Approximation Theory, 2010
ISBN 978-3-642-11219-5

Vol. 252. Cengiz Kahraman, Mesut Yavuz (Eds.)
Production Engineering and Management under Fuzziness, 2010
ISBN 978-3-642-12051-0

Vol. 253. Badredine Arfi
Linguistic Fuzzy Logic Methods in Social Sciences, 2010
ISBN 978-3-642-13342-8

Vol. 254. Weldon A. Lodwick, Janusz Kacprzyk (Eds.)
Fuzzy Optimization, 2010
ISBN 978-3-642-13934-5

Vol. 255. Zongmin Ma, Li Yan (Eds.)
Soft Computing in XML Data Management, 2010
ISBN 978-3-642-14009-9

Vol. 256. Robert Jeansoulin, Odile Papini, Henri Prade, and Steven Schockaert (Eds.)
Methods for Handling Imperfect Spatial Information, 2010
ISBN 978-3-642-14754-8

Vol. 257. Salvatore Greco, Ricardo Alberto Marques Pereira, Massimo Squillante, Ronald R. Yager, and Janusz Kacprzyk (Eds.)
*Preferences and Decisions,*2010
ISBN 978-3-642-15975-6

Vol. 258. Jorge Casillas and Francisco José Martínez López
Marketing Intelligent Systems Using Soft Computing, 2010
ISBN 978-3-642-15605-2

Vol. 259. Alexander Gegov
Fuzzy Networks for Complex Systems, 2010
ISBN 978-3-642-15599-4

Alexander Gegov

Fuzzy Networks for Complex Systems

A Modular Rule Base Approach

Springer

Author

Alexander Gegov
University of Portsmouth
School of Computing
Buckingham Building
Portsmouth PO1 3HE
United Kingdom
E-mail: alexander.gegov@port.ac.uk

Additional material to this book can be downloaded from http://extra.springer.com

ISBN 978-3-642-15599-4 e-ISBN 978-3-642-15600-7

DOI 10.1007/978-3-642-15600-7

Studies in Fuzziness and Soft Computing ISSN 1434-9922

Library of Congress Control Number: 2010933870

Typeset & Cover Design: Scientific Publishing Services Pvt. Ltd., Chennai, India.

Printed on acid-free paper

9 8 7 6 5 4 3 2 1

springer.com

To the architects of United Europe

Preface

There is only one limit to fundamental research – the sky.

This book introduces the novel concept of a fuzzy network. In particular, it describes further developments of some results from its predecessor book on Complexity Management in Fuzzy Systems, published in 2007 in the Springer Series in Studies in Fuzziness and Soft Computing.

The book contents build on a number of special presentations made by the author at international scientific events in the recent years. These presentations include an invited lecture at the EPSRC International Summer School in Complexity Science in 2007, tutorials at the IEEE International Conferences on Fuzzy Systems in 2007 and 2010, tutorials at the IEEE International Conferences on Intelligent Systems in 2008 and 2010, a tutorial at the IFSA World Congress in 2009 as well as plenary lectures at the WSEAS International Conferences on Fuzzy Systems in 2008 and Artificial Intelligence in 2009.

The notion of complexity has recently become a serious challenge to scientific research in a multi-disciplinary context. For example, it is quite common to find complex systems in biology, cosmology, engineering, computing, finance and other areas. However, the understanding of complex systems is often a difficult task.

There are two main aspects of complexity – quantitative and qualitative. The quantitative aspect is usually associated with a large scale of an entity or a large number of elements within this entity. The qualitative aspect is often characterised by uncertainty about data, information or knowledge that relates to an entity.

A natural way of coping with quantitative complexity is to use the concept of a general network. The latter consists of nodes and connections whereby the nodes represent the elements of an entity and the connections reflect the interactions among these elements. In this case, the scale of the entity is reflected by the overall size of the network whereas the number of elements is given by the number of nodes.

An obvious way of dealing with qualitative complexity is to use the concept of a fuzzy network. The latter consists of nodes and connections whereby the nodes are fuzzy systems and the connections reflect the interactions among these fuzzy systems. In this case, the uncertainty about data, information or knowledge related to an entity are reflected by the rule bases of the corresponding fuzzy systems and the underlying fuzzy logic.

In the context of the considerations made above, a fuzzy network represents a natural counterpart of a neural network. Both neural networks and fuzzy networks are computational intelligence based networks with nodes and connections.

However, the nodes in a neural network are represented by neurons whereas the nodes in a fuzzy network are represented by rule bases.

The author would like to thank Mathworks for including this book in their Book Programme and for providing a free individual licence of Matlab and the Fuzzy Logic Toolbox. These software products have been used for validating some of the theoretical results on fuzzy networks.

The author would also like to thank the Springer Series Editor Prof Janusz Kacprzyk for the useful comments on the draft contents of the book. His feedback has been very helpful for the subsequent improvements made to the final version.

The author is very indebted to the Springer Editorial Assistant Heather King for the cooperation on the editorial aspects of the book. Her kind help from the very start of the writing process until the final submission is gratefully acknowledged.

The author is very thankful to Annette Wilson, Head of School of Computing at the University of Portsmouth, for her managerial support with regard to the book. Her cooperation in keeping the teaching and administration duties of the author within reasonable bounds has helped for the timely publication of the book.

The author is also very thankful to the PhD students Nedyalko Petrov and Emil Gegov from the University of Portsmouth and the University of Brunel for validating some of the theoretical results from this book in the Matlab software environment. Without their help the book would have only a theoretical focus.

The author would like to acknowledge the visiting research fellowships granted to him in the past by the Alexander von Humboldt Foundation and the European Commission. The associated research visits to the Universities of Duisburg and Wuppertal in Germany as well as the Delft University of Technology in the Netherlands have laid down the early foundations for some of the ideas presented in the book.

The author would like to thank his wife, parents and sister for their spiritual support during his work on this book. Without their support the writing process would have been more difficult and more time consuming.

Finally, the author would like to thank his friend Diana Koleva for her help during the proofreading process, his colleagues from the University of Portsmouth rock band Discovery for the musical-turned-scientific inspiration over the years as well as his favorite bands and music channels for the entertainment during the typing process.

June 2010, Portsmouth, UK Alexander Gegov

Contents

Abbreviations

AGG – aggregation
AI – accuracy index
APP – application
CFS – chained fuzzy system
DEF – deffuzification
EI – efficiency index
FI – feasibility index
FID – fuzzification inference defuzzification
FN – fuzzy network
FNN – fuzzy neural network
FUZ – fuzzification
HFS – hierarchical fuzzy system
IMP – implication
NFS – networked fuzzy system
SFS – standard fuzzy system
TI – transparency index

Chapter 1
Introduction

1.1 Attributes of Systemic Complexity

Processes that are the subject of studies by humans are often referred to as systems. In this context, the term system has a fairly general meaning that can be associated either with a process operating on its own or under some human intervention. In this book, a system is associated with a process operating on its own.

Processes are usually studied by humans for the purpose of modelling, simulation and control which are aimed at managing these processes for the benefit of humans. In this book, modelling, simulation and control of processes is referred to as system management. In a more general context, system management may also include other activities such as diagnosis, classification and recognition.

Complexity is quite a versatile feature of existing systems in that it can not be described by a single definition. However, complexity is usually associated with a number of attributes such as nonlinearity, uncertainty, dimensionality and structure which make the management of systems with these attributes more difficult.

Therefore, the complexity of a given system can be accounted for by listing the complexity related attributes that are to be found in this system. In a more general setting, other complexity related attributes may also be considered, as for example, multiple levels of abstraction or multiple modes of operation.

1.2 Complexity Management by Fuzzy Logic

The effectiveness of system management depends on the way in which the outputs in a system are influenced by its inputs. These influences may be described by nonlinear functional mappings that are usually referred to as nonlinearity. The latter represents a serious challenge to system management due to the difficulties in handling nonlinear mappings for modelling, simulation and control.

The effectiveness of system management also depends on the way in which the relevant data, information and knowledge are handled. The latter may be characterised by imprecision, incompleteness, vagueness or ambiguity. These characteristics are often referred to collectively as uncertainty and they may lead to potential problems in safety-critical situations. The causes for uncertainty can be imprecise measurement devices, faulty sensors, noisy communication channels, subjective expert knowledge and others. As a whole, uncertainty in data,

A. Gegov: Fuzzy Networks for Complex Systems, STUDFUZZ 259, pp. 1–3.
springerlink.com

information and knowledge represents another serious challenge to system management as it may not be possible to obtain a solution within acceptable safety bounds.

Another factor that influences the effectiveness of system management is the way in which dimensionality is accounted for. The larger the number of inputs in a system, the more difficult it is to deal with it due to potential problems in time-critical situations. Such problems may compromise the reliability of the system as it may not be possible to obtain a solution within reasonable time scales.

A factor that also influences the effectiveness of system management is the way in which structure is reflected. In this context, many systems consist of subsystems that interact with each other. The ability to reflect this structure explicitly is a key to understanding how a system operates and a prerequisite for improving its functionality.

In general, fuzzy logic has proved itself as a powerful tool for dealing with nonlinearity and uncertainty. In this context, the concept of fuzziness is very suitable for approximating strong nonlinearity in terms of non-linearisable and non-analytical functional mappings between system inputs and outputs [22, 29, 38, 42, 71, 96, 126, 141, 144, 145, 146, 152, 153, 174]. Fuzziness is also quite suitable for reflecting non-probabilistic uncertainty such as imprecision, incompleteness, vagueness and ambiguity [27, 44, 49, 54, 58, 86, 101, 118, 154, 164, 167].

However, fuzzy logic has not been very effective in dealing with dimensionality and structure. In this respect, dimensionality is usually associated with the number of fuzzy rules which is an exponential function of the number of system inputs and the number of linguistic terms per input [18, 19, 20, 26, 28, 134, 156, 157]. As far as structure is concerned, it is often used to describe interacting modules which can not be taken into account explicitly due to the black box nature of the fuzzy rules [87, 163, 172, 175].

1.3 Description of Book Chapters

This book consists of ten chapters. The current chapter discusses complexity as a systemic feature and the ability of fuzzy systems to handle different attributes of complexity. Chapter 2 reviews several types of fuzzy systems in the context of systemic complexity, including systems with single, multiple and networked rule bases. Chapter 3 introduces the novel concept of fuzzy networks by means of formal models such as if-then rules and integer tables, Boolean matrices and binary relations, grid and interconnections structures, incidence and adjacency matrices as well as block schemes and topological expressions. Chapter 4 presents basic operations on nodes in fuzzy networks, including merging and splitting in horizontal, vertical and output context. Chapter 5 describes some structural properties of node operations in fuzzy networks such as associativity of merging and variability of splitting in horizontal, vertical and output context. Chapter 6 illustrates some advanced operations on nodes in fuzzy networks, including node transformation in the context of input augmentation, output permutation and feedback equivalence as well as node identification in the context of horizontal merging, vertical merging and output merging. Chapters 7-8 show the application of the theoretical results from Chapters 4-6 in feedforward fuzzy networks with single or multiple levels or layers and in feedback fuzzy networks with single or multiple local or global feedback. Chapter 9 gives an overall evaluation of fuzzy networks by means of assessment of structural complexity, composition of hierarchical fuzzy systems, decomposition of standard fuzzy systems, indicators of model performance and applications for case studies. The last chapter highlights the theoretical significance, the application areas and the methodological impact of fuzzy networks as well as the philosophical aspects of the book contents.

Chapter 2
Types of Fuzzy Systems

2.1 Introduction to Fuzzy Systems

A fuzzy system is described by input-output if-then rules in the form of a rule base. The inputs and the outputs take values which are linguistic terms such as 'small', 'big', 'low' and 'high'. In this context, the inputs and their linguistic terms in the 'if' part of the rule base are called 'antecedents' whereas the outputs and their linguistic terms in the 'then' part of the rule base are called 'consequents'.

A fuzzy system with two inputs x_1, x_2 taking linguistic terms from the set {small, big} and two outputs y_1, y_2 taking linguistic terms from the set {low, high} can be described in a detailed rule base form by Eqs.(2.1)-(2.4).

If x_1 is small and/or x_2 is small, then y_1 is high and y_2 is high (2.1)

If x_1 is small and/or x_2 is big, then y_1 is high and y_2 is low (2.2)

If x_1 is big and/or x_2 is small, then y_1 is low and y_2 is high (2.3)

If x_1 is big and/or x_2 is big, then y_1 is low and y_2 is low (2.4)

Depending on the type of the logical connections 'and/or' among antecedents, the latter can be conjunctive or disjunctive. However, consequents are always conjunctive as any multiple-output fuzzy system can be represented equivalently as a conjunctive set of single-output fuzzy systems. As far as the fuzzy rules are concerned, they are usually disjunctive as it is unreasonable to assume that all rules in a fuzzy system can be applied simultaneously.

A fuzzy system with conjunctive antecedents that has two inputs x_1, x_2 and two outputs y_1, y_2 taking linguistic terms from any admissible sets, can be described in a compact rule base form by Eq.(2.5).

If x_1 and x_2, then y_1 and y_2 (2.5)

A fuzzy system with disjunctive antecedents that has two inputs x_1, x_2 and two outputs y_1, y_2 taking linguistic terms from any admissible sets, can be described in a compact rule base form by Eq.(2.6).

If x_1 or x_2, then y_1 and y_2 (2.6)

A. Gegov: Fuzzy Networks for Complex Systems, STUDFUZZ 259, pp. 5–12.
springerlink.com

A multiple-output fuzzy system with two inputs x_1, x_2 and two outputs y_1, y_2 taking linguistic terms from any admissible sets can be described in a compact rule base form by Eq.(2.7).

$$\text{If } x_1 \text{ and/or } x_2, \text{ then } y_1 \text{ and } y_2 \tag{2.7}$$

The multiple-output fuzzy system above can be represented equivalently as a conjunctive set of two single-output fuzzy systems which are described in a compact rule base form by Eqs.(2.8)-(2.9).

$$\text{If } x_1 \text{ and/or } x_2, \text{ then } y_1 \tag{2.8}$$

$$\text{If } x_1 \text{ and/or } x_2, \text{ then } y_2 \tag{2.9}$$

The operation of a fuzzy system is characterised by a Fuzzification-Inference-Defuzzification (FID) sequence which is applied for each output. During this sequence, the crisp values of inputs are first fuzzified into fuzzy values in terms of fuzzy membership degrees, then these fuzzy membership degrees are mapped to fuzzy values of the output in terms of a fuzzy membership function, and finally, this fuzzy membership function is defuzzified into a single crisp value for the output. In the case of multiple outputs, the fuzzification and the initial part of the inference may be applied only once whereas the other parts of the inference and the defuzzification must be applied separately for each output.

The fuzzification stage in a FID sequence is based on the fuzzy membership functions of the inputs. Each of these functions represents a mathematical description of a linguistic term for an input. Depending on the shape of the input fuzzy membership functions, different fuzzification formulas are applied.

For the fuzzy system described by Eqs.(2.1)-(2.4), the fuzzification stage can be described by a function that maps the crisp values x_{1C}, x_{2C} of the inputs x_1, x_2 to their fuzzy membership degrees x_{1F}^{i}, x_{2F}^{i}, i=1,4 in each rule. This function is denoted as FUZ and is given for the two inputs by Eqs.(2.10)-(2.11).

$$FUZ_1\ (x_{1C}) = (x_{1F}^{1}, x_{1F}^{2}, x_{1F}^{3}, x_{1F}^{4}) \tag{2.10}$$

$$FUZ_2\ (x_{2C}) = (x_{2F}^{1}, x_{2F}^{2}, x_{2F}^{3}, x_{2F}^{4}) \tag{2.11}$$

The inference stage in a FID sequence consists of three substages - application, implication and aggregation. The application substage maps the fuzzy membership degrees of the inputs in each rule to a single membership degree that represents the firing strength for this rule. The latter is mapped by the implication substage to a fuzzy membership function for the output in this rule that represents an amended image of the original membership function. The aggregation substage maps the amended fuzzy membership functions for all rules to a single fuzzy membership function for the output that represents the whole rule base.

For the fuzzy system described by Eqs.(2.1)-(2.4), the application substage of the inference stage can be described by a function that maps the input fuzzy membership degrees x_{1F}^{i}, x_{2F}^{i}, i=1,4 in each rule to the firing strength x_F^{i}, i=1,4 for this rule. This function is denoted as APP and is given for the four rules by Eqs.(2.12)-(2.15).

$$APP_1\ (x_{1F}^{1}, x_{2F}^{1}) = x_F^{1} \tag{2.12}$$

$$APP_2\ (x_{1F}^{2}, x_{2F}^{2}) = x_F^{2} \tag{2.13}$$

$$APP_3\ (x_{1F}^{3}, x_{2F}^{3}) = x_F^{3} \tag{2.14}$$

$$APP_4\ (x_{1F}^{4}, x_{2F}^{4}) = x_F^{4} \tag{2.15}$$

For the fuzzy system described by Eqs.(2.1)-(2.4), the implication substage of the inference stage can be described by a function that maps the firing strength x_F^{i}, i=1,4 for each rule to fuzzy membership functions y_{1F}^{i}, y_{2F}^{i}, i=1,4 in each rule for the outputs y_1, y_2. This function is denoted as IMP and is given for the two outputs by Eqs.(2.16)-(2.17).

$$IMP_1\ (x_F^{1}, x_F^{2}, x_F^{3}, x_F^{4}) = (y_{1F}^{1}, y_{1F}^{2}, y_{1F}^{3}, y_{1F}^{4}) \tag{2.16}$$

$$IMP_2\ (x_F^{1}, x_F^{2}, x_F^{3}, x_F^{4}) = (y_{2F}^{1}, y_{2F}^{2}, y_{2F}^{3}, y_{2F}^{4}) \tag{2.17}$$

For the fuzzy system described by Eqs.(2.1)-(2.4), the aggregation substage of the inference stage can be described by a function that maps the output fuzzy membership functions y_{1F}^{i}, y_{2F}^{i}, i=1,4 in each rule to a single fuzzy membership function y_{jF}, j=1,2 for each output. This function is denoted as AGG and is given for the two outputs by Eqs.(2.18)-(2.19).

$$AGG_1\ (y_{1F}^{1}, y_{1F}^{2}, y_{1F}^{3}, y_{1F}^{4}) = y_{1F} \tag{2.18}$$

$$AGG_2\ (y_{2F}^{1}, y_{2F}^{2}, y_{2F}^{3}, y_{2F}^{4}) = y_{2F} \tag{2.19}$$

The defuzzification stage in a FID sequence is based on the fuzzy membership functions of the outputs. Each of these functions represents a mathematical description of a linguistic term for an output. Depending on the shape of the output fuzzy membership functions, different defuzzification formulas are applied.

For the fuzzy system described by Eqs.(2.1)-(2.4), the defuzzification stage can be described by a function that maps the single fuzzy membership function y_{jF}, j=1,2 for each output to crisp value y_{jC}, j=1,2 of this output. This function is denoted as DEF and is given for the two outputs by Eqs.(2.20)-(2.21).

$$DEF_1\ (y_{1F}) = y_{1C} \tag{2.20}$$

$$DEF_2\ (y_{2F}) = y_{2C} \tag{2.21}$$

The FID sequence described by Eqs.(2.10)-(2.21) is used for a Mamdani fuzzy system where both the inputs and the outputs are presented by linguistic terms. A variation of this FID sequence is used for a Sugeno fuzzy system where the inputs are also presented by linguistic terms but the crisp values of the outputs are presented as linear functions of the crisp values of the inputs.

2.2 Systems with Single Rule Base

The most common type of fuzzy system has a single rule base [5, 11, 53, 123, 127, 137]. This type of system is usually referred to as Standard Fuzzy System (SFS). The latter is characterised by a black-box nature whereby the inputs are mapped directly to the outputs without the consideration of any intermediate variables. The operation of a SFS is based on a single FID sequence.

A Mamdani SFS with 'r' rules, 'm' inputs $x_1 \ldots x_m$ taking linguistic terms from the input sets $\{A_{11},\ldots,A_{1r}\},\ldots,\{A_{m1},\ldots,A_{mr}\}$ and 'n' outputs $y_1 \ldots y_n$ taking linguistic terms from the output sets $\{B_{11},\ldots,B_{1r}\},\ldots,\{B_{n1},\ldots,B_{nr}\}$ can be described in a generic rule base form by Eq.(2.22).

If x_1 is A_{11} and/or ... and x_m is A_{m1}, then y_1 is B_{11} and ... and y_n is B_{n1} (2.22)

....

If x_1 is A_{1r} and/or ... and/or x_m is A_{mr}, then y_1 is B_{1r} and ... and y_n is B_{nr}

A SFS is usually quite accurate for output modelling as it reflects the simultaneous influence of all inputs on the output. However, this accuracy depends on the feasibility of the model which may be compromised due to the difficulties in reflecting the simultaneous influence of a large number of inputs. Also, the efficiency and transparency of a SFS deteriorate with the increase of the number of inputs and the number of linguistic terms per input. The main cause for this deterioration is the number of fuzzy rules which is an exponential function of the number of inputs and their linguistic terms. Therefore, as the number of rules increases, it not only takes longer to simulate the output but it is also more difficult to understand how the output is affected by the inputs.

2.3 Systems with Multiple Rule Bases

Another type of fuzzy system has multiple rule bases [15, 76, 81, 95, 105, 129, 136, 140, 155, 162, 166, 173, 177]. This type of system is often described by cascaded rule bases and its most common forms are referred to as Chained Fuzzy System (CFS) or Hierarchical Fuzzy System (HFS). The latter are characterised by a white-box nature whereby the inputs are mapped to the outputs by means of some intermediate variables. The operation of a CFS/HFS is based on multiple FID sequences whereby each intermediate variable links the FID sequences for two adjacent rule bases.

A CFS may have an arbitrary structure in terms of subsystems and the connections among them [12, 66, 74]. In this case, each subsystem represents an individual rule base as the one described by Eq.(2.22) whereas each interaction is represented by an intermediate variable linking a pair of adjacent rule bases. This intermediate variable is identical with an output from the first rule base and an input to the second rule base in the pair. The presentation of connections among multiple rule bases is discussed in Sect.2.4 in the context of networked rule bases which are a special type of multiple rule bases.

A CFS is usually used as a detailed presentation of a SFS for the purpose of improving transparency by explicitly taking into account all subsystems and the interactions among them. Also, some efficiency is gained because of the smaller number of inputs to the individual rule bases. The same positive effect is observed for feasibility which is enhanced by the ability to reflect better the simultaneous influence of the reduced number of inputs to the individual rule bases. However, some accuracy is lost due to the accumulation of errors as a result of the repetitive application of fuzzification, inference and defuzzification within the multiple FID sequences.

A HFS is a special type of a CFS that has a specific structure [2, 4, 13, 24, 25, 30, 36, 39, 64, 70, 75, 79, 90, 94, 139, 143, 151, 169]. Each subsystem in a HFS has two inputs and one output. Some intermediate variables represent identical mappings which may propagate across parts of the system.

A HFS is usually used as a simplified presentation of a SFS for the purpose of improving efficiency and transparency. Efficiency is improved by the reduction of the overall number of rules which is a linear function of the number of inputs to the subsystems and the number of linguistic terms per input. Transparency is also improved by explicitly taking into all subsystems and the interactions among them. The same applies to feasibility which is facilitated by the small number of inputs to the individual rule bases. However, these improvements are at the expense of losing accuracy for the same reason as in the case of a CFS.

2.4 Systems with Networked Rule Bases

A third type of fuzzy system has networked rule bases. This is a novel type of system that has been introduced recently in [50]. This system is referred to as Networked Fuzzy System (NFS) and it is characterised by a white-box nature whereby the inputs are mapped to the outputs by means of some intermediate variables. Each subsystem in a NFS is represented by a node whereas the interactions among subsystems are the connections among these nodes.

A Mamdani NFS with 'p×q' nodes $\{N_{11}\ldots N_{p1}\},\ldots,\{N_{1q}\ldots N_{pq}\}$, 'p×q' node inputs $\{x_{11}\ldots x_{p1}\},\ldots,\{x_{1q}\ldots x_{pq}\}$ taking linguistic terms from any admissible input sets, 'p×q' node outputs $\{y_{11}\ldots y_{p1}\},\ldots,\{y_{1q}\ldots y_{pq}\}$ taking linguistic terms from any admissible output sets, 'p' horizontal levels and 'q' vertical layers in the underlying grid structure for this NFS can be described by Eq.(2.23).

$$\begin{array}{lll} & \text{Layer 1} \ldots\ldots\ldots\ldots\ldots \text{Layer q} & (2.23) \\ \text{Level 1} & N_{11}(x_{11}, y_{11}) \ldots\ldots\ldots\ldots N_{1q}(x_{1q}, y_{1q}) & \\ \ldots\ldots\ldots\ldots & \ldots\ldots\ldots\ldots\ldots\ldots\ldots\ldots\ldots\ldots & \\ \text{Level p} & N_{p1}(x_{p1}, y_{p1}) \ldots\ldots\ldots\ldots N_{pq}(x_{pq}, y_{pq}) & \end{array}$$

The grid structure in Eq.(2.23) specifies the location of nodes as well as their inputs and outputs. In this case, each input and output can be either a scalar or a vector. The levels in this grid structure represent a spatial hierarchy of the nodes in terms of subordination in space whereas the layers represent a temporal hierarchy in terms of consecutiveness in time.

However, Eq.(2.23) does not give any information about the connections among the nodes in the grid structure. Such information is contained in the inter-connection structure in Eq.(2.24) whereby the 'p×(q-1)' node connections $\{z_{11,12} \ldots z_{p1,p2}\}, \ldots, \{z_{1q-1,1q} \ldots z_{pq-1,pq}\}$ take linguistic terms from the admissible sets for the associated node outputs and inputs.

$$\begin{array}{lll} & \text{Layer 1} \ldots\ldots\ldots\ldots\ldots \text{Layer q-1} & (2.24) \\ \text{Level 1} & z_{11,12}=y_{11}=x_{12} \ldots\ldots\ldots\ldots z_{1q-1,1q}=y_{1q-1}=x_{1q} & \\ \ldots\ldots\ldots\ldots & \ldots\ldots\ldots\ldots\ldots\ldots\ldots\ldots\ldots\ldots & \\ \text{Level p} & z_{p1,p2}=y_{p1}=x_{p2} \ldots\ldots\ldots\ldots z_{pq-1,pq}=y_{pq-1}=x_{pq} & \end{array}$$

For consistency, the NFS system described by Eqs.(2.23)-(2.24) assumes the existence of a node in each location of the underlying grid structure. However, not all locations in a NFS have to be populated by nodes. For simplicity, all connections in the NFS system above are among nodes in the same level and consecutive layers. However, some connections in a NFS may be among nodes in different levels or non-consecutive layers.

A NFS is a hybrid between a SFS and a CFS/HFS. On one hand, the structure of a NFS is similar to the structure of a CFS/HFS due to the explicit presentation of subsystems and the interactions among them. On the other hand, the operation of a NFS resembles the operation of a SFS due to the possibility of simplifying the original multiple rule bases to a linguistically equivalent single rule base. This simplification is based on a linguistic composition approach that is central to a NFS and is introduced in more detail further in this book.

Being a hybrid concept, a NFS potentially has some of the advantages and the disadvantages of a SFS and a CFS/HFS. In this respect, on the positive side, a NFS could be as feasible and transparent as a CFS/HFS due to the original multiple rule base presentation. However, on the negative side, a NFS could also be as efficient as a SFS due to the equivalent single rule base presentation. As far as accuracy is concerned, a NFS could have either advantages or disadvantages in relation to a SFS or a CFS/HFS. For example, a NFS could be more accurate than a CFS/HFS due to the single application of a FID sequence but less accurate than a SFS due to the approximation effect of the linguistic composition approach.

2.5 Comparison of Fuzzy Systems

The attributes of systemic complexity discussed in this chapter can be handled with a different level of success by the fuzzy systems introduced earlier in the chapter. For example, a SFS is usually suitable for representing nonlinearity and uncertainty but may have problems with dimensionality and structure. A CFS/HFS is often suitable for representing nonlinearity, dimensionality and structure but may experience problems with uncertainty. A NFS should be suitable for nonlinearity, uncertainty and structure but may exhibit problems with dimensionality.

Properties of fuzzy systems such as feasibility, accuracy, efficiency and transparency are directly related to attributes of systemic complexity such as nonlinearity, uncertainty, dimensionality and structure. In this respect, nonlinearity represents a challenge for feasibility because it is more difficult to reflect simultaneous nonlinear influence of the inputs on the output [21, 45, 59, 82, 102, 124, 135, 138]. Also, uncertainty is a challenge for accuracy as it is harder to build an accurate model under uncertainty [69, 78, 93, 97, 100, 108, 148, 160 170, 171]. Furthermore, dimensionality represents an obstacle to efficiency as it is more difficult to reduce the amount of computations in a FID sequence for a large number of rules [1, 32, 51, 56, 57, 65, 68, 73, 80, 109, 117, 128, 142, 161]. And finally, structure is an obstacle to transparency as it is harder to understand the behaviour of a black-box model with no interactions among subsystems [16, 35, 43, 46, 62, 63, 67, 83, 91, 107, 116, 149, 178].

The relationships between the different attributes of systemic complexity and the associated properties of fuzzy systems are summarised in Table 2.1.

Table 2.1 Systemic complexity attributes and fuzzy system properties

Systemic complexity attribute	Fuzzy system property
Nonlinearity	Feasibility
Uncertainty	Accuracy
Dimensionality	Efficiency
Structure	Transparency

This book has three main objectives. The first objective is to introduce a detailed theoretical framework for NFSs. The second objective is to demonstrate this framework as a modelling methodology for different types of NFSs. The third objective is to compare NFSs with SFSs and CFSs/HFSs in a number of application case studies.

For clarity and simplicity, a NFS is referred to as a Fuzzy Network (FN) further in this book. As a concept, a FN is different from a Fuzzy Neural Network (FNN). Although there may be some similarities between a FN and a FNN, the nodes in the latter can be fuzzy logical neurons but not fuzzy rule bases [9, 23, 34, 48, 60, 72, 85, 92, 98, 106, 115, 119, 176].

The book focuses on Mamdani fuzzy systems, described by conjunctive antecedents, disjunctive rules, multiple outputs and any type of FID sequences in terms of the fuzzification, inference and defuzzification stages. As far as the application, implication and aggregation substages of the inference stage are concerned, they can be also of any type. These choices are based on the fact that Mamdani fuzzy systems are not only widely used and but are also well suited to the linguistic composition approach used in the book.

The next chapter introduces some basic concepts from the theoretical framework for FNs. In particular, it discusses several types of formal models for FNs.

Chapter 3
Formal Models for Fuzzy Networks

3.1 Introduction to Formal Models

A formal model is usually characterised by mathematical formalisms. Their purpose is to avoid ambiguities that may be found in an informal model. For example, formal models in software engineering are used for specifying uniquely requirements to software products. Such models often make use of different areas of mathematics that has proved itself as a valuable tool for formal modelling.

This chapter introduces several types of formal models for FNs. Some of these models have been around for sometime and are therefore well known to the fuzzy community [6, 7, 33, 37, 41, 52, 61, 77, 88, 111, 112, 113, 121, 130]. However, other models are quite novel in terms of their application for fuzzy systems and FNs. These novel models have been used in some areas of mathematics, computing and engineering but they have been introduced only recently as formal models of networked rule bases in FNs.

As FNs represent an extension of fuzzy systems, some of the formal models for FNs are fairly basic and similar to the formal models for fuzzy systems. However, other formal models are more advanced and different from the formal models for fuzzy systems in that they can reflect connections among nodes in networked rule bases and facilitate the simplification of such rule bases to a linguistically equivalent single rule base. Apart from that, the more advanced formal models for FNs usually represent compressed images of these networks in terms of nodes and connections whereby only the most essential information is preserved and all unnecessary details are removed.

3.2 If-then Rules and Integer Tables

If-then rules and integer tables are known formal models for fuzzy systems. These models can represent nodes in a FN without the connections and are used here as a bridge between fuzzy systems and FNs.

A FN is considered which has four nodes N_{11}, N_{12}, N_{21}, N_{22} located within two levels and two layers. Initially, the nodes are assumed to be isolated and they can be described by the if-then rules given in Eqs.(3.1)-(3.12).

A. Gegov: Fuzzy Networks for Complex Systems, STUDFUZZ 259, pp. 13–21.
springerlink.com

Rule 1 for N_{11} : If x_{11} is small, then y_{11} is low (3.1)

Rule 2 for N_{11} : If x_{11} is medium, then y_{11} is high (3.2)

Rule 3 for N_{11} : If x_{11} is big, then y_{11} is average (3.3)

Rule 1 for N_{12} : If x_{12} is low, then y_{12} is moderate (3.4)

Rule 2 for N_{12} :If x_{12} is average, then y_{12} is heavy (3.5)

Rule 3 for N_{12} : If x_{12} is high, then y_{12} is light (3.6)

Rule 1 for N_{21} : If x_{21} is small, then y_{21} is average (3.7)

Rule 2 for N_{21} : If x_{21} is medium, then y_{21} is low (3.8)

Rule 3 for N_{21} : If x_{21} is big, then y_{21} is high (3.9)

Rule 1 for N_{22} : If x_{22} is low, then y_{22} is heavy (3.10)

Rule 2 for N_{22} : If x_{22} is average, then y_{22} is light (3.11)

Rule 3 for N_{22} : If x_{22} is high, then y_{22} is moderate (3.12)

The four nodes above can also be described by integer tables whereby each row in the table represents a rule. In this case, the linguistic terms 'small', 'low' and 'light' are represented by '1', the linguistic terms 'medium', 'average' and 'moderate' are represented by '2' whereas the linguistic terms 'big', 'high' and 'heavy' are represented by '3'. These integer tables are given in Tables 3.1-3.4.

Table 3.1 Integer table for node N_{11}

Linguistic terms for input x_{11}	Linguistic terms for output y_{11}
1	1
2	3
3	2

Table 3.2 Integer table for node N_{12}

Linguistic terms for input x_{12}	Linguistic terms for output y_{12}
1	2
2	3
3	1

Table 3.3 Integer table for node N_{21}

Linguistic terms for input x_{21}	Linguistic terms for output y_{21}
1	2
2	1
3	3

Table 3.4 Integer table for node N_{22}

Linguistic terms for input x_{22}	Linguistic terms for output y_{22}
1	3
2	1
3	2

If-then rules and integer tables as the ones presented above are very suitable for formal modelling of fuzzy systems with a single rule base such as SFSs. However, they are not quite suitable for formal modelling of fuzzy systems with multiple or networked rule bases. This is due to the fact that if-then rules and integer tables can not take into account any connections among nodes in networked rule bases. Also, if-then rules and integer tables do not lend themselves easily to manipulation for the purpose of simplifying networked rule bases to a linguistically equivalent single rule base using the linguistic composition approach.

3.3 Boolean Matrices and Binary Relations

Boolean matrices and binary relations are novel formal models for fuzzy systems that can represent nodes in a FN. Similarly to if-then rules and integer tables, these models can represent nodes without the connections.

A Boolean matrix compresses the information from a rule base that is represented by a node. In this case, the row and column labels of the Boolean matrix are all possible permutations of the positive integers representing the linguistic terms of the inputs and the outputs from the integer table for this rule base. The elements of the Boolean matrix are either '0's or '1's whereby each '1' reflects a rule from the rule base.

The rule bases represented by the isolated nodes N_{11}, N_{12}, N_{21}, N_{22} from Eqs.(3.1)-(3.12) can be described by the Boolean matrices in Eqs.(3.13)-(3.16).

N_{11}: (3.13)

x_{11} \ y_{11}	1	2	3
1	1	0	0
2	0	0	1
3	0	1	0

N_{12}: (3.14)

x_{12} \ y_{12}	1	2	3
1	0	1	0
2	0	0	1
3	1	0	0

N_{21}: (3.15)

x_{21} \ y_{21}	1	2	3
1	0	1	0
2	1	0	0
3	0	0	1

N_{22}: (3.16)

x_{22} \ y_{22}	1	2	3
1	0	0	1
2	1	0	0
3	0	1	0

A binary relation compresses further the information from the Boolean matrix of a rule base that is represented by a node. In this case, the pairs in the relation are the permutations of positive integers representing the linguistic terms of the inputs and the outputs from the row and column labels of the Boolean matrix which correspond to a rule. Therefore, each pair in the binary relation reflects a rule from the rule base.

The rule bases represented by the isolated nodes N_{11}, N_{12}, N_{21}, N_{22} from Eqs.(3.1)-(3.12) can be described by the binary relations in Eqs.(3.17)-(3.20).

$$N_{11}: \ \{(1, 1), (2, 3)\ (3, 2)\} \tag{3.17}$$

$$N_{12}: \ \{(1, 2), (2, 3)\ (3, 1)\} \tag{3.18}$$

$$N_{21}: \ \{(1, 2), (2, 1)\ (3, 3)\} \tag{3.19}$$

$$N_{22}: \ \{(1, 3), (2, 1)\ (3, 2)\} \tag{3.20}$$

Boolean matrices and binary relations as the ones presented above are very suitable for formal modelling of fuzzy systems with multiple or networked rule bases. In particular, they are well suited for formal modelling of FNs at a lower level of abstraction whereby detailed input-output mappings are specified for isolated individual nodes. Besides this, Boolean matrices and binary relations work well with other formal models which can take into account connections among nodes in FNs. A more detailed treatment of this matter is presented further in the book.

3.4 Grid and Interconnection Structures

Grid and interconnection structures are novel formal models that represent compressed images of the overall structure of a FN. These models describe the location of nodes and the connections among them. The models are introduced briefly in Chapter 2 and are discussed in more detail here.

The rule bases represented by the isolated nodes N_{11}, N_{12}, N_{21}, N_{22} from Eqs.(3.1)-(3.12) can be described by the grid structure in Eq.(3.21).

	Layer 1	Layer 2	(3.21)
Level 1	$N_{11}(x_{11}, y_{11})$	$N_{12}(x_{12}, y_{12})$	
Level 2	$N_{21}(x_{21}, y_{21})$	$N_{22}(x_{22}, y_{22})$	

The grid structure above with two levels and two layers is a formal model for a FN with a node set $\{N_{11}, N_{12}, N_{11}, N_{22}\}$, an input set $\{x_{11}, x_{12}, x_{21}, x_{22}\}$ and an output set $\{y_{11}, y_{12}, y_{21}, y_{22}\}$. This grid structure specifies the location of nodes as well as their inputs and outputs but it can not take into account any connections among nodes.

Therefore, it is assumed that the nodes N_{11}, N_{12}, N_{21}, N_{22} from Eqs.(3.1)-(3.12) are not isolated anymore and their connections are described by the connection set is$\{z_{11,12}, z_{21,22}\}$. In this case, the first connection is assumed to be identical with the output from N_{11} and the input to N_{12} whereas the second connection is assumed to be identical with the output from N_{21} and the input to N_{22}. These connections can be taken into account by the interconnection structure with two levels and one layer in Eq.(3.22).

	Layer 1	(3.22)
Level 1	$z_{11,12}=y_{11}=x_{12}$	
Level 2	$z_{21,22}=y_{21}=x_{22}$	

Grid and interconnection structures as the ones presented above are quite suitable for formal modelling of fuzzy systems with multiple or networked rule bases. In particular, these structures are well suited for formal modelling of FNs at a higher level of abstraction whereby only locations, inputs, outputs and connections for individual nodes are specified.

Grid and interconnection structures describe FNs at overall network level. They work quite well with Boolean matrices and binary relations which describe FNs at individual node level. However, grid and interconnection structures do not lend themselves easily to manipulation for the purpose of simplifying networked rule bases to a linguistically equivalent single rule base using the linguistic composition approach. For this reason, these structures are not considered further in this book.

3.5 Incidence and Adjacency Matrices

Incidence and adjacency matrices are other novel formal models that represent compressed images of the overall structure of a FN. Similarly to grid and interconnection structures, these models describe the location of nodes and the connections among them. In this case, it is necessary to introduce a start node S for virtual generation of inputs and an end node E for virtual receipt of outputs. These two additional nodes make possible the transformation of a FN into an equivalent graph which can then be analytically described by an incidence matrix and an adjacency matrix.

An incidence matrix describes the interactions among individual rule bases whereby the row labels designate the rule bases and the column labels designate the associated interactions. Each existing connection between an ordered pair of nodes is represented by a (1,-1) pair of elements in the incidence matrix whereas missing connections are represented by zeros. For each connection, the '1' element in the associated pair specifies the outgoing node and the '-1' element specifies the incoming node.

The FN from Eqs.(3.21)-(3.22) can be described by the incidence matrix in Eq.(3.23).

(3.23)

Node \ Connection	x_{11}	$z_{11,12}$	y_{12}	x_{21}	$z_{21,22}$	y_{22}
S	1	0	0	1	0	0
N_{11}	-1	1	0	0	0	0
N_{12}	0	-1	1	0	0	0
N_{21}	0	0	0	-1	1	0
N_{22}	0	0	0	0	-1	1
E	0	0	-1	0	0	-1

An adjacency matrix presents the information from an incidence matrix in a slightly different way. In this case, both the row and column labels in the adjacency matrix designate the rule bases. Each existing connection between an ordered pair of nodes is represented by a '1' whereas missing connections are represented by a '0'. As opposed to an incidence matrix which specifies explicitly each connection and the associated pair of nodes, the adjacency matrix specifies only which pairs of nodes are linked by a connection.

The FN from Eqs.(3.21)-(3.22) can be described by the adjacency matrix in Eq.(3.24).

(3.24)

Node	S	N_{11}	N_{12}	N_{21}	N_{22}	E
Node						
S	0	1	0	1	0	0
N_{11}	0	0	1	0	0	0
N_{12}	0	0	0	0	0	1
N_{21}	0	0	0	0	1	0
N_{22}	0	0	0	0	0	1
E	0	0	0	0	0	0

Incidence and adjacency matrices as the ones presented above are quite suitable for formal modelling of fuzzy systems with multiple or networked rule bases. In particular, they are well suited for formal modelling of FNs at a higher level of abstraction whereby only inputs, outputs and connections for individual nodes are specified.

Like grid and interconnection structures, incidence and adjacency matrices describe FNs at overall network level. They work quite well with Boolean matrices and binary relations which describe FNs at individual node level. However, incidence and adjacency matrices do not lend themselves easily to manipulation for the purpose of simplifying networked rule bases to a linguistically equivalent single rule base using the linguistic composition approach. For this reason, these matrices are not considered further in this book.

3.6 Block Schemes and Topological Expressions

Block schemes and topological expressions are advanced novel formal models that represent compressed images of the overall structure of a FN. Similarly to grid and interconnection structures, these models describe the location of nodes and the connections among them. In this case, the subscripts of each node specify its location in the network whereby the first subscript gives the level number and the second subscript gives the layer number. Besides this, block scheme and topological expressions specify all inputs, outputs and connections with respect to the nodes.

The four-node FN from Eqs.(3.21)-(3.22) can be described by the block scheme in Fig.3.1.

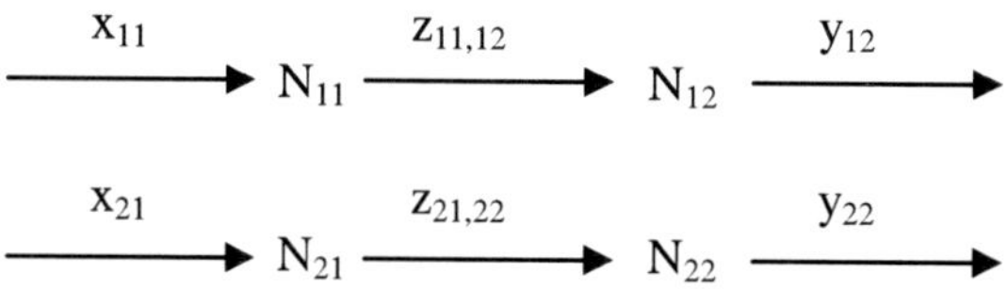

Fig. 3.1 Block scheme for a four-node FN

The arrows in the block scheme above designate the input set $\{x_{11}, x_{21}\}$ for the nodes in the first layer and the output set $\{y_{12}, y_{22}\}$ for the nodes in the second layer. Also, the arrows designate the connection set $\{z_{11,12}, z_{21,22}\}$ for connected pairs of nodes whereby for each pair of nodes the first node is in the first layer and the second node is in the second layer.

The FN from Eqs.(3.21)-(3.22) can be described by the topological expression in Eq.(3.25).

$$\{[N_{11}](x_{11} \mid z_{11,12})\ H\ [N_{12}](z_{11,12} \mid y_{12})\}V\{[N_{21}](x_{21} \mid z_{21,22})\ H\ [N_{22}](z_{21,22} \mid y_{22})\} \quad (3.25)$$

Each node in the topological expression above is placed within a pair of square brackets '[ ]'. The inputs and the outputs for each node are placed within a pair of simple brackets '( )' right after the node. In this case, the inputs are separated from the outputs by a vertical slash '|'. Nodes in sequence are designated by the symbol 'H' for horizontal relative location whereas nodes in parallel are designated by the symbol 'V' for vertical relative location. The higher priority of horizontal relative location with respect to vertical relative location in Eq.(3.25) is specified by pairs of curly brackets '{ }'.

Block schemes and topological expressions as the ones presented above are very suitable for formal modelling of fuzzy systems with multiple or networked rule bases. In particular, they are well suited for formal modelling of FNs at a higher level of abstraction whereby only inputs, outputs and connections for individual nodes are specified.

Like grid and interconnection structures, block schemes and topological expressions describe FNs at overall network level. They work very well with Boolean matrices and binary relations which describe FNs at individual node level. Besides this, block schemes and topological expressions lend themselves easily to manipulation for the purpose of simplifying networked rule bases to a linguistically equivalent single rule base using the linguistic composition approach. A more detailed treatment of this matter is presented further in the book.

3.7 Comparison of Formal Models

The formal models introduced in this chapter can be used with a different level of success for different types of FNs. For example, if-then rules and integer tables are mainly suitable for formal modelling of nodes in fairly simple and disconnected FNs. However, Boolean matrices and binary relations are suitable for formal modelling of nodes in more complex and connected FNs. Grid and interconnection structures are suitable for formal modelling of whole FNs, but mainly in a static context and alongside Boolean matrices or binary relations. The same characteristics can be attributed to incidence and adjacency matrices under the assumption that the associated FNs do not change their structure. However, when networked rule bases are simplified to a linguistically equivalent single rule base, the associated FNs change their structure during this process and the most suitable formal models for this purpose are block schemes and topological expressions. The latter are suitable for formal modelling of whole FNs in a dynamic context and alongside Boolean matrices or binary relations.

The characteristics of different types of formal models in terms of their ability to handle nodes, connections and dynamics in FNs are summarised in Table 3.5.

Table 3.5 Characteristics of formal models in relation to FNs

Formal model	Nodes	Connections	Dynamics
If-then rules	Yes	No	No
Integer tables	Yes	No	No
Boolean matrices	Yes	No	Yes
Binary relations	Yes	No	Yes
Grid structures	No	Yes	No
Interconnection structures	No	Yes	No
Incidence matrices	No	Yes	No
Adjacency matrices	No	Yes	No
Block schemes	No	Yes	Yes
Topological expressions	No	Yes	Yes

This book focuses on Boolean matrices and binary relations for formal modelling of nodes. As far as formal modelling of connections is concerned, the focus is on block schemes and topological expressions. The choice of these formal models is justified by their ability to handle dynamics in FNs and thereby to facilitate the intended use of the linguistic composition approach in the book.

The next chapter introduces more basic concepts from the theoretical framework for FNs. In particular, it discusses several types of basic operations in FNs.

Chapter 4
Basic Operations in Fuzzy Networks

4.1 Introduction to Basic Operations

The process of simplifying networked rule bases to a linguistically equivalent single rule base is central to the linguistic composition approach used in this book. This approach is based on several basic operations on nodes which can be either binary or unary in that they can be applied to a pair of nodes or a single node. In this respect, each binary operation has a unary counterpart that is an inverse operation with respect to the binary operation. For simplicity, all operations are illustrated with examples of nodes with scalar inputs, outputs and intermediate variables but their extension to the vector case is straightforward.

Some of the basic operations are to be found in mathematics and are therefore well known whereas others are quite novel in terms of the underlying theory and have been introduced only recently. The operations make use of Boolean matrices or binary relations as formal models for FNs at node level. These formal models lend themselves easily to manipulation in the context of the linguistic composition approach. Therefore, the basic operations can be viewed as elementary building blocks for the simplification of an arbitrarily complex FN to a fuzzy system.

4.2 Horizontal Merging of Nodes

Horizontal merging is a binary operation that can be applied to a pair of sequential nodes, i.e. nodes located in the same level of a FN. This operation merges the operand nodes from the pair into a single product node. The operation can be applied when the output from the first node is fed forward as an input to the second node in the form of an intermediate variable. In this case, the product node has the same input as the input to the first operand node and the same output as the output from the second operand node whereas the intermediate variable does not appear in the product node.

When Boolean matrices are used as formal models for the operand nodes, the horizontal merging operation is identical with Boolean matrix multiplication. The latter is similar to conventional matrix multiplication whereby each arithmetic multiplication is replaced by a ‘minimum’ operation and each arithmetic addition is replaced by a ‘maximum’ operation. In this case, the row labels of the product matrix are the same as the row labels of the first operand matrix whereas the

A. Gegov: Fuzzy Networks for Complex Systems, STUDFUZZ 259, pp. 23–50.
springerlink.com

column labels of the product matrix are the same as the column labels of the second operand matrix.

The horizontal merging operation can also be applied in the context of binary relations when such relations are used as formal models for the operand nodes. In this case, the horizontal merging operation is identical with standard relational composition.

Example 4.1

This example considers the sequential operand nodes N_{11} and N_{12} located in the first level of the four-node FN from Fig.3.1. These nodes are described there by the Boolean matrices in Eqs.(3.13)-(3.14) and the binary relations in Eqs.(3.17)-(3.18). The connections among these nodes are given by the interconnection structure in Eq.(3.22). In this context, nodes N_{11} and N_{12} represent a two-node FN that is a subnetwork of the four-node FN. This two-node FN can be described by the block-scheme in Fig.4.1 and the topological expression in Eq.(4.1).

Fig. 4.1 Two-node FN with operand nodes N_{11} and N_{12}

$$[N_{11}]\ (x_{11} \mid z_{11,12}) * [N_{12}]\ (z_{11,12} \mid y_{12}) \tag{4.1}$$

The use of the symbol '*' in Fig.4.1 and Equation (4.1) implies that the horizontal merging operation can be applied to the operand nodes N_{11} and N_{12}. In this context, the use of the symbol '*' makes valid the precondition for horizontal merging of nodes N_{11} and N_{12}.

The horizontal merging of the operand nodes N_{11} and N_{12} results into a single product node N_{11*12} which represents a simplified image of the two-node FN in the form of a one-node FN. The latter can be described by the block scheme in Fig.4.2 and the topological expression in Eq.(4.2).

Fig. 4.2 One-node FN with product node N_{11*12}

$$[N_{11*12}]\ (x_{11} \mid y_{12}) \tag{4.2}$$

The use of the symbol '*' in Fig.4.2 and Eq.(4.2) implies that the application of the horizontal merging operation has resulted in the product node N_{11*12}. This is justifiable due to the disappearance of the intermediate variable $z_{11,12}$ as well as to the fact that the input x_{11} to the product node is the same the input to the first operand node and the output y_{12} from the product node is the same as the output from the second operand node. In this context, the use of the symbol '*' makes valid the postcondition for the formation of node N_{11*12} as a result of horizontal merging.

Finally, the product node N_{11*12} can be described by the Boolean matrix in Eq.(4.3) and the binary relation in Eq.(4.4).

N_{11*12}: (4.3)

x_{11} \ y_{12}	1	2	3
1	0	1	0
2	1	0	0
3	0	0	1

$$N_{11*12}: \{(1, 2), (2, 1)\ (3, 3)\} \tag{4.4}$$

Example 4.2

This example considers the sequential operand nodes N_{21} and N_{22} located in the second level of the four-node FN from Fig.3.1. These nodes are described there by the Boolean matrices in Eqs.(3.15)-(3.16) and the binary relations in Eqs.(3.19)-(3.20). The connections among these nodes are given by the interconnection structure in Eq.(3.22). In this context, nodes N_{21} and N_{22} represent a two-node subnetwork of the four-node FN. The latter can be described by the block-scheme in Fig.4.3 and the topological expression in Eq.(4.5).

Fig. 4.3 Two-node FN with operand nodes N_{21} and N_{22}

$$[N_{21}]\ (x_{21} \mid z_{21,22})\ *\ [N_{22}]\ (z_{21,12} \mid y_{22}) \tag{4.5}$$

The use of the symbol '*' in Fig.4.3 and Eq.(4.5) implies that the horizontal merging operation can be applied to the operand nodes N_{21} and N_{22}. In this context, the use of the symbol '*' makes valid the precondition for horizontal merging of nodes N_{21} and N_{22}.

The horizontal merging of the operand nodes N_{21} and N_{22} results into a single product node N_{21*22} which represents a simplified image of the two-node FN in the form of a one-node FN. This one-node FN can be described by the block-scheme in Fig.4.4 and the topological expression in Eq.(4.6).

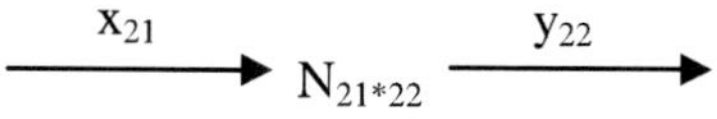

Fig. 4.4 One-node FN with product node N_{21*22}

$[N_{21*22}]\ (x_{21} \mid y_{22})$ (4.6)

The use of the symbol '*' in Fig.4.4 and Eq.(4.6) implies that the application of the horizontal merging operation has resulted in the product node N_{21*22}. This is justifiable due to the disappearance of the intermediate variable $z_{21,22}$ as well as to the fact that the input x_{21} to the product node is the same the input to the first operand node and the output y_{22} from the product node is the same as the output from the second operand node. In this context, the use of the symbol '*' makes valid the postcondition for the formation of node N_{21*22} as a result of horizontal merging.

Finally, the product node N_{21*22} can be described by the Boolean matrix in Eq.(4.7) and the binary relation in Eq.(4.8).

N_{21*22}: (4.7)

x_{21} \ y_{22}	1	2	3
1	1	0	0
2	0	0	1
3	0	1	0

N_{21*22}: {(1, 1), (2, 3) (3, 2)} (4.8)

4.3 Horizontal Splitting of Nodes

Horizontal splitting is a unary operation that can be applied to a single node in a FN. The operation splits an operand node into a pair of sequential product nodes whereby the input to the first product node is the same as the input to the operand node and the output from the second product node is the same as the output from the operand node. In this case, an intermediate variable appears between the product nodes in the form of an output from the first product node that is fed forwarded as an input to the second product node.

When a Boolean matrix is used as a formal model for the operand node, the horizontal splitting operation is identical with Boolean matrix factorisation. The latter is similar to conventional matrix factorisation whereby the multiplication of the two factors gives the matrix that has been factorised. In this case, the row labels of the first product matrix are the same as the row labels of the operand matrix whereas the column labels of the second product matrix are the same as the column labels of the operand matrix. As any Boolean matrix can be factorised into the same matrix and an identity matrix of appropriate dimension, the horizontal splitting operation can always be applied at least in this trivial context.

The horizontal splitting operation can also be applied in the context of binary relations when such a relation is used as a formal model for the operand node. In this case, the horizontal splitting operation is identical with standard relational decomposition. As any binary relation can be decomposed into the same relation and an identity relation of appropriate cardinality, the horizontal splitting operation can always be applied at least in this trivial context.

Example 4.3

This example considers an operand node $N_{11/12}$ that is the same as the product node N_{11*12} from Example 4.1. This node is described there by the Boolean matrix in Eq.(4.3) and the binary relation in Eq.(4.4). In this context, the node N_{11*12} represents a simplified image of a two-node FN in the form of a one-node FN. The latter can be described by the block-scheme in Fig.4.5 and the topological expression in Eq.(4.9).

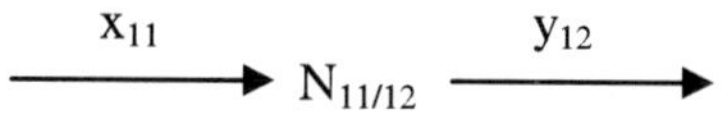

Fig. 4.5 One-node FN with operand node $N_{11/12}$

$$[N_{11/12}]\ (x_{11} \mid y_{12}) \tag{4.9}$$

The use of the symbol '/' in Fig.4.5 and Equation (4.1) implies that the horizontal splitting operation can be applied to the operand node $N_{11/12}$. In this context, the use of the symbol '/' makes valid the precondition for horizontal splitting of node $N_{11/12}$.

The horizontal splitting of the operand node $N_{11/12}$ results into a pair of sequential product nodes N_{11} and N_{12} which represent a complexified image of the one-node FN in the form of a two-node FN. The latter can be described by the block-scheme in Fig.4.6 and the topological expression in Eq.(4.10).

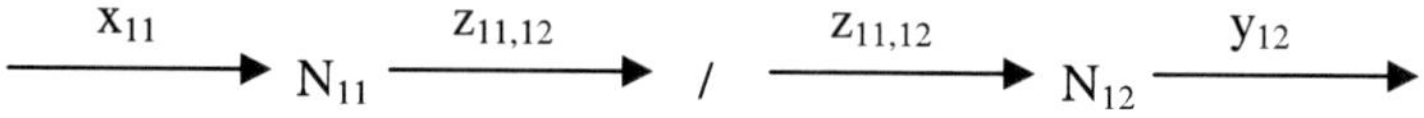

Fig. 4.6 Two-node FN with product nodes N_{11} and N_{12}

$$[N_{11}]\ (x_{11} \mid z_{11,12})\ /\ [N_{12}]\ (z_{11,12} \mid y_{12}) \tag{4.10}$$

The use of the symbol '/' in Fig.4.6 and Eq.(4.10) implies that the application of the horizontal splitting operation has resulted in the product nodes N_{11} and N_{12}. This is justifiable due to the appearance of the intermediate variable $z_{11,12}$ as well as to the fact that the input x_{11} to the first product node is the same as the input to the operand node and the output y_{12} from the second product node is the same as the output from the operand node. In this context, the use of the symbol '/' makes valid the postcondition for the formation of nodes N_{11} and N_{12} as a result of horizontal splitting.

Finally, the product nodes N_{11} and N_{12} can be described by the Boolean matrices in Eqs.(3.13)-(3.14) and the binary relations in Eqs.(3.17)-(3.18). However, this solution is not unique due to the existence of a trivial solution with an identity node as one of the two product nodes. For example, the trivial solution with an identity node N_{I3} of dimension 3×3 as a first product node can be described by the block-scheme in Fig.4.7, the topological expression in Eq.(4.11), the Boolean matrices in Eqs.(4.12)-(4.13) and the binary relations in Eqs. (4.14)-(4.15).

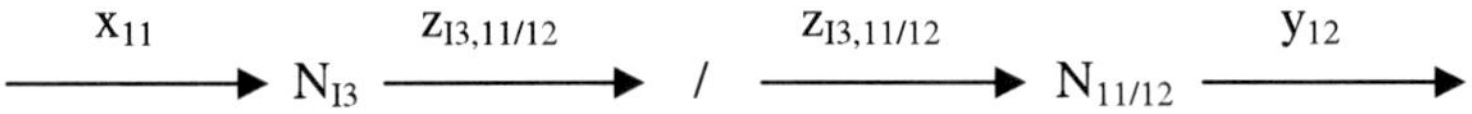

Fig. 4.7 Two-node FN with product nodes N_{I3} and $N_{11/12}$

$$[N_{I3}]\ (x_{11} \mid z_{11/12,I3})\ /\ [N_{11/12}]\ (z_{11/12,I3} \mid y_{12}) \tag{4.11}$$

N_{I3}: (4.12)

x_{11} \ $z_{I3,11/12}$	1	2	3
1	1	0	0
2	0	1	0
3	0	0	1

$N_{11/12}$: (4.13)

$z_{I3,11/12}$ \ y_{12}	1	2	3
1	0	1	0
2	1	0	0
3	0	0	1

$$N_{I3}: \{(1, 1), (2, 2)\ (3, 3)\} \tag{4.14}$$

$$N_{11/12}: \{(1, 2), (2, 1)\ (3, 3)\} \tag{4.15}$$

Example 4.4

This example considers an operand node $N_{21/22}$ that is the same as the product node N_{21*22} from Example 4.2. This node is described there by the Boolean matrix in Eq.(4.7) and the binary relation in Eq.(4.8). In this context, the node N_{21*22} represents a simplified image of a two-node FN in the form of a one-node FN. The latter can be described by the block-scheme in Fig.4.8 and the topological expression in Eq.(4.16).

Fig. 4.8 One-node FN with operand node $N_{21/22}$

$$[N_{21/22}]\ (x_{21} \mid y_{22}) \tag{4.16}$$

The use of the symbol '/' in Fig.4.8 and Equation (4.16) implies that the horizontal splitting operation can be applied to the operand node $N_{21/22}$. In this context, the use of the symbol '/' makes valid the precondition for horizontal splitting of node $N_{21/22}$.

The horizontal splitting of the operand node $N_{21/22}$ results into a pair of sequential product nodes N_{21} and N_{22} which represent a complexified image of the one-node FN in the form of a two-node FN. The latter can be described by the block-scheme in Fig.4.9 and the topological expression in Eq.(4.17).

$$\xrightarrow{x_{21}} N_{21} \xrightarrow{z_{21,22}} / \xrightarrow{z_{21,22}} N_{22} \xrightarrow{y_{22}}$$

Fig. 4.9 Two-node FN with product nodes N_{21} and N_{22}

$$[N_{21}]\ (x_{21} \mid z_{21,22})\ /\ [N_{22}]\ (z_{21,22} \mid y_{22}) \qquad (4.17)$$

The use of the symbol '/' in Fig.4.9 and Eq.(4.17) implies that the application of the horizontal splitting operation has resulted in the product nodes N_{21} and N_{22}. This is justifiable due to the appearance of the intermediate variable $z_{21,22}$ as well as to the fact that the input x_{21} to the first product node is the same the input to the operand node and the output y_{22} from the second product node is the same as the output from the operand node. In this context, the use of the symbol '/' makes valid the postcondition for the formation of nodes N_{21} and N_{22} as a result of horizontal splitting.

Finally, the product nodes N_{21} and N_{22} can be described by the Boolean matrices in Eqs.(3.15)-(3.16) and the binary relations in Eqs.(3.19)-(3.20). However, this solution is not unique due to the existence of a trivial solution with an identity node as one of the two product nodes. For example, the trivial solution with an identity node N_{I3} of dimension 3X3 as a second product node can be described by the block-scheme in Fig.4.10, the topological expression in Eq.(4.18), the Boolean matrices in Eqs.(4.19)-(4.20) and the binary relations in Eqs.(4.21)-(4.22).

$$\xrightarrow{x_{21}} N_{21/22} \xrightarrow{z_{21/22,I3}} / \xrightarrow{z_{21/22,I3}} N_{I3} \xrightarrow{y_{22}}$$

Fig. 4.10 Two-node FN with product nodes $N_{21/22}$ and N_{I3}

$$[N_{21/22}]\ (x_{21} \mid z_{21/22,I3})\ /\ [N_{I3}]\ (z_{21/22,I3} \mid y_{22}) \qquad (4.18)$$

$N_{21/22}$: (4.19)

x_{21} \ $z_{21/22,I3}$	1	2	3
1	1	0	0
2	0	0	1
3	0	1	0

N_{I3}: (4.20)

$z_{21/22,I3}$ \ y_{22}	1	2	3
1	1	0	0
2	0	1	0
3	0	0	1

$$N_{11/12}: \{(1, 1), (2, 3) (3, 2)\} \tag{4.21}$$

$$N_{I3}: \{(1, 1), (2, 2) (3, 3)\} \tag{4.22}$$

4.4 Vertical Merging of Nodes

Vertical merging is a binary operation that can be applied to a pair of parallel nodes, i.e. nodes located in the same layer of a FN. This operation merges the operand nodes from the pair into a single product node. In this case, the inputs to the product node represent the union of the inputs to the operand nodes whereas the outputs from the product node represent the union of the outputs from the operand nodes. The operation of vertical merging can always be applied due to the ability to concatenate the inputs and the outputs of any two parallel nodes.

When Boolean matrices are used as formal models for the operand nodes, the vertical merging operation is like an expansion of the first operand matrix along its rows and columns. In particular, the product matrix is obtained by expanding each non-zero element from the first operand matrix to a block that is the same as the second operand matrix and by expanding each zero element from the first operand matrix to a zero block of the same dimension as the second operand matrix. In this case, the row labels of the product matrix are all possible permutations of row labels of the operand matrices whereas the column labels of the product matrix are all permutations of column labels of the operand matrices.

The vertical merging operation can also be applied in the context of binary relations when such relations are used as formal models for the operand nodes. In this case, the vertical merging operation represents a special type of relational composition in the form of a modified Cartesian product that is applied independently to the first and the second elements from the pairs of the operand relations.

Example 4.5

This example considers the parallel operand nodes N_{11} and N_{21} located in the first layer of the four-node FN from Fig.3.1. These nodes are described there by the

Boolean matrices and the binary relations in Eq.(3.13), Eq.(3.15), Eq.(3.17) and Eq.(3.19). The connections of these nodes with the nodes in the second layer of this FN are given by the interconnection structure in Eq.(3.22). In this context, nodes N_{11} and N_{21} represent a two-node subnetwork of this FN. This two-node FN can be described by the block-scheme in Fig.4.11 and the topological expression in Eq.(4.23).

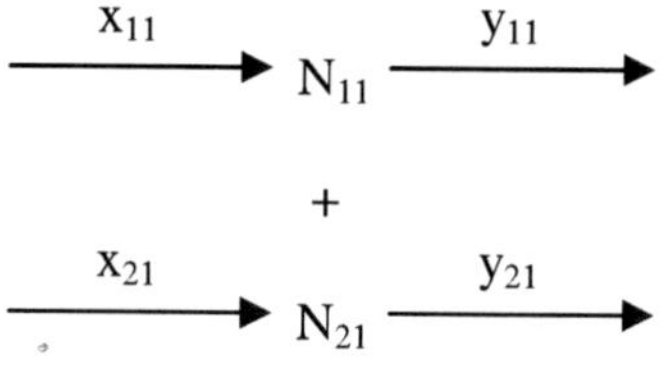

Fig. 4.11 Two-node FN with operand nodes N_{11} and N_{21}

$$[N_{11}]\ (x_{11} \,|\, y_{11}) + [N_{21}]\ (x_{21} \,|\, y_{21}) \tag{4.23}$$

The use of the symbol '+' in Fig.4.11 and Equation (4.23) implies that the vertical merging operation can be applied to the operand nodes N_{11} and N_{21}. In this context, the use of the symbol '+' confirms the validity of the precondition for vertical merging of nodes N_{11} and N_{21}.

The vertical merging of the operand nodes N_{11} and N_{21} results into a single product node N_{11+21} which represents a simplified image of the two-node FN in the form of a one-node FN. The latter can be described by the block scheme in Fig.4.12 and the topological expression in Eq.(4.24).

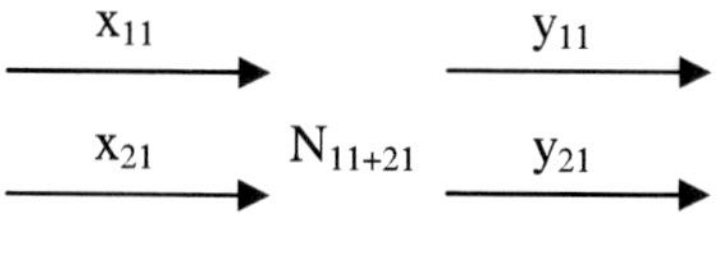

Fig. 4.12 One-node FN with product node N_{11+21}

$$[N_{11+12}]\ (x_{11}, x_{21} \,|\, y_{11}, y_{21}) \tag{4.24}$$

The use of the symbol '+' in Fig.4.12 and Eq.(4.24) implies that the application of the vertical merging operation has resulted in the product node N_{11+12}. This is justifiable due to the concatenation of the inputs to the operand nodes as inputs x_{11}, x_{21} to the product node and the concatenation of the outputs from the operand nodes as outputs y_{11}, y_{21} from the product node. In this context, the use of the symbol '+' makes valid the postcondition for the formation of node N_{11+21} as a result of vertical merging.

Finally, the product node N_{11+21} can be described by the Boolean matrix in Eq.(4.25) and the binary relation in Eq.(4.26).

N_{11+12}: (4.25)

x_{11}, x_{21} \ y_{11}, y_{21}	11	12	13	21	22	23	31	32	33
11	0	1	0	0	0	0	0	0	0
12	1	0	0	0	0	0	0	0	0
13	0	0	1	0	0	0	0	0	0
21	0	0	0	0	0	0	0	1	0
22	0	0	0	0	0	0	1	0	0
23	0	0	0	0	0	0	0	0	1
31	0	0	0	0	1	0	0	0	0
32	0	0	0	1	0	0	0	0	0
33	0	0	0	0	0	1	0	0	0

$$N_{11+12}: \{(11, 12), (12, 11), (13, 13), (21, 32), (22, 31), (23, 33), (31, 22), (32, 21), (33, 23)\} \quad (4.26)$$

Example 4.6

This example considers the parallel operand nodes N_{12} and N_{22} located in the second layer of the four-node FN from Fig.3.1. These nodes are described there by the Boolean matrices and the binary relations in Eq.(3.14), Eq.(3.16), Eq.(3.18) and Eq.(3.20). The connections of these nodes with the nodes in the first layer of this FN are given by the interconnection structure in Eq.(3.22). In this context, nodes N_{12} and N_{22} represent a two-node subnetwork of this FN. This two-node FN can be described by the block-scheme in Fig.4.13 and the topological expression in Eq.(4.27).

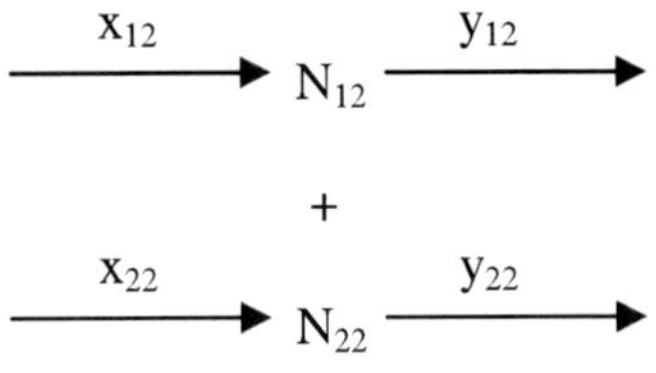

Fig. 4.13 Two-node FN with operand nodes N_{12} and N_{22}

$$[N_{12}]\,(x_{12} \,|\, y_{12}) + [N_{22}]\,(x_{22} \,|\, y_{22}) \quad (4.27)$$

The use of the symbol '+' in Fig.4.13 and Equation (4.27) implies that the vertical merging operation can be applied to the operand nodes N_{12} and N_{22}. In this context, the use of the symbol '+' confirms the validity of the precondition for vertical merging of nodes N_{12} and N_{22}.

The vertical merging of the operand nodes N_{12} and N_{22} results into a single product node N_{12+22} which represents a simplified image of the two-node FN in the form of a one-node FN. The latter can be described by the block scheme in Fig.4.14 and the topological expression in Eq.(4.28).

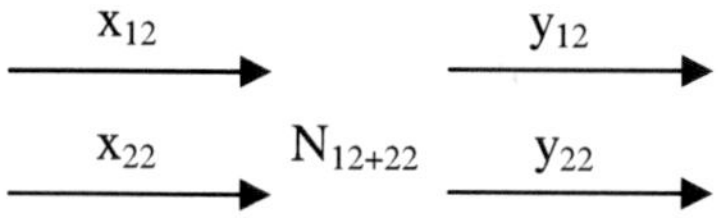

Fig. 4.14 One-node FN with product node N_{12+22}

$$[N_{12+22}]\ (x_{12}, x_{22} \mid y_{12}, y_{22}) \tag{4.28}$$

The use of the symbol '+' in Fig.4.18 and Eq.(4.28) implies that the application of the vertical merging operation has resulted in the product node N_{12+22}. This is justifiable due to the concatenation of the inputs to the operand nodes as inputs x_{12}, x_{22} to the product node and the concatenation of the outputs from the operand nodes as outputs y_{12}, y_{22} from the product node. In this context, the use of the symbol '+' makes valid the postcondition for the formation of node N_{12+22} as a result of vertical merging.

Finally, the product node N_{12+22} can be described by the Boolean matrix in Eq.(4.29) and the binary relation in Eq.(4.30).

N_{12+22}: (4.29)

x_{12}, x_{22} \ y_{12}, y_{22}	11	12	13	21	22	23	31	32	33
11	0	0	0	0	0	1	0	0	0
12	0	0	0	1	0	0	0	0	0
13	0	0	0	0	1	0	0	0	0
21	0	0	0	0	0	0	0	0	1
22	0	0	0	0	0	0	1	0	0
23	0	0	0	0	0	0	0	1	0
31	0	0	1	0	0	0	0	0	0
32	1	0	0	0	0	0	0	0	0
33	0	1	0	0	0	0	0	0	0

$$N_{12+22}: \{(11, 23), (12, 21), (13, 22), (21, 33), (22, 31), (23, 32), (31, 13), (32, 11), (33, 12)\} \tag{4.30}$$

4.5 Vertical Splitting of Nodes

Vertical splitting is a unary operation that can be applied to a single node in a FN. This operation splits an operand node into a pair of parallel product nodes whereby the union of the inputs to the product nodes represents the inputs to the operand node and the union of the outputs from the product nodes represents the outputs from the operand node.

When a Boolean matrix is used as a formal model for the operand node, the vertical splitting operation is like a contraction of this matrix with respect to its rows and columns. This contraction can be applied if the operand matrix contains some identical non-zero matrix blocks, i.e. two-dimensional blocks with the same elements such that not all of these elements are '0's. Also, the operand matrix must contain some identical zero matrix blocks of the same dimension as the non-zero matrix blocks. It is obvious from these considerations that the vertical splitting operation can not always be applied because not any Boolean matrix representing a multiple-input-multiple-output node contains only identical non-zero matrix blocks and zero blocks of the same dimension.

The product matrices are obtained by contracting each non-zero block from the operand matrix to a '1' in the first product matrix and each zero block from the operand matrix to a '0' in the first product matrix whereas the second product matrix is the same as the non-zero block from the operand matrix. In this case, the row labels of the first product matrix are the terms in the permutations representing the row labels of the operand matrix whereas the column labels of the second product matrix are the terms in the permutations representing the column labels of the operand matrix.

The vertical splitting operation can also be applied in the context of binary relations when such a relation is used as a formal model for the operand node. In this case, the vertical splitting operation represents a special type of relational decomposition in the form of an inverse modified Cartesian product that is applied independently to the first and the second elements from the pairs of the operand relation. It is obvious again from these considerations that the vertical splitting operation can not always be applied because not any binary relation representing a multiple-input-multiple-output node can be decomposed by means of an inverse modified Cartesian product.

Example 4.7

This example considers an operand node $N_{11\text{-}21}$ that is the same as the product node N_{11+21} from Example 4.5. This node is described there by the Boolean matrix in Eq.(4.25) and the binary relation in Eq.(4.26). In this context, the node $N_{11\text{-}21}$ represents a simplified image of a two-node FN in the form of a one-node FN. The latter can be described by the block-scheme in Fig.4.15 and the topological expression in Eq.(4.31).

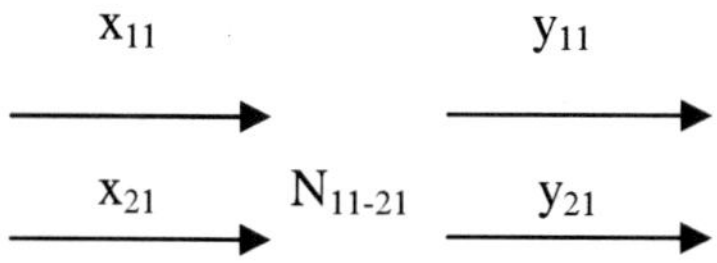

Fig. 4.15 One-node FN with operand node $N_{11\text{-}21}$

$$[N_{11\text{-}21}]\ (x_{11}, x_{21} \mid y_{11}, y_{21}) \tag{4.31}$$

The use of the symbol '-' in Fig.4.15 and Equation (4.31) implies that the vertical splitting operation can be applied to the operand node $N_{11\text{-}12}$. In this context, the use of the symbol '-' makes valid the precondition for vertical splitting of node $N_{11\text{-}21}$.

The vertical splitting of the operand node $N_{11\text{-}12}$ results into a pair of parallel product nodes N_{11} and N_{21} which represent a complexified image of the one-node FN in the form of a two-node FN. The latter can be described by the block-scheme in Fig.4.16 and the topological expression in Eq.(4.32).

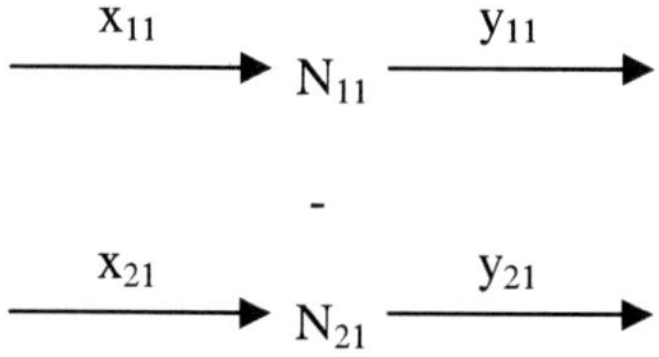

Fig. 4.16 Two-node FN with product nodes N_{11} and N_{21}

$$[N_{11}]\ (x_{11} \mid y_{11}) - [N_{21}]\ (x_{21} \mid y_{21}) \tag{4.32}$$

The use of the symbol '-' in Fig.4.16 and Eq.(4.32) implies that the application of the vertical splitting operation has resulted in the product nodes N_{11} and N_{21}. This is justifiable due to the deconcatenation of the inputs to the operand node as inputs x_{11}, x_{21} to the product nodes and the deconcatenation of the outputs from the operand node as outputs y_{11}, y_{21} from the product nodes. In this context, the use of the symbol '-' makes valid the postcondition for the formation of nodes N_{11} and N_{21} as a result of vertical splitting.

Finally, the product nodes N_{11} and N_{21} can be described by the Boolean matrices and the binary relations in Eq.(3.13), Eq.(3.15), Eq.(3.17) and Eq.(3.19).

Example 4.8

This example considers an operand node $N_{12\text{-}22}$ that is the same as the product node N_{12+22} from Example 4.6. This node is described there by the Boolean matrix in Eq.(4.29) and the binary relation in Eq.(4.30). In this context, the node $N_{12\text{-}22}$ represents a simplified image of a two-node FN in the form of a one-node FN. The latter can be described by the block-scheme in Fig.4.17 and the topological expression in Eq.(4.31).

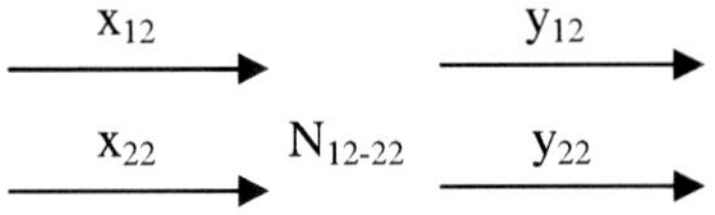

Fig. 4.17 One-node FN with operand node N_{12-22}

$$[N_{12-22}]\ (x_{12}, x_{22} \mid y_{12}, y_{22}) \tag{4.33}$$

The use of the symbol '-' in Fig.4.17 and Equation (4.33) implies that the vertical splitting operation can be applied to the operand node N_{12-22}. In this context, the use of the symbol '-' makes valid the precondition for vertical splitting of node N_{12-22}.

The vertical splitting of the operand node N_{12-22} results into a pair of parallel product nodes N_{12} and N_{22} which represent a complexified image of the one-node FN in the form of a two-node FN. The latter can be described by the block-scheme in Fig.4.18 and the topological expression in Eq.(4.34).

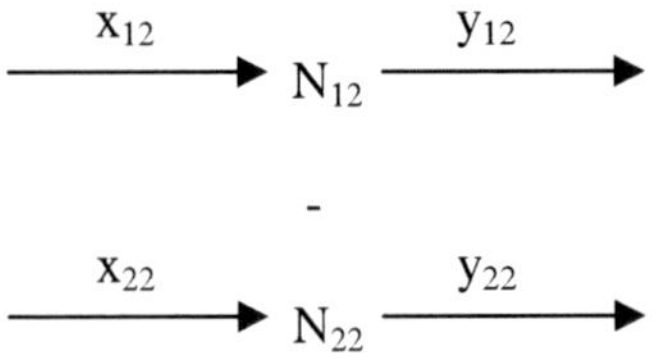

Fig. 4.18 Two-node FN with product nodes N_{12} and N_{22}

$$[N_{12}]\ (x_{12} \mid y_{12})\ -\ [N_{22}]\ (x_{22} \mid y_{22}) \tag{4.34}$$

The use of the symbol '-' in Fig.4.18 and Eq.(4.32) implies that the application of the vertical splitting operation has resulted in the product nodes N_{12} and N_{22}. This is justifiable due to the deconcatenation of the inputs to the operand node as inputs x_{12}, x_{22} to the product nodes and the deconcatenation of the outputs from the operand node as outputs y_{12}, y_{22} from the product nodes. In this context, the use of the symbol '-' makes valid the postcondition for the formation of nodes N_{12} and N_{22} as a result of vertical splitting.

Finally, the product nodes N_{12} and N_{22} can be described by the Boolean matrices and the binary relations in Eq.(3.14), Eq.(3.16), Eq.(3.18) and Eq.(3.20).

4.6 Output Merging of Nodes

Output merging is a binary operation that can be applied to a pair of parallel nodes with common inputs. This operation merges the operand nodes from the pair into a single product node. In this case, the inputs to the product node are the same as the common inputs to the operand nodes whereas the outputs from the product node represent the union of the outputs from the operand nodes. The operation of output merging can always be applied due to the ability to concatenate the outputs of any two parallel nodes with common inputs.

When Boolean matrices are used as formal models for the operand nodes, the output merging operation is like an expansion of the first operand matrix along its columns. In particular, the product matrix is obtained by expanding each non-zero element from the first operand matrix to a row-block that is the same as the corresponding row of the second operand matrix and by expanding each zero element from the first operand matrix to a zero row-block of the same dimension as the rows of the second product matrix. In this case, the row labels of the product matrix are the same as the identical row labels of the operand matrices whereas the column labels of the product matrix are all possible permutations of column labels of the operand matrices.

The output merging operation can also be applied in the context of binary relations when such relations are used as formal models for the operand nodes. In this case, the output merging operation represents a special type of relational composition in the form of a partially modified Cartesian product that is applied only to the second elements from the pairs of the operand relations whereas the first elements remain unchanged.

Example 4.9

This example considers the parallel operand nodes N_{11} and N_{21} located in the first layer of the four-node FN from Fig.3.1 in a modified context. In particular, the two independent inputs x_{11} and x_{21} to these nodes are replaced by a common input $x_{11,21}$. The nodes are described by the Boolean matrices and the binary relations in Eq.(3.13), Eq.(3.15), Eq.(3.17) and Eq.(3.19). The connections of these nodes with the nodes in the second layer of this FN are given by the interconnection structure in Eq.(3.22). In this context, the nodes N_{11} and N_{21} represent a modified two-node subnetwork of this FN. This two-node FN can be described by the block-scheme in Fig.4.19 and the topological expression in Eq.(4.35).

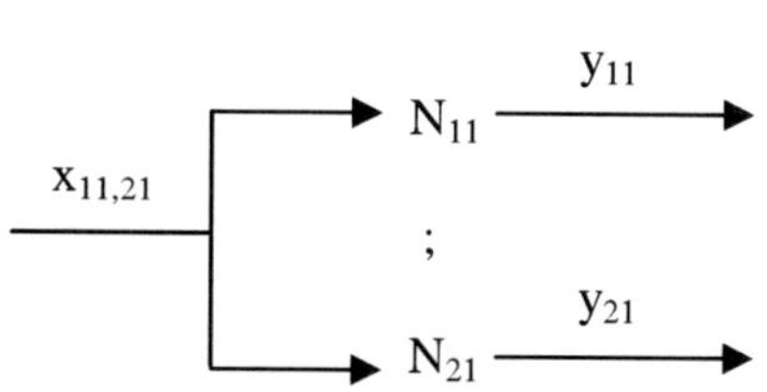

Fig. 4.19 Two-node FN with operand nodes N_{11}, N_{21} and common input

$[N_{11}]\ (x_{11,21} \mid y_{11})\ ;\ [N_{21}]\ (x_{11,21} \mid y_{21})$ (4.35)

The use of the symbol ';' in Fig.4.19 and Equation (4.35) implies that the output merging operation can be applied to the operand nodes N_{11} and N_{21}. In this context, the use of the symbol ';' confirms the validity of the precondition for output merging of nodes N_{11} and N_{21}.

The output merging of the operand nodes N_{11} and N_{21} results into a single product node $N_{11;21}$ which represents a simplified image of the two-node FN in the form of a one-node FN. The latter can be described by the block scheme in Fig.4.20 and the topological expression in Eq.(4.36).

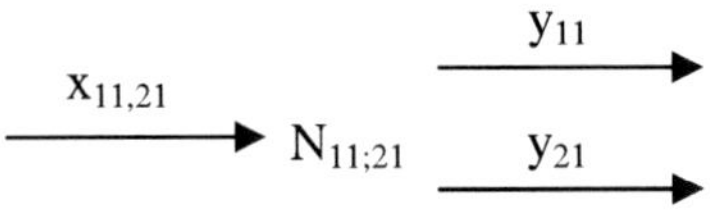

Fig. 4.20 One-node FN with product node $N_{11;21}$

$[N_{11;12}]\ (x_{11,21} \mid y_{11}, y_{21})$ (4.36)

The use of the symbol ';' in Fig.4.20 and Eq.(4.36) implies that the application of the output merging operation has resulted in the product node $N_{11;12}$. This is justifiable due to the concatenation of the outputs from the operand nodes as outputs y_{11}, y_{21} from the product node while preserving the common input to the operand nodes as an input $x_{11,21}$ to the product node. In this context, the use of the symbol ';' makes valid the postcondition for the formation of node $N_{11;21}$ as a result of output merging.

Finally, the product node $N_{11;21}$ can be described by the Boolean matrix in Eq.(4.37) and the binary relation in Eq.(4.38).

$N_{11;21}$: (4.37)

$x_{11,21}$ \ y_{11}, y_{21}	11	12	13	21	22	23	31	32	33
1	0	1	0	0	0	0	0	0	0
2	0	0	0	0	0	0	1	0	0
3	0	0	0	0	0	1	0	0	0

$N_{11;21}: \{(1, 12), (2, 31), (3, 23)\}$ (4.38)

Example 4.10

This example considers the parallel operand nodes N_{12} and N_{22} located in the second layer of the four-node FN from Fig.3.1 in a modified context. In particular, the two independent inputs x_{12} and x_{22} to these nodes are replaced by a common input $x_{12,22}$. The nodes are described by the Boolean matrices and the binary relations in Eq.(3.14), Eq.(3.16), Eq.(3.18) and Eq.(3.20). In this context, nodes N_{12}

and N_{22} represent a modified two-node subnetwork of this FN. This two-node FN can be described by the block-scheme in Fig.4.21 and the topological expression in Eq.(4.39).

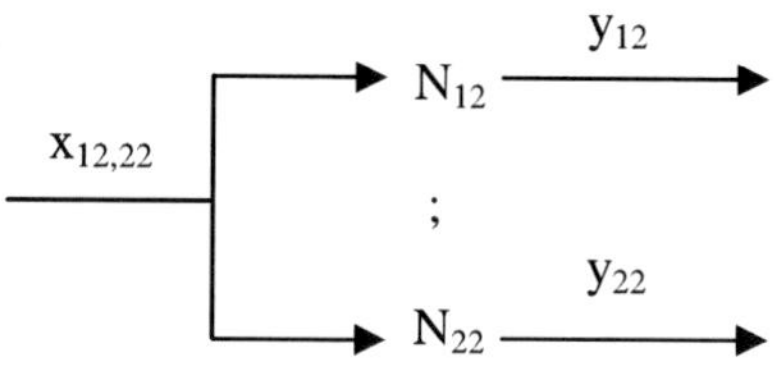

Fig. 4.21 Two-node FN with operand nodes N_{12}, N_{22} and common input

$$[N_{12}]\ (x_{12,22} \mid y_{12})\ ;\ [N_{21}]\ (x_{12,22} \mid y_{22}) \tag{4.39}$$

The use of the symbol ';' in Fig.4.21 and Equation (4.39) implies that the output merging operation can be applied to the operand nodes N_{12} and N_{22}. In this context, the use of the symbol ';' confirms the validity of the precondition for output merging of nodes N_{12} and N_{22}.

The output merging of the operand nodes N_{12} and N_{22} results into a single product node $N_{12;22}$ which represents a simplified image of the two-node FN in the form of a one-node FN. The latter can be described by the block scheme in Fig.4.22 and the topological expression in Eq.(4.40).

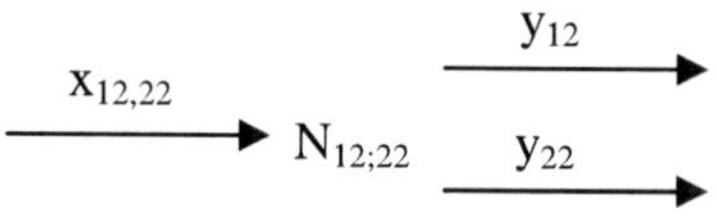

Fig. 4.22 One-node FN with product node $N_{12;22}$

$$[N_{12;22}]\ (x_{12,22} \mid y_{12},\ y_{22}) \tag{4.40}$$

The use of the symbol ';' in Fig.4.22 and Eq.(4.40) implies that the application of the output merging operation has resulted in the product node $N_{12;22}$. This is justifiable due to the concatenation of the outputs from the operand nodes as outputs y_{12}, y_{22} to the product node while preserving the common input to the operand nodes as input $x_{12,22}$ to the product node. In this context, the use of the symbol ';' makes valid the postcondition for the formation of node $N_{12;22}$ as a result of output merging.

Finally, the product node $N_{12;22}$ can be described by the Boolean matrix in Eq.(4.41) and the binary relation in Eq.(4.42).

$N_{12;22}$: (4.41)

$x_{12,22}$ \ y_{12}, y_{22}	11	12	13	21	22	23	31	32	33
1	0	0	0	0	0	1	0	0	0
2	0	0	0	0	0	0	1	0	0
3	0	1	0	0	0	0	0	0	0

$$N_{12;22}: \{(1, 23), (2, 31), (3, 12)\} \quad (4.42)$$

4.7 Output Splitting of Nodes

Output splitting is a unary operation that can be applied to a single node in a FN. This operation splits an operand node into a pair of parallel product nodes whereby the inputs to the product nodes are the same as the inputs to the operand node and the union of the outputs from the product nodes represents the outputs from the operand node.

When a Boolean matrix is used as a formal model for the operand node, the output splitting operation is like a contraction of this matrix along its columns. This operation can be applied if the operand matrix contains some not necessarily identical non-zero row blocks of the same dimension, i.e. one-dimensional blocks with the same number of elements such that not all of these elements are '0's. Also, the operand matrix must contain some identical zero row blocks of the same dimension as the non-zero row blocks. It is obvious from these considerations that the output splitting operation can always be applied because any Boolean matrix representing a multiple-output node contains some not necessarily identical non-zero row blocks of the same dimension and some identical zero blocks with the same dimension as the non-zero blocks.

The product matrices are obtained by contracting each non-zero block from the operand matrix to a '1' in the first product matrix and each zero block from the operand matrix to a '0' in the first product matrix whereas the rows of the second product matrix are the same as the corresponding non-zero blocks from the operand matrix. In this case, the row labels of both product matrices are the same as the row labels of the operand matrix whereas the column labels of the product matrices are the terms in the permutations representing the column labels of the operand matrix.

The output splitting operation can also be applied in the context of binary relations when such a relation is used as a formal model for the operand node. In this case, the output splitting operation represents a special type of relational decomposition in the form of an inverse partially modified Cartesian product that is applied to the second elements from the pairs of the operand relation whereas the first elements remain unchanged. It is obvious again from these considerations that the output splitting operation can always be applied because any binary relation representing a multiple-output node can be decomposed by means of an inverse partially modified Cartesian product.

Example 4.11

This example considers an operand node $N_{11:21}$ that is assumed to be the same as the product node $N_{11;21}$ from Example 4.9. This node is described there by the Boolean matrix in Eq.(4.37) and the binary relation in Eq.(4.38). In this context, the node $N_{11:21}$ represents a simplified image of a two-node FN in the form of a one-node FN. The latter can be described by the block-scheme in Fig.4.23 and the topological expression in Eq.(4.43).

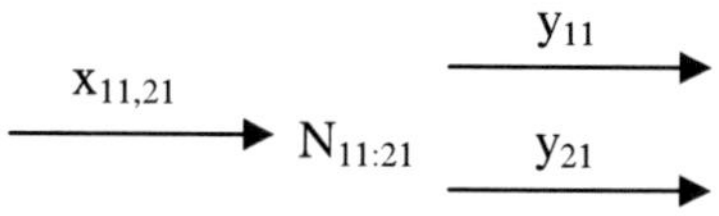

Fig. 4.23 One-node FN with operand node $N_{11:21}$

$$[N_{11:21}]\ (x_{11,21} \mid y_{11},\ y_{21}) \tag{4.43}$$

The use of the symbol ':' in Fig.4.23 and Equation (4.43) implies that the output splitting operation can be applied to the operand node $N_{11:21}$. In this context, the use of the symbol ':' makes valid the precondition for output splitting of node $N_{11:21}$.

The output splitting of the operand node $N_{11:21}$ results into a pair of parallel product nodes N_{11} and N_{21} which represent a complexified image of the one-node FN in the form of a two-node FN. The latter can be described by the block-scheme in Fig.4.24 and the topological expression in Eq.(4.44).

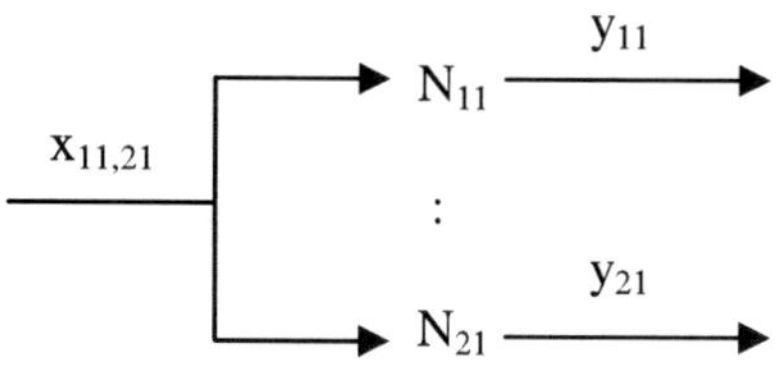

Fig. 4.24 Two-node FN with product nodes N_{11}, N_{21} and common input

$$[N_{11}]\ (x_{11,21} \mid y_{11}) : [N_{21}]\ (x_{11,21} \mid y_{21}) \tag{4.44}$$

The use of the symbol ':' in Fig.4.24 and Eq.(4.44) implies that the application of the output splitting operation has resulted in the product nodes N_{11} and N_{21}. This is justifiable due the deconcatenation of the outputs from the operand node as outputs y_{11}, y_{21} from the product nodes while preserving the input to the operand node as a common input $x_{11,21}$ to the product nodes.

Finally, the product nodes N_{11} and N_{21} can be described by the Boolean matrices and the binary relations in Eq.(3.13), Eq.(3.15), Eq.(3.17) and Eq.(3.19).

Example 4.12

This example considers an operand node $N_{12:22}$ that is assumed to be the same as the product node $N_{12;22}$ from Example 4.10. This node is described there by the Boolean matrix in Eq.(4.41) and the binary relation in Eq.(4.42). In this context, the node $N_{12:22}$ represents a simplified image of a two-node FN in the form of a one-node FN. The latter can be described by the block-scheme in Fig.4.25 and the topological expression in Eq.(4.45).

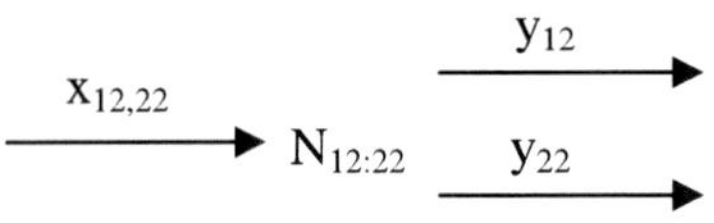

Fig. 4.25 One-node FN with operand node $N_{12:22}$

$$[N_{12:22}]\ (x_{11,21} \mid y_{12}, y_{22}) \tag{4.45}$$

The use of the symbol ':' in Fig.4.25 and Equation (4.45) implies that the output splitting operation can be applied to the operand node $N_{12:22}$. In this context, the use of the symbol ':' makes valid the precondition for output splitting of node $N_{12:22}$.

The output splitting of the operand node $N_{12:22}$ results into a pair of parallel product nodes N_{12} and N_{22} which represent a complexified image of the one-node FN in the form of a two-node FN. The latter can be described by the block-scheme in Fig.4.26 and the topological expression in Eq.(4.46).

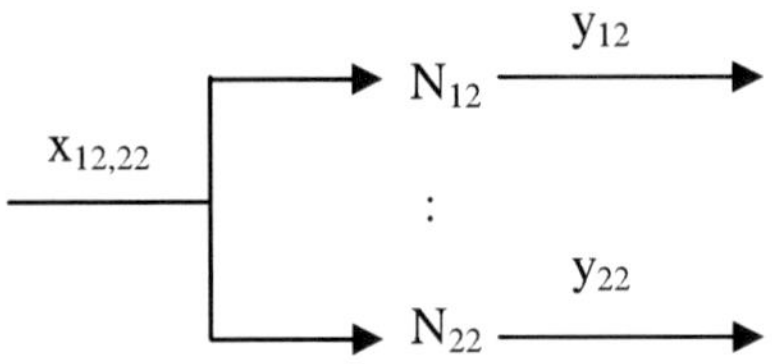

Fig. 4.26 Two-node FN with product nodes N_{12}, N_{22} and common input

$$[N_{12}]\ (x_{12,22} \mid y_{12}) : [N_{22}]\ (x_{12,22} \mid y_{22}) \tag{4.46}$$

The use of the symbol ':' in Fig.4.26 and Eq.(4.46) implies that the application of the output splitting operation has resulted in the product nodes N_{12} and N_{22}. This is justifiable due the deconcatenation of the outputs from the operand node as outputs y_{12}, y_{22} from the product nodes while preserving the input to the operand node as a common input $x_{12,22}$ to the product nodes.

Finally, the product nodes N_{12} and N_{22} can be described by the Boolean matrices and the binary relations in Eq.(3.14), Eq.(3.16), Eq.(3.18) and Eq.(3.20).

4.8 Combined Operations on Nodes

The basic operations on nodes introduced in the preceding sections of this chapter are all atomic, i.e. the operations are applied on their own. However, each of these operations can also be applied in a combined context with other operations. Unless there are brackets specifying the order of the atomic operations, these operations are applied from left to right for merging and from right to left for splitting.

This section introduces some combined operations. For simplicity, these operations are assumed to include only one permutation of any two atomic operations of the same type, e.g. merging or splitting of nodes. However, the extension of these operations to more complex combined operations including other permutations of any two atomic operations or more than two atomic operations of any type is straightforward.

Similarly to atomic operations, combined operations can be described by block schemes and topological expressions at network level as well as by Boolean matrices and binary relations at node level. For simplicity, each combined operation is described here only by block schemes and topological expressions whereby the associated Boolean matrices and binary relations are assumed to be embedded implicitly in the description.

Also, block schemes and topological expressions are used for combined operations in a modified context in this section. In this respect, nodes may occupy more than one location in a particular level and have no subscripts indicating their location in the grid structure. These modifications are aimed at facilitating the understanding of the fairly complex nature of combined operations.

For consistency, all combined operations are presented by three stages. The first stage describes the initial state of the combined operation with the operand nodes before the application of any atomic operations. The second stage describes an intermediate state of the operation with some temporary nodes after the application of some atomic operations. The third stage describes the final state of the combined operation with the product nodes after the application of all atomic operations.

Example 4.13

This example considers a combined operation of horizontal-vertical merging that is applied to three operand nodes A, B and C. First, node A is horizontally merged with node B into a temporary node A*B. Then, node A*B is vertically merged with node C into a product node A*B+C. The three states of this combined operation are described by the block schemes and the topological expressions in Figs.4.27-4.29 and Eqs.(4.47)-(4.49).

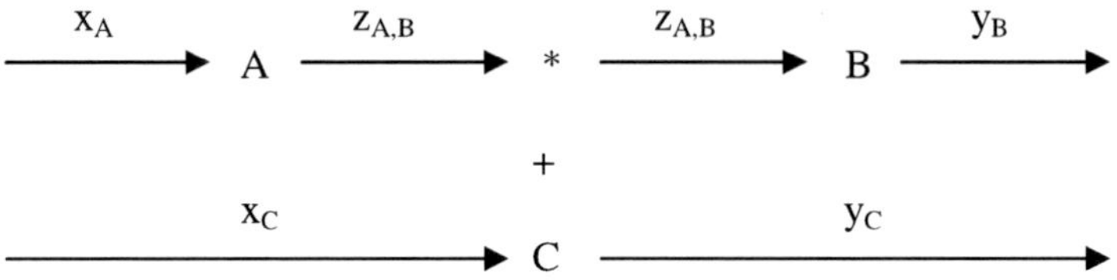

Fig. 4.27 Initial state for horizontal-vertical merging of nodes A, B and C

$$[A]\ (x_A \mid z_{A,B}) * [B]\ (z_{A,B} \mid y_B) + [C]\ (x_C \mid y_C) \tag{4.47}$$

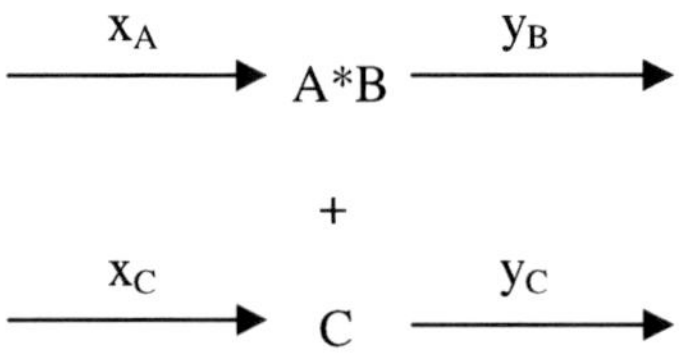

Fig. 4.28 Intermediate state for horizontal-vertical merging of nodes A, B and C

$$[A*B]\ (x_A \mid y_B) + [C]\ (x_C \mid y_C) \tag{4.48}$$

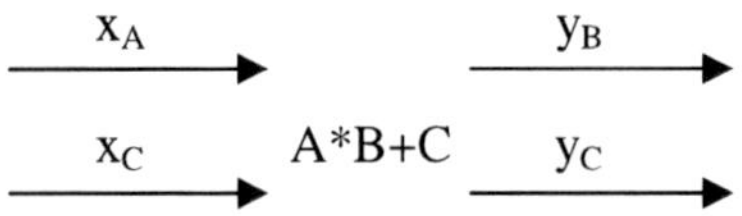

Fig. 4.29 Final state for horizontal-vertical merging of nodes A, B and C

$$[A*B+C]\ (x_A, x_C \mid y_B, y_C\ ) \tag{4.49}$$

Example 4.14

This example considers a combined operation of vertical-horizontal splitting that is applied to an operand node A/B-C. First, this node is vertically split into a temporary node A/B and a product node C. Then, node A/B is horizontally split into product nodes A and B. The three states of this combined operation are described by the block schemes and the topological expressions in Figs.4.30-4.32 and Eqs.(4.50)-(4.52).

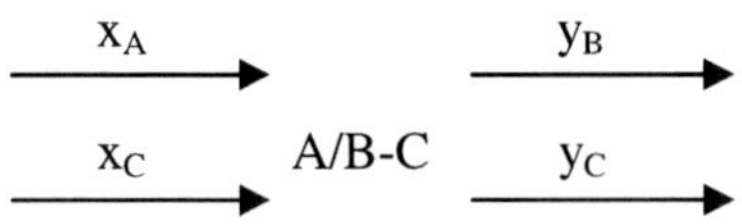

Fig. 4.30 Initial state for vertical-horizontal splitting of node A/B-C

$$[A/B\text{-}C]\ (x_A, x_C \mid y_B, y_C) \tag{4.50}$$

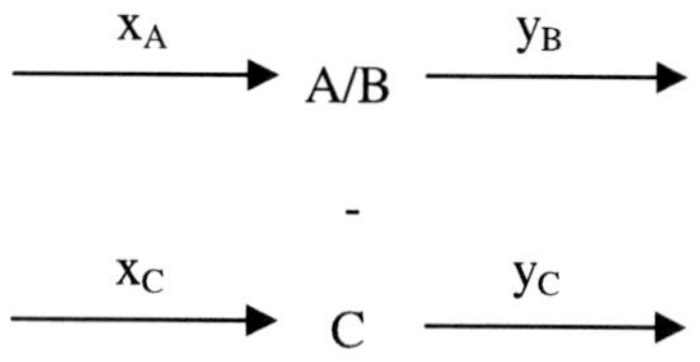

Fig. 4.31 Intermediate state for vertical-horizontal splitting of node A/B-C

$$[A/B]\ (x_A \mid y_B)\ \text{-}\ [C]\ (x_C \mid y_C) \tag{4.51}$$

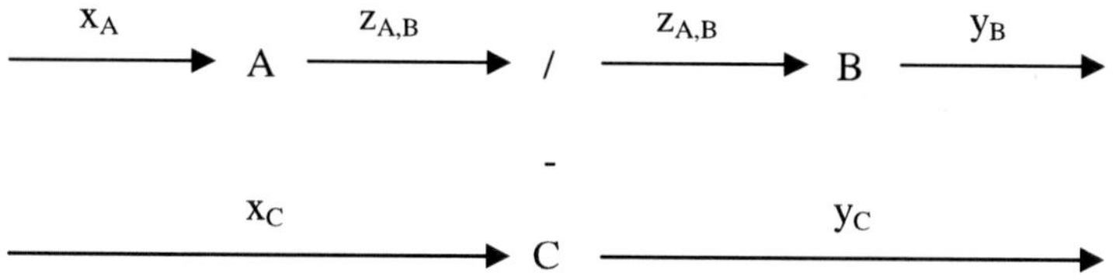

Fig. 4.32 Final state for vertical-horizontal splitting of node A/B-C

$$[A]\ (x_A \mid z_{A,B})\ /\ [B]\ (z_{A,B} \mid y_B)\ \text{-}\ [C]\ (x_C \mid y_C) \tag{4.52}$$

Example 4.15

This example considers a combined operation of horizontal-output merging that is applied to three operand nodes A, B and C. First, node A is horizontally merged with node B into a temporary node A*B. Then, node A*B is output merged with node C into a product node A*B;C. The three states of this combined operation are described by the block schemes and the topological expressions in Figs.4.33-4.35 and Eqs.(4.53)-(4.55).

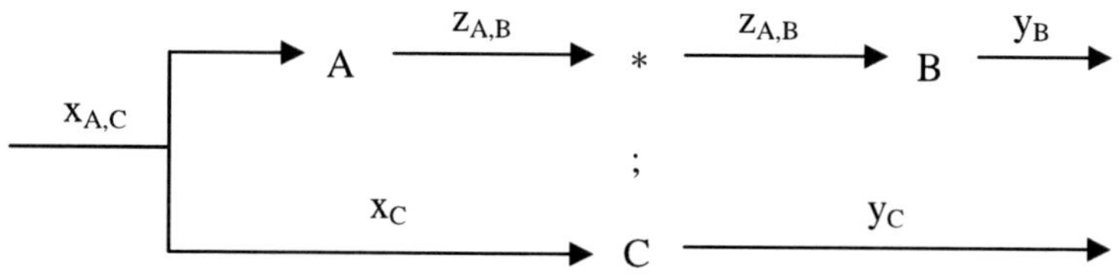

Fig. 4.33 Initial state for horizontal-output merging of nodes A, B and C

$$[A]\ (x_{A,C} \mid z_{A,B})\ *\ [B]\ (z_{A,B} \mid y_B)\ ;\ [C]\ (x_C \mid y_C) \tag{4.53}$$

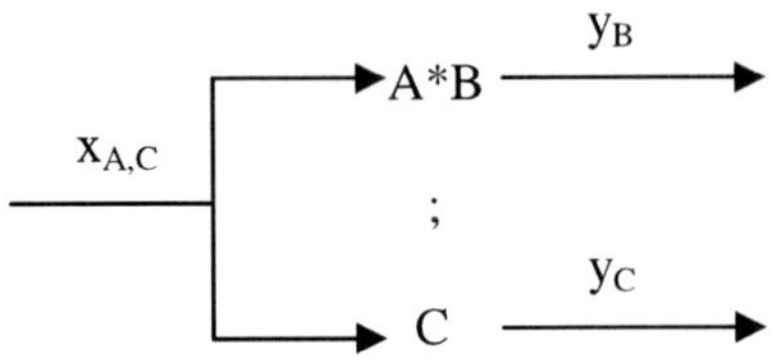

Fig. 4.34 Intermediate state for horizontal-output merging of nodes A, B and C

$$[A*B]\ (x_{A,C} \mid y_B)\ ;\ [C]\ (x_{A,C} \mid y_C) \tag{4.54}$$

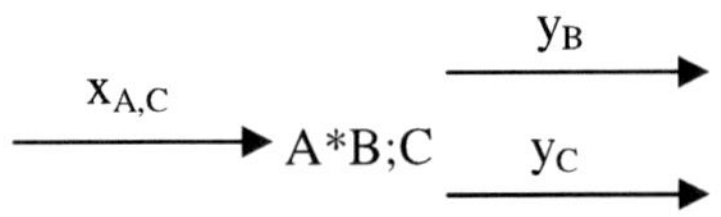

Fig. 4.35 Final state for horizontal-output merging of nodes A, B and C

$$[A*B;C]\ (x_{A,C} \mid y_B,\ y_C) \tag{4.55}$$

Example 4.16

This example considers a combined operation of output-horizontal splitting that is applied to an operand node A/B:C. First, this node is output split into a temporary node A/B and a product node C. Then, node A/B is horizontally split into product nodes A and B. The three states of this combined operation are described by the block schemes and the topological expressions in Figs.4.36-4.38 and Eqs.(4.56)-(4.58).

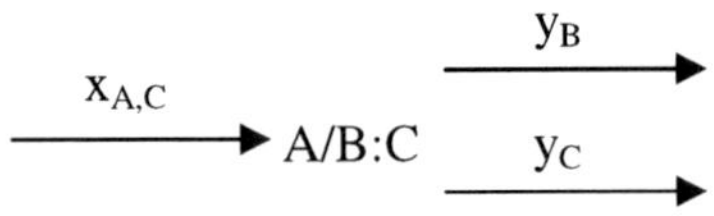

Fig. 4.36 Initial state for output-horizontal splitting of node A/B:C

$$[A/B:C]\ (x_{A,C} \mid y_B,\ y_C) \tag{4.56}$$

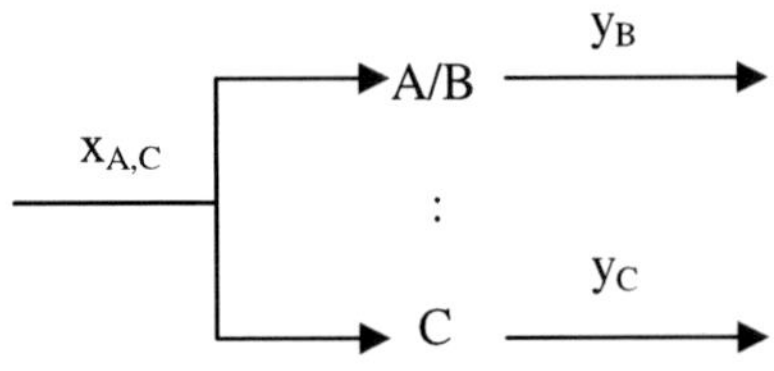

Fig. 4.37 Intermediate state for output-horizontal splitting of node A/B:C

$$[A/B]\ (x_{A,C} \mid y_B) : [C]\ (x_{A,C} \mid y_C) \tag{4.57}$$

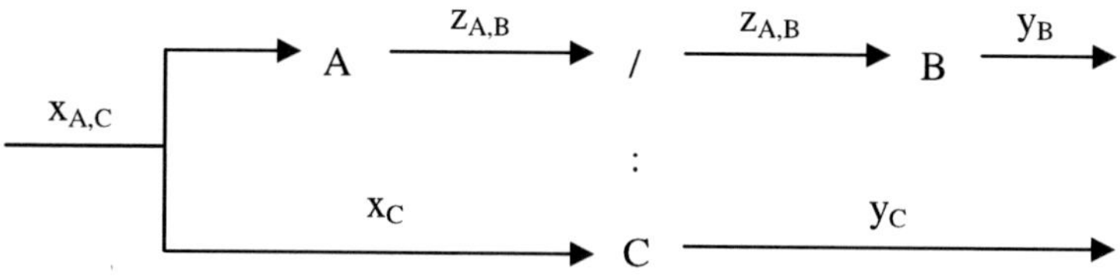

Fig. 4.38 Final state for output-horizontal splitting of node A/B:C

$$[A]\ (x_{A,C} \mid z_{A,B})\ /\ [B]\ (z_{A,B} \mid y_B) : [C]\ (x_C \mid y_C) \tag{4.58}$$

Example 4.17

This example considers a combined operation of vertical-output merging that is applied to three operand nodes A, B and C. First, node A is vertically merged with node B into a temporary node A+B. Then, node A+B is output merged with node C into a product node A+B;C. The three states of this combined operation are described by the block schemes and the topological expressions in Figs.4.39-4.41 and Eqs.(4.59)-(4.61).

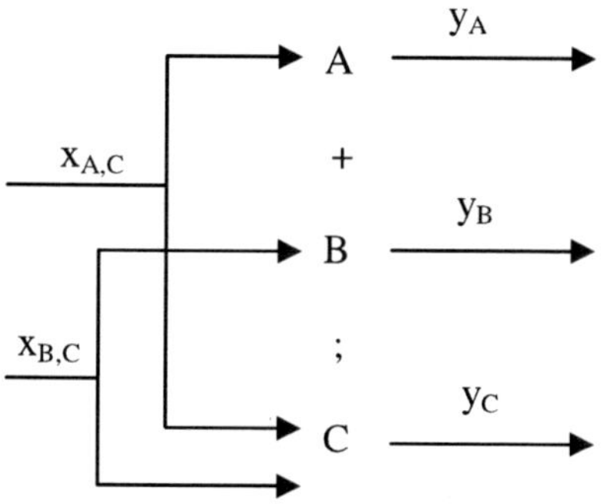

Fig. 4.39 Initial state for vertical-output merging of nodes A, B and C

$$[A]\ (x_{A,C} \mid y_A) + [B]\ (x_{B,C} \mid y_B)\ ;\ [C]\ (x_{A,C},\ x_{B,C} \mid y_C) \tag{4.59}$$

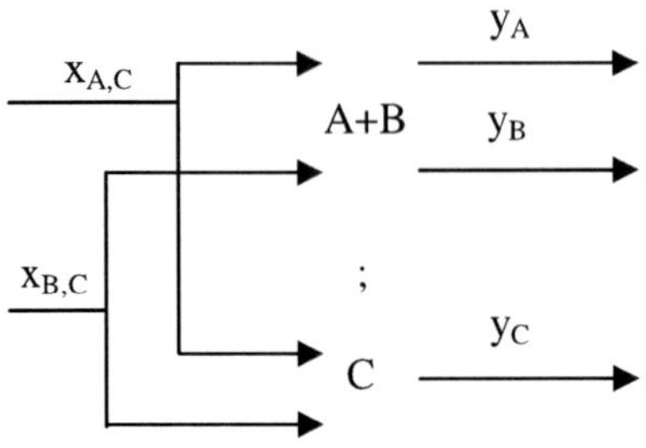

Fig. 4.40 Intermediate state for vertical-output merging of nodes A, B and C

$$[A+B]\ (x_{A,C},\ x_{B,C}\ |\ y_A,\ y_B)\ ;\ [C]\ (x_{A,C},\ x_{B,C}\ |\ y_C) \tag{4.60}$$

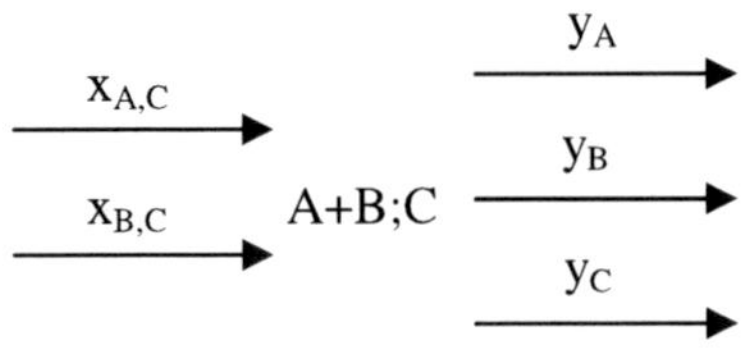

Fig. 4.41 Final state for vertical-output merging of nodes A, B and C

$$[A+B;C]\ (x_{A,C},\ x_{B,C}\ |\ y_A,\ y_B,\ y_C) \tag{4.61}$$

Example 4.18

This example considers a combined operation of output-vertical splitting that is applied to an operand node A-B:C. First, this node is output split into a temporary node A-B and a product node C. Then, node A-B is vertically split into product nodes A and B. The three states of this combined operation are described by the block schemes and the topological expressions in Figs.4.42-4.44 and Eqs. (4.62)-(4.64).

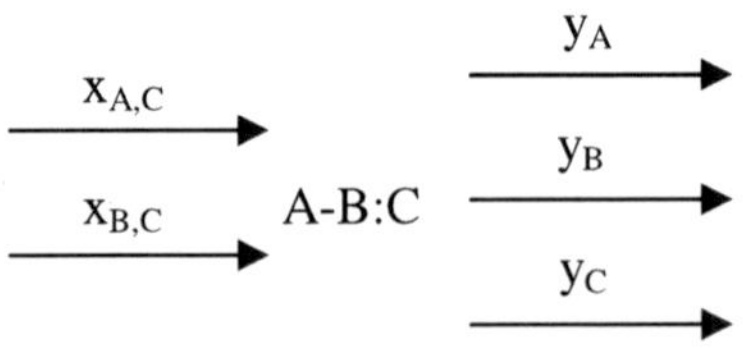

Fig. 4.42 Initial state for output-vertical splitting of node A-B:C

$$[A-B:C]\ (x_{A,C},\ x_{B,C}\ |\ y_A,\ y_B,\ y_C) \tag{4.62}$$

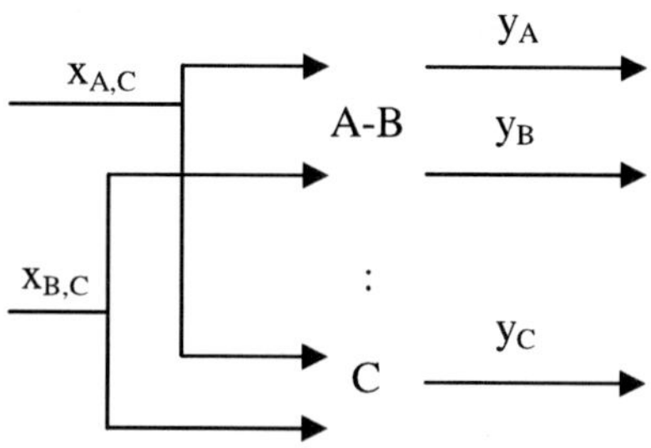

Fig. 4.43 Intermediate state for output-vertical splitting of node A-B:C

$$[A\text{-}B]\ (x_{A,C}, x_{B,C} \mid y_A, y_B) : [C]\ (x_{A,C}, x_{B,C} \mid y_C) \tag{4.63}$$

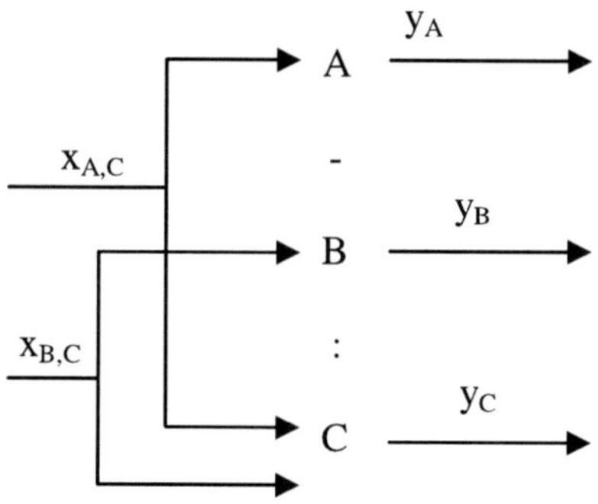

Fig. 4.44 Final state for output-vertical splitting of node A-B:C

$$[A]\ (x_{A,C} \mid y_A) - [B]\ (x_{B,C} \mid y_B) : [C]\ (x_{A,C}, x_{B,C} \mid y_C) \tag{4.64}$$

4.9 Comparison of Basic Operations

The basic operations introduced in this chapter are central to the linguistic composition approach used in the book. This applies particularly to merging operations which are aimed at composing the networked rule bases within a FN into a linguistically equivalent single rule base for a fuzzy system. On the contrary, splitting operations are aimed at decomposing a single rule base for a fuzzy system into linguistically equivalent networked rule bases within a FN. However, in some cases splitting operations may facilitate merging operations in the context of the linguistic composition approach and this is shown by some examples further in this book.

The solution to most types of basic operations always exists. The only exception in this respect is the operation of vertical splitting that may not have a solution. However, only merging operations have a unique solution whereas splitting operations usually have multiple solutions.

The characteristics of solutions to different types of basic operations in FNs are summarised in Table 4.1.

Table 4.1 Solution characteristics for basic operations in FNs

Basic operation	Composition	Existence	Uniqueness
Horizontal merging	Yes	Yes	Yes
Horizontal splitting	No	Yes	No
Vertical merging	Yes	Yes	Yes
Vertical splitting	No	No	No
Output merging	Yes	Yes	Yes
Output splitting	No	Yes	No

The next chapter introduces some advanced concepts from the theoretical framework for FNs. In particular, it discusses several structural properties of basic operations in FNs.

Chapter 5
Structural Properties of Basic Operations

5.1 Introduction to Structural Properties

The basic operations introduced in Chapter 4 can be applied to fairly simple FNs with only a pair of nodes or a single node. However, an arbitrarily complex FN may have a large number of nodes whereby all of them have to be manipulated for the purpose of using the linguistic composition approach. Therefore, it is important to know how the basic operations can be applied in this more realistic and complex context.

A key to the solution of the above problem are some structural properties of basic operations. These properties make the manipulation of nodes within the structure of an arbitrarily complex FN very flexible. In this respect, each property related to a merging operation has a counterpart related to the corresponding splitting operation for this merging operation. In this case, the property related to the splitting operation has an inverse effect with respect to its counterpart related to the merging operation. The structural properties are similar to some properties of mathematical operations. However, these properties are novel in that they are all related to operations within a FN which is a novel concept.

All structural properties are proved and illustrated with examples of nodes with scalar inputs, outputs and intermediate variables but the extension of these proofs and examples to the vector case is straightforward. The proofs and the examples are based on the use of Boolean matrices or binary relations as formal models for FNs at node level as these formal models lend themselves easily to manipulation in the context of the linguistic composition approach. Therefore, the structural properties can be viewed as the glue that makes the elementary building blocks for the simplification of an arbitrarily complex FN to a fuzzy system, i.e. the basic operations on nodes, stick together.

5.2 Associativity of Horizontal Merging

Associativity is a property related to the operation of horizontal merging when the latter is applied to three sequential nodes for the purpose of merging them into a single node. In particular, this property allows the merging of three operand nodes A, B and C into a product node A*B*C to take place as a sequence of two binary merging operations that can be applied either from left to right or from right to left. The property can be applied when the output from the first node A is fed

A. Gegov: Fuzzy Networks for Complex Systems, STUDFUZZ 259, pp. 51–108.
springerlink.com

forward as an input to the second node B in the form of an intermediate variable and the output from the second node B is fed forward as an input to the third node C as another intermediate variable. In this case, the product node A*B*C has the same input as the input to the first operand node A and the same output as the output from the third operand node C whereas the two intermediate variables do not appear in the product node.

Proof 5.1

It has to be proved here that the operation of horizontal merging is associative in accordance with Eq.(5.1). In this case, the horizontal merging of any three operand nodes A, B and C from left to right should be equivalent to their horizontal merging from right to left.

$$(A*B)*C = A*(B*C) = A*B*C \tag{5.1}$$

The proof is based on the use of binary relations as formal models for the operand nodes A, B and C, as shown in Eqs. (5.2)-(5.4). In this case, the elements of the relational pairs are denoted by the letter a in A, the letters a and c in B, and the letter c in C. For simplicity, all pairs in the middle relation B are assumed to be composable with pairs from the left relation A and the right relation C. For this reason, the first and the second element of each pair in B are denoted by a and c, respectively, and not by b.

$$A = \{(a_1^1, a_2^1),\ldots,(a_1^p, a_2^p)\} \tag{5.2}$$

$$B = \{(a_2^1, c_1^1),\ldots,(a_2^1, c_1^q),\ldots,(a_2^p, c_1^1),\ldots,(a_2^p, c_1^q)\} \tag{5.3}$$

$$C = \{(c_1^1, c_2^1),\ldots,(c_1^q, c_2^q)\} \tag{5.4}$$

The first and the second element of any relational pair in A and C are denoted by the subscripts ‘1’ and ‘2’, respectively. However, the superscripts for the first and the second element of any relational pair in A and C are identical as they indicate the corresponding number for each pair. In particular, the relation A has ‘p’ pairs and the relation C has ‘q’ pairs. The subscripts for the first and the second element of any relational pair in B are ‘2’ and ‘1’, respectively. This is due to the requirement for left and right composability of B, i.e. the first element of each pair in B must be identical with a second element of a pair in A whereas the second element of each pair in B must be identical with a first element of a pair in C. In this case, the superscripts for the elements of the relational pairs in B don’t have to be identical and the relation B has ‘p.q’ pairs.

The horizontal composition of the operand relations A and B gives the temporary relation A*B, as shown in Eq.(5.5).

$$A*B = \{(a_1^1, c_1^1),\ldots,(a_1^1, c_1^q),\ldots,(a_1^p, c_1^1),\ldots,(a_1^p, c_1^q)\} \tag{5.5}$$

Further on, the horizontal composition of the temporary relation A*B and the operand relation C gives the product relation (A*B)*C, as shown in Eq.(5.6).

$$(A*B)*C = \{(a_1^{\,1}, c_2^{\,1}),\ldots,(a_1^{\,1}, c_2^{\,q}),\ldots,(a_1^{\,p}, c_2^{\,1}),\ldots,(a_1^{\,p}, c_2^{\,q})\} \tag{5.6}$$

On the other hand, the horizontal composition of the operand relations B and C gives the temporary relation B*C, as shown in Eq.(5.7).

$$B*C = \{(a_2^{\,1}, c_2^{\,1}),\ldots,(a_2^{\,1}, c_2^{\,q}),\ldots,(a_2^{\,p}, c_2^{\,1}),\ldots,(a_2^{\,p}, c_2^{\,q})\} \tag{5.7}$$

In this case, the horizontal composition of the operand relation A and the temporary relation B*C gives the product relation A*(B*C). As the latter is identical with the product relation (A*B)*C from Eq.(5.6), this implies the validity of Eq.(5.1) and concludes the proof.

Example 5.1

This example considers a FN with three sequential operand nodes. The first and the second node N_{11} and N_{12} are taken from the four-node FN in Fig.3.1. These two nodes are described there by the Boolean matrices in Eqs.(3.13)-(3.14) and the binary relations in Eqs.(3.17)-(3.18) whereas the third node N_{13} is described by the Boolean matrix in Eq.(5.8) and the binary relation in Eq.(5.9). The connections among the three nodes are given by Eqs.(5.10)-(5.11).

N_{13}: (5.8)

x_{13} \ y_{13}	1	2	3
1	0	0	1
2	0	1	0
3	1	0	0

$$N_{13}: \ \{(1, 3), (2, 2)\ (3, 1)\} \tag{5.9}$$

$$z_{11,12} = y_{11} = x_{12} \tag{5.10}$$

$$z_{12,13} = y_{12} = x_{13} \tag{5.11}$$

The nodes N_{11}, N_{12} and N_{13} represent a three-node FN that is an extended sub-network of the four-node FN from Fig.3.1. This three-node FN can be described by the block-scheme in Fig.5.1 and the topological expression in Eq.(5.12).

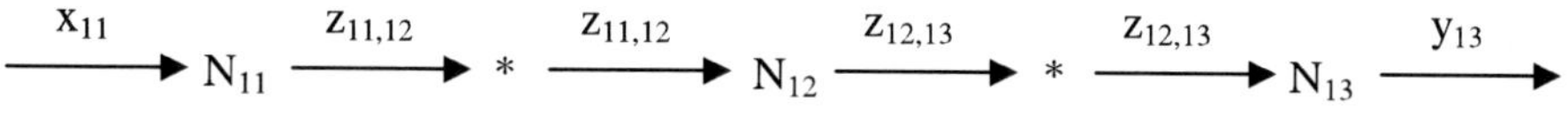

Fig. 5.1 FN with operand nodes N_{11}, N_{12} and N_{13}

$[N_{11}]\ (x_{11} \mid z_{11,12}) * [N_{12}]\ (z_{11,12} \mid z_{12,13}) * [N_{13}]\ (z_{12,13} \mid y_{13})$ (5.12)

The horizontal merging of the first operand node N_{11} and the second operand node N_{12} from Fig.5.1 results into a temporary node N_{11*12} that is connected to the right with the third operand node N_{13} in the form of a two-node FN. The latter can be described by the block scheme in Fig.5.2 and the topological expression in Eq.(5.13).

Fig. 5.2 FN with temporary node N_{11*12} and operand node N_{13}

$[N_{11*12}]\ (x_{11} \mid z_{11*12,13}) * [N_{13}]\ (z_{11*12,13} \mid y_{13})$ (5.13)

Further on, the horizontal merging of the temporary node N_{11*12} and the operand node N_{13} from Fig.5.2 results into a product node $N_{(11*12)*13}$ in the form of a one-node FN. The latter can be described by the block scheme in Fig.5.3 and the topological expression in Eq.(5.14).

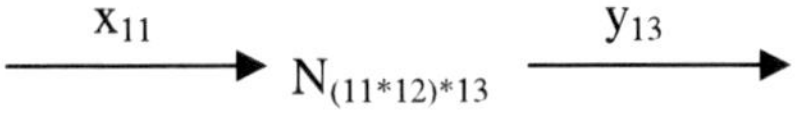

Fig. 5.3 FN with product node $N_{(11*12)*13}$

$[N_{(11*12)*13}]\ (x_{11} \mid y_{13})$ (5.14)

As a result, the product node $N_{(11*12)*13}$ is described by the Boolean matrix in Eq.(5.15) and the binary relation in Eq.(5.16).

$N_{(11*12)*13}$: (5.15)

x_{11} \ y_{13}	1	2	3
1	0	1	0
2	0	0	1
3	1	0	0

$N_{(11*12)*13}$: {(1, 2), (2, 3) (3, 1)} (5.16)

On the other hand, the horizontal merging of the second operand node N_{12} and the third operand node N_{13} from Fig.5.1 results into a temporary node N_{12*13} that is connected to the left with the first operand node N_{11} in the form of a two-node FN. The latter can be described by the block scheme in Fig.5.4 and the topological expression in Eq.(5.17).

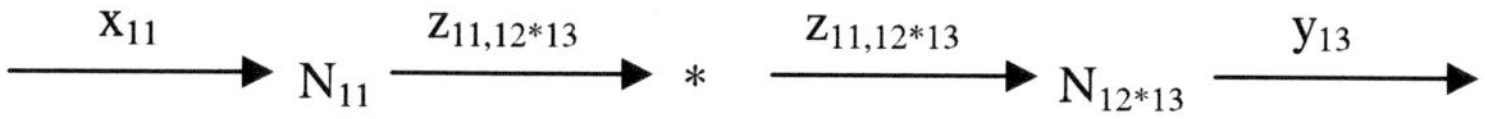

Fig. 5.4 FN with operand node N_{11} and temporary node N_{12*13}

$$[N_{11}]\ (x_{11} \mid z_{11,12*13}) * [N_{12*13}]\ (z_{11,12*13} \mid y_{13}) \tag{5.17}$$

Further on, the horizontal merging of the operand node N_{11} and the temporary node N_{12*13} from Fig.5.4 results into a product node $N_{11*(12*13)}$ in the form of a one-node FN. The latter can be described by the block scheme in Fig.5.5 and the topological expression in Eq.(5.18).

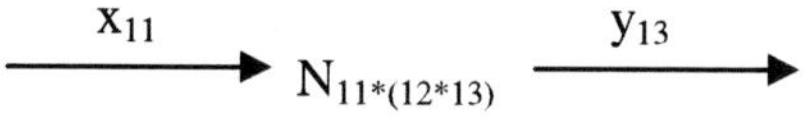

Fig. 5.5 FN with product node $N_{11*(12*13)}$

$$[N_{11*(12*13)}]\ (x_{11} \mid y_{13}) \tag{5.18}$$

In this case, the product node $N_{11*(12*13)}$ is also described by the Boolean matrix in Eq.(5.15) and the binary relation in Eq.(5.16). Therefore, $N_{11*(12*13)}$ is identical with $N_{(11*12)*13}$ and this identity is defined by Eq.(5.19) in accordance with Eq.(5.1) from Proof 5.1.

$$N_{(11*12)*13} = N_{11*(12*13)} = N_{11*12*13} \tag{5.19}$$

Example 5.2

This example considers a FN with three sequential operand nodes. The first and the second node N_{21} and N_{22} are taken from the four-node FN in Fig.3.1. These two nodes are described there by the Boolean matrices in Eqs.(3.15)-(3.16) and the binary relations in Eqs.(3.19)-(3.20) whereas the third node N_{23} is described by the Boolean matrix in Eq.(5.20) and the binary relation in Eq.(5.21). The connections among the three nodes are given by Eqs.(5.22)-(5.23).

$$N_{23}: \quad \begin{array}{cccc} & y_{23} & & \\ x_{23} & 1 & 2 & 3 \\ 1 & 0 & 0 & 1 \\ 2 & 0 & 1 & 0 \\ 3 & 1 & 0 & 0 \end{array} \tag{5.20}$$

$$N_{23}: \ \{(1, 3), (2, 2)\ (3, 1)\} \tag{5.21}$$

$$z_{21,22} = y_{21} = x_{22} \tag{5.22}$$

$$z_{22,23} = y_{22} = x_{23} \tag{5.23}$$

The nodes N_{21}, N_{22} and N_{23} represent a three-node FN that is an extended sub-network of the four-node FN from Fig.3.1. This three-node FN can be described by the block-scheme in Fig.5.6 and the topological expression in Eq.(5.24).

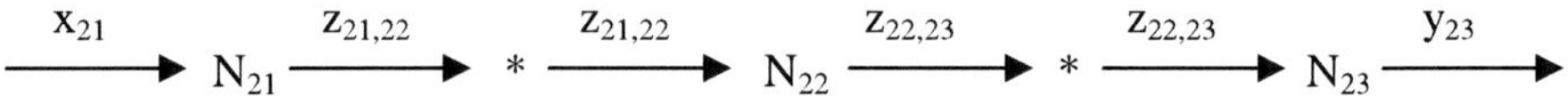

Fig. 5.6 FN with operand nodes N_{21}, N_{22} and N_{23}

$$[N_{21}]\,(x_{21}\,|\,z_{21,22}) * [N_{22}]\,(z_{21,22}\,|\,z_{22,23}) * [N_{23}]\,(z_{22,23}\,|\,y_{23}) \tag{5.24}$$

The horizontal merging of the first operand node N_{21} and the second operand node N_{22} from Fig.5.1 results into a temporary node N_{21*22} that is connected to the right with the third operand node N_{23} in the form of a two-node FN. The latter can be described by the block scheme in Fig.5.7 and the topological expression in Eq.(5.25).

Fig. 5.7 FN with temporary node N_{21*22} and operand node N_{23}

$$[N_{21*22}]\,(x_{21}\,|\,z_{21*22,23}) * [N_{23}]\,(z_{21*22,23}\,|\,y_{23}) \tag{5.25}$$

Further on, the horizontal merging of the temporary node N_{21*22} and the operand node N_{23} from Fig.5.7 results into a product node $N_{(21*22)*23}$ in the form of a one-node FN. The latter can be described by the block scheme in Fig.5.8 and the topological expression in Eq.(5.26).

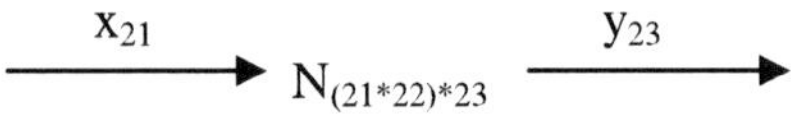

Fig. 5.8 FN with product node $N_{(21*22)*23}$

$$[N_{(21*22)*23}]\,(x_{21}\,|\,y_{23}) \tag{5.26}$$

As a result, the product node $N_{(21*22)*23}$ is described by the Boolean matrix in Eq.(5.27) and the binary relation in Eq.(5.28).

$N_{(21*22)*23}$: (5.27)

x_{21} \ y_{23}	1	2	3
1	0	0	1
2	1	0	0
3	0	1	0

$$N_{(21*22)*23}: \{(1, 3), (2, 1) (3, 2)\} \quad (5.28)$$

On the other hand, the horizontal merging of the second operand node N_{22} and the third operand node N_{23} from Fig.5.6 results into a temporary node N_{22*23} that is connected to the left with the first operand node N_{21} in the form of a two-node FN. The latter can be described by the block scheme in Fig.5.9 and the topological expression in Eq.(5.29).

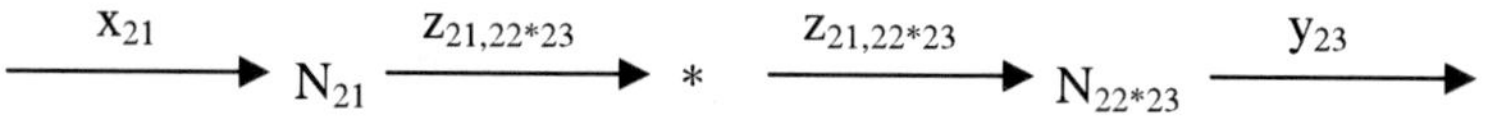

Fig. 5.9 FN with operand node N_{21} and temporary node N_{22*23}

$$[N_{21}]\ (x_{21} \mid z_{21,22*23}) * [N_{22*23}]\ (z_{21,22*23} \mid y_{23}) \quad (5.29)$$

Further on, the horizontal merging of the operand node N_{21} and the temporary node N_{22*23} from Fig.5.9 results into a product node $N_{21*(22*23)}$ in the form of a one-node FN. The latter can be described by the block scheme in Fig.5.10 and the topological expression in Eq.(5.30).

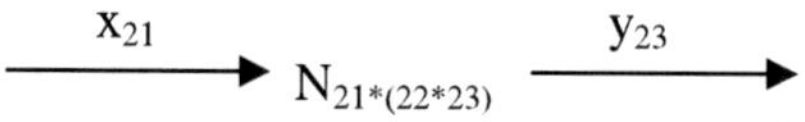

Fig. 5.10 FN with product node $N_{21*(22*23)}$

$$[N_{21*(22*23)}]\ (x_{21} \mid y_{23}) \quad (5.30)$$

In this case, the product node $N_{21*(22*23)}$ is also described by the Boolean matrix in Eq.(5.27) and the binary relation in Eq.(5.28). Therefore, $N_{21*(22*23)}$ is identical with $N_{(21*22)*23}$ and this identity is defined by Eq.(5.31) in accordance with Eq.(5.1) from Proof 5.1.

$$N_{(21*22)*23} = N_{21*(22*23)} = N_{21*22*23} \quad (5.31)$$

5.3 Variability of Horizontal Splitting

Variability is a property related to the operation of horizontal splitting when the latter is applied to a single node with the purpose of splitting it into three sequential nodes. In particular, this property allows the splitting of an operand node A/B/C into three product nodes A, B and C to take place as a sequence of two unary splitting operations that can be applied either from left to right or from right to left. The property always holds as any node can be split at least trivially into itself and two identical identity nodes. This trivial case follows from an extension of the horizontal splitting operation whereby an identity product node can be further split into two nodes that are both identical with this node. In either the trivial or the general case, the input to the first product node A is the same as the input to the operand node A/B/C and the output from the third product node C is the same as the output from the operand node. Also, two intermediate variables appear between the product nodes whereby one of them is between A and B and the other one is between B and C.

Proof 5.2

It has to be proved here that the operation of horizontal splitting is variable in accordance with Eq.(5.32). In this case, the horizontal splitting of any single operand node A/B/C from left to right should be equivalent to its horizontal splitting from right to left.

$$A, (B/C) = (A/B), C = A, B, C \tag{5.32}$$

The proof is based on the use of a binary relation as a formal model for the operand node A/B/C, as shown in Eq.(5.33). In this case, the first and the second element of each pair in A/B/C are denoted by a and c, respectively.

$$A/B/C = \{(a_1^1, c_2^1),\ldots,(a_1^1, c_2^q),\ldots,(a_1^p, c_2^1),\ldots,(a_1^p, c_2^q)\} \tag{5.33}$$

The first and the second element of any relational pair in A/B/C are denoted by the subscripts ‘1’ and ‘2’, respectively. In this case, the superscripts for the elements of the relational pairs in A/B/C don’t have to be identical and the relation A/B/C has ‘p.q’ pairs.

The horizontal decomposition of the operand relation A/B/C from left to right gives the product relation A in Eq.(5.34) and the temporary relation B/C in Eq.(5.35).

$$A = \{(a_1^1, a_2^1),\ldots,(a_1^p, a_2^p)\} \tag{5.34}$$

$$B/C = \{(a_2^1, c_2^1),\ldots,(a_2^1, c_2^q),\ldots,(a_2^p, c_2^1),\ldots,(a_2^p, c_2^q)\} \tag{5.35}$$

Further on, the horizontal decomposition of the temporary relation B/C gives the product relations B and C, as shown in Eqs.(5.36)-(5.37).

$$B = \{(a_2^1, c_1^1),\ldots,(a_2^1, c_1^q),\ldots,(a_2^p, c_1^1),\ldots,(a_2^p, c_1^q)\} \tag{5.36}$$

$$C = \{(c_1^1, c_2^1),\ldots,(c_1^q, c_2^q)\} \tag{5.37}$$

On the other hand, the horizontal decomposition of the operand relation A/B/C from right to left gives the temporary relation A/B in Eq.(5.38) and the product relation C which is identical with the product relation from Eq.(5.37).

$$A/B = \{(a_1^1, c_1^1),\ldots,(a_1^1, c_1^q),\ldots,(a_1^p, c_1^1),\ldots,(a_1^p, c_1^q)\} \tag{5.38}$$

In this case, the horizontal decomposition of the temporary relation A/B gives the product relations A and B. As the latter are also identical with the product relations from Eq.(5.34) and Eq.(5.36), this implies the validity of Eq.(5.32) and concludes the proof.

Example 5.3

This example considers a one-node FN located in the first level of a larger FN. This one-node FN has a single operand node $N_{11/12/13}$ that is described by the Boolean matrix in Eq.(5.15) and the binary relation in Eq.(5.16). The one-node FN can be described by the block-scheme in Fig.5.11 and the topological expression in Eq.(5.39).

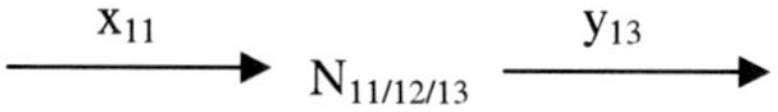

Fig. 5.11 FN with operand node $N_{11/12/13}$

$$[N_{11/12/13}]\ (x_{11} \mid y_{13}) \tag{5.39}$$

The horizontal splitting of the operand node $N_{11/12/13}$ from left to right results into a product node N_{11} that is connected to the right with a temporary node $N_{12/13}$ by means of an intermediate variable $z_{11,12/13}$. In this case, the node N_{11} can be described by the Boolean matrix in Eq.(3.13) and the binary relation Eq.(3.17) whereas the node $N_{12/13}$ can be described by the Boolean matrix in Eq.(3.15) and the binary relation in Eq.(3.19).

The overall result of the above operation is a two-node FN that can be described by the block scheme in Fig.5.12 and the topological expression in Eq.(5.40).

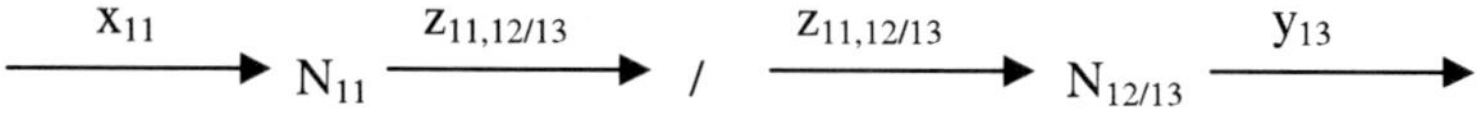

Fig. 5.12 FN with product node N_{11} and temporary node $N_{12/13}$

$$[N_{11}]\ (x_{11} \mid z_{11,12/13})\ /\ [N_{12/13}]\ (z_{11,12/13} \mid y_{13}) \tag{5.40}$$

Further on, the horizontal splitting of the temporary node $N_{12/13}$ results into two product nodes N_{12} and N_{13} that are connected with each other by an intermediate variable $z_{12,13}$ whereby the other intermediate variable $z_{11,12/13}$ is renamed as $z_{11,12}$ for consistency. In this case, the node N_{12} can be described by the Boolean matrix in Eq.(3.14) and the binary relation Eq.(3.18) whereas the node N_{13} can be described by the Boolean matrix in Eq.(5.8) and the binary relation in Eq.(5.9).

The overall result of the above two operations is a three-node FN that can be described by the block scheme in Fig.5.13 and the topological expression in Eq.(5.41).

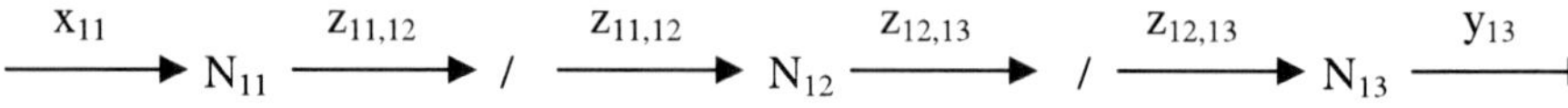

Fig. 5.13 FN with product nodes N_{11}, N_{12} and N_{13}

$$[N_{11}]\ (x_{11} \mid z_{11,12}) \ /\ [N_{12}]\ (z_{11,12} \mid z_{12,13}) \ /\ [N_{13}]\ (z_{12,13} \mid y_{13}) \tag{5.41}$$

On the other hand, the horizontal splitting of the operand node $N_{11/12/13}$ from right to left results into a product node N_{13} that is connected to the left with a temporary node $N_{11/12}$ by means of an intermediate variable $z_{11/12,13}$. In this case, the node N_{13} can be described by the Boolean matrix in Eq.(5.8) and the binary relation Eq.(5.9) whereas the node $N_{11/12}$ can be described by the Boolean matrix in Eq.(4.3) and the binary relation in Eq.(4.4).

The overall result of the above operation is a two-node FN that can be described by the block scheme in Fig.5.14 and the topological expression in Eq.(5.42).

Fig. 5.14 FN with temporary node N_{11*12} and operand node N_{13}

$$[N_{11/12}]\ (x_{11} \mid z_{12/13,13}) \ /\ [N_{13}]\ (z_{12/13,13} \mid y_{13}) \tag{5.42}$$

Further on, the horizontal splitting of the temporary node $N_{11/12}$ results into two product nodes N_{11} and N_{12} that are connected with each other by an intermediate variable $z_{11,12}$ whereby the other intermediate variable $z_{11,12/13}$ is renamed as $z_{12/13}$ for consistency. In this case, the node N_{11} can be described by the Boolean matrix in Eq.(3.13) and the binary relation in Eq.(3.17) whereas the node N_{12} can be described by the Boolean matrix in Eq.(3.14) and the binary relation in Eq.(3.18).

The overall result of the above two operations is a three-node FN that can be described by the block scheme in Fig.5.13 and the topological expression in Eq.(5.41).

Therefore, the horizontal splitting of the operand node $N_{11/12/13}$ results into an identical set of product nodes $\{N_{11}, N_{12}, N_{13}\}$ for both cases of left-to-right and right-to-left splitting. This identity is defined by Eq.(5.43) in accordance with Eq.(5.32) from Proof 5.2.

$$N_{11}, N_{12/13} = N_{11/12}, N_{13} = N_{11}, N_{12}, N_{13} \tag{5.43}$$

Example 5.4

This example considers a one-node FN located in the second level of a larger FN. This one-node FN has a single operand node $N_{21/22/23}$ that is described by the Boolean matrix in Eq.(5.27) and the binary relation in Eq.(5.28). The one-node FN can be describe``d by the block-scheme in Fig.5.15 and the topological expression in Eq.(5.44).

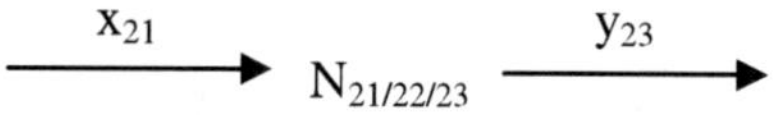

Fig. 5.15 FN with operand node $N_{21/22/23}$

$$[N_{21/22/23}]\ (x_{21} \mid y_{23}) \tag{5.44}$$

The horizontal splitting of the operand node $N_{21/22/23}$ from left to right results into a product node N_{21} that is connected to the right with a temporary node $N_{22/23}$ by means of an intermediate variable $z_{21,22/23}$. In this case, the node N_{21} can be described by the Boolean matrix in Eq.(3.15) and the binary relation Eq.(3.19) whereas the node $N_{22/23}$ can be described by the Boolean matrix in Eq.(3.13) and the binary relation in Eq.(3.17).

The overall result of the above operation is a two-node FN that can be described by the block scheme in Fig.5.16 and the topological expression in Eq.(5.45).

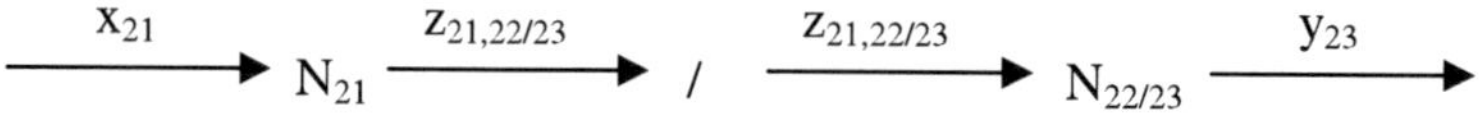

Fig. 5.16 FN with product node N_{21} and temporary node $N_{22/23}$

$$[N_{21}]\ (x_{21} \mid z_{21,22/23})\ /\ [N_{22/23}]\ (z_{21,22/23} \mid y_{23}) \tag{5.45}$$

Further on, the horizontal splitting of the temporary node $N_{22/23}$ results into two product nodes N_{22} and N_{23} that are connected with each other by an intermediate variable $z_{22,23}$ whereby the other intermediate variable $z_{21,22/23}$ is renamed as $z_{21,22}$ for consistency. In this case, the node N_{22} can be described by the Boolean matrix in Eq.(3.16) and the binary relation Eq.(3.20) whereas the node N_{13} can be described by the Boolean matrix in Eq.(5.20) and the binary relation in Eq.(5.21).

The overall result of the above two operations is a three-node FN that can be described by the block scheme in Fig.5.17 and the topological expression in Eq.(5.46).

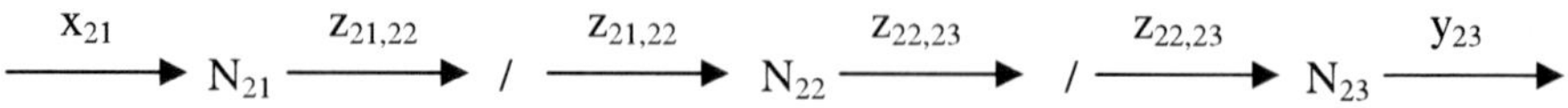

Fig. 5.17 FN with product nodes N_{21}, N_{22} and N_{23}

$$[N_{21}]\ (x_{21} \mid z_{21,22})\ /\ [N_{22}]\ (z_{21,22} \mid z_{22,23})\ /\ [N_{23}]\ (z_{22,23} \mid y_{23}) \tag{5.46}$$

On the other hand, the horizontal splitting of the operand node $N_{21/22/23}$ from right to left results into a product node N_{23} that is connected to the left with a temporary node $N_{21/22}$ by means of an intermediate variable $z_{21/22,23}$. In this case, the node N_{23} can be described by the Boolean matrix in Eq.(5.20) and the binary relation Eq.(5.21) whereas the node $N_{21/22}$ can be described by the Boolean matrix in Eq.(4.7) and the binary relation in Eq.(4.8).

The overall result of the above operation is a two-node FN that can be described by the block scheme in Fig.5.18 and the topological expression in Eq.(5.47).

Fig. 5.18 FN with temporary node $N_{21/22}$ and operand node N_{23}

$$[N_{21/22}]\ (x_{21} \mid z_{21/22,23})\ /\ [N_{23}]\ (z_{21,22/23} \mid y_{23}) \tag{5.47}$$

Further on, the horizontal splitting of the temporary node $N_{21/22}$ results into two product nodes N_{21} and N_{22} that are connected with each other by an intermediate variable $z_{21,22}$ whereby the other intermediate variable $z_{21,22/23}$ is renamed as $z_{22,23}$ for consistency. In this case, the node N_{21} can be described by the Boolean matrix in Eq.(3.15) and the binary relation in Eq.(3.19) whereas the node N_{22} can be described by the Boolean matrix in Eq.(3.16) and the binary relation in Eq.(3.20).

The overall result of the above two operations is a three-node FN that can be described by the block scheme in Fig.5.17 and the topological expression in Eq.(5.46).

Therefore, the horizontal splitting of the operand node $N_{21/22/23}$ results into an identical set of product nodes $\{N_{21}, N_{22}, N_{23}\}$ for both cases of left-to-right and right-to-left splitting. This identity is defined by Eq.(5.48) in accordance with Eq.(5.32) from Proof 5.2.

$$N_{21}, N_{22/23} = N_{21/22}, N_{23} = N_{21}, N_{22}, N_{23} \tag{5.48}$$

5.4 Associativity of Vertical Merging

Associativity is a property related to the operation of vertical merging when the latter is applied to three parallel nodes for the purpose of merging them into a single node. In particular, this property allows the merging of three operand nodes A, B and C into a product node A+B+C to take place as a sequence of two binary merging operations that can be applied either from top to bottom or from bottom to top. The property can be applied when the inputs to and the outputs from each of the three nodes A, B and C are self-standing. In this case, the input set to the product node A+B+C is the union of the inputs to the operand nodes A, B and C whereas the output set from the product node is the union of the outputs from the operand nodes.

Proof 5.3

It has to be proved here that the operation of vertical merging is associative in accordance with Eq.(5.49). In this case, the horizontal merging of any three operand nodes A, B and C from top to bottom should be equivalent to their vertical merging from bottom to top.

$$(A+B)+C = A+(B+C) = A+B+C \tag{5.49}$$

The proof is based on the use of binary relations as formal models for the operand nodes A, B and C, as shown in Eqs. (5.50)-(5.52). In this case, the elements of the relational pairs are denoted by the letter a in A, the letter b in B and the letter c in C.

$$A = \{(a_1^1, a_2^1),\ldots,(a_1^p, a_2^p)\} \tag{5.50}$$

$$B = \{(b_1^1, b_2^1),\ldots,(b_1^q, b_2^q)\} \tag{5.51}$$

$$C = \{(c_1^1, c_2^1),\ldots,(c_1^r, c_2^r)\} \tag{5.52}$$

The first and the second element of any relational pair in A, B and C are denoted by the subscripts '1' and '2', respectively. However, the superscripts for the first and the second element of any relational pair in A, B and C are identical as they indicate the corresponding number for each pair. In particular, the relation A has 'p' pairs, the relation B has 'q' pairs and the relation C has 'r' pairs.

The vertical composition of the operand relations A and B gives the temporary relation A+B, as shown in Eq.(5.53).

$$A+B = \{(a_1^1 b_1^1, a_2^1 b_2^1),\ldots,(a_1^1 b_1^q, a_2^1 b_2^q),\ldots, \\ (a_1^p b_1^1, a_2^p b_2^1),\ldots,(a_1^p b_1^q, a_2^p b_2^q)\} \tag{5.53}$$

Further on, the vertical composition of the temporary relation A+B and the operand relation C gives the product relation (A+B)+C, as shown in Eq.(5.54).

$$(A+B)+C = \{(a_1^1 b_1^1 c_1^1, a_2^1 b_2^1 c_2^1),\ldots,(a_1^1 b_1^1 c_1^r, a_2^1 b_2^1 c_2^r),\ldots, \\ (a_1^1 b_1^q c_1^1, a_2^1 b_2^q c_2^1),\ldots,(a_1^1 b_1^q c_1^r, a_2^1 b_2^q c_2^r),\ldots, \tag{5.54}$$

$$(a_1^p b_1^1 c_1^1,\ a_2^p b_2^1 c_2^1),\ldots,(a_1^p b_1^1 c_1^r,\ a_2^p b_2^1 c_2^r),\ldots,$$

$$(a_1^p b_1^q c_1^1,\ a_2^p b_2^q c_2^1),\ldots,(a_1^p b_1^q c_1^r,\ a_2^p b_2^q c_2^r)\}$$

On the other hand, the vertical composition of the operand relations B and C gives the temporary relation B+C, as shown in Eq.(5.55).

$$B+C = \{(b_1^1 c_1^1,\ b_2^1 c_2^1),\ldots,(b_1^1 c_1^r,\ b_2^1 c_2^r),\ldots, \quad (5.55)$$

$$(b_1^q c_1^1,\ b_2^q c_2^1),\ldots,(b_1^q c_1^r,\ b_2^q c_2^r)\}$$

In this case, the vertical composition of the operand relation A and the temporary relation B+C gives the product relation A+(B+C). As the latter is identical with the product relation (A+B)+C from Eq.(5.54), this implies the validity of Eq.(5.49) and concludes the proof.

Example 5.5

This example considers a FN with three parallel operand nodes. The first and the second node N_{11} and N_{21} are taken from the four-node FN in Fig.3.1. These two nodes are described there by the Boolean matrices and the binary relations in Eqs.(3.13), (3.15), (3.17) and (3.19) whereas the third node N_{31} is described by the Boolean matrix in Eq.(5.56) and the binary relation in Eq.(5.57).

N_{31} : (5.56)

x_{31} \ y_{31}	1	2	3
1	0	0	1
2	0	1	0
3	1	0	0

$$N_{31}:\ \{(1, 3), (2, 2)\ (3, 1)\} \quad (5.57)$$

The nodes N_{11}, N_{21} and N_{31} represent a three-node FN that is an extended sub-network of the four-node FN from Fig.3.1. This three-node FN can be described by the block-scheme in Fig.5.19 and the topological expression in Eq.(5.58).

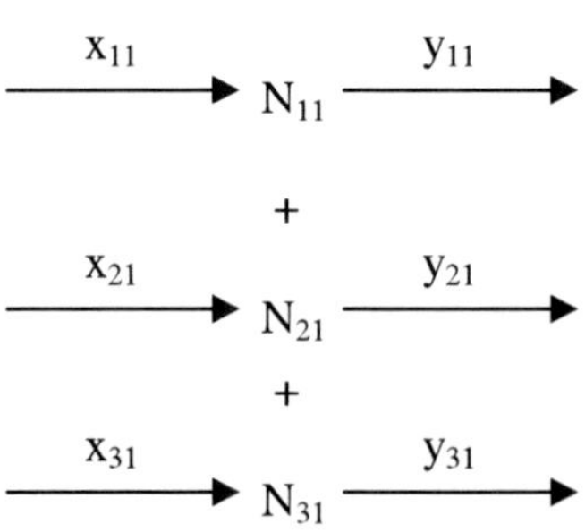

Fig. 5.19 FN with operand nodes N_{11}, N_{21} and N_{31}

$[N_{11}]\ (x_{11} \mid y_{11}) + [N_{21}]\ (x_{21} \mid y_{21}) + [N_{31}]\ (x_{31} \mid y_{31})$ (5.58)

The vertical merging of the first operand node N_{11} and the second operand node N_{21} from Fig.5.19 results into a temporary node N_{11+21} that can be described by the Boolean matrix in Eq.(4.25) and the binary relation in Eq.(4.26). This temporary node is connected at the bottom with the third operand node N_{31} in the form of a two-node FN. The latter can be described by the block scheme in Fig.5.20 and the topological expression in Eq.(5.59).

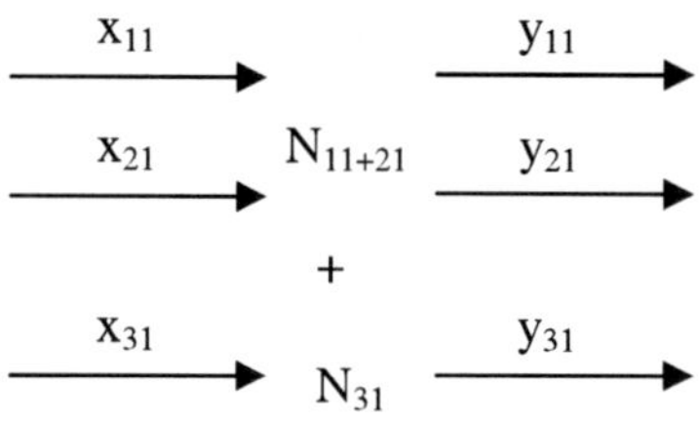

Fig. 5.20 FN with temporary node N_{11+21} and operand node N_{31}

$[N_{11+21}]\ (x_{11}, x_{21} \mid y_{11}, y_{21}) + [N_{31}]\ (x_{31} \mid y_{31})$ (5.59)

Further on, the vertical merging of the temporary node N_{11+21} and the operand node N_{31} from Fig.5.20 results into a product node $N_{(11+21)+31}$ in the form of a one-node FN. The latter can be described by the block scheme in Fig.5.21 and the topological expression in Eq.(5.60).

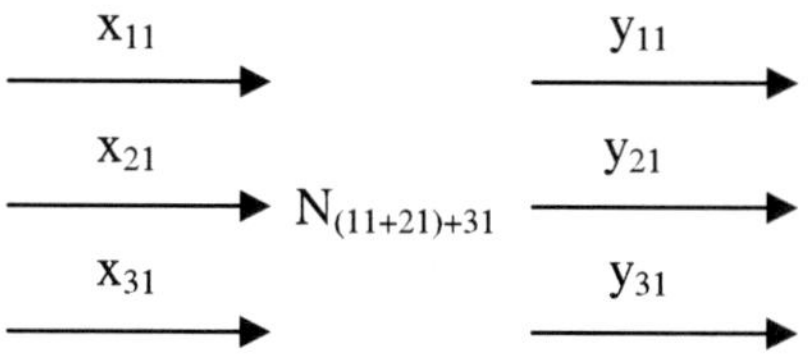

Fig. 5.21 FN with product node $N_{(11+21)+31}$

$[N_{(11+21)+31}]\ (x_{11}, x_{21}, x_{31} \mid y_{11}, y_{21}, y_{31})$ (5.60)

As a result, the product node $N_{(11+21)+31}$ is described by the Boolean matrix in Eq.(5.61) and the binary relation in Eq.(5.62). In this case, the labels and the elements of the Boolean matrix are represented by a compact notation. In particular, each of the capital letters A, B, C, D, E, F, G, H, I stands for three sequential rows and columns as indicated in brackets, 1_3 denotes the square Boolean matrix from Eq.(5.56) and 0_3 denotes a zero Boolean matrix of dimension 3×3.

$N_{(11+21)+31}$: (5.61)

x_{11}, x_{21}, x_{31} \ y_{11}, y_{21}, y_{31}	A	B	C	D	E	F	G	H	I
A (111-113)	0_3	1_3	0_3	0_3	0_3	0_3	0_3	0_3	0_3
B (121-123)	1_3	0_3	0_3	0_3	0_3	0_3	0_3	0_3	0_3
C (131-133)	0_3	0_3	1_3	0_3	0_3	0_3	0_3	0_3	0_3
D (211-213)	0_3	0_3	0_3	0_3	0_3	0_3	0_3	1_3	0_3
E (221-223)	0_3	0_3	0_3	0_3	0_3	0_3	1_3	0_3	0_3
F (231-133)	0_3	0_3	0_3	0_3	0_3	0_3	0_3	0_3	1_3
G (311-313)	0_3	0_3	0_3	0_3	1_3	0_3	0_3	0_3	0_3
H (321-323)	0_3	0_3	0_3	1_3	0_3	0_3	0_3	0_3	0_3
I (331-333)	0_3	0_3	0_3	0_3	0_3	1_3	0_3	0_3	0_3

$N_{(11+21)+31}$: {(111, 123), (112, 122), (113, 121),
(121, 113), (122, 112), (123, 111),
(131, 133), (132, 132), (133, 131),
(211, 323), (212, 322), (213, 321),
(221, 313), (222, 312), (223, 311),
(231, 333), (232, 332), (233, 331),
(311, 223), (312, 222), (313, 221),
(321, 213), (322, 212), (323, 211),
(331, 233), (332, 232), (333, 231)} (5.62)

On the other hand, the vertical merging of the second operand node N_{21} and the third operand node N_{31} from Fig.5.19 results into a temporary node N_{21+31} that can be described by the Boolean matrix in Eq.(5.63) and the binary relation in Eq.(5.64). This temporary node is connected at the top with the first operand node N_{11} in the form of a two-node FN. The latter can be described by the block scheme in Fig.5.22 and the topological expression in Eq.(5.65).

N_{21+31}: (5.63)

x_{21}, x_{31} \ y_{21}, y_{31}	11	12	13	21	22	23	31	32	33
11	0	0	0	0	0	1	0	0	0
12	0	0	0	0	1	0	0	0	0
13	0	0	0	1	0	0	0	0	0
21	0	0	1	0	0	0	0	0	0
22	0	1	0	0	0	0	0	0	0
23	1	0	0	0	0	0	0	0	0
31	0	0	0	0	0	0	0	0	1
32	0	0	0	0	0	0	0	1	0
33	0	0	0	0	0	0	1	0	0

N_{21+31}: {(11, 23), (12, 22), (13, 21),
(21, 13), (22, 12), (23, 11),
(31, 33), (32, 32), (33, 31)} (5.64)

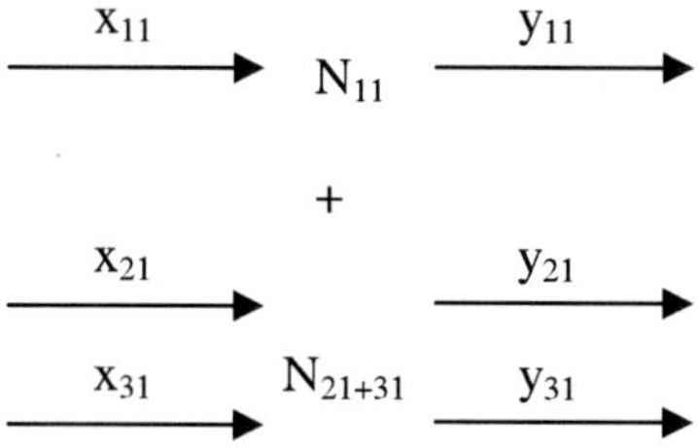

Fig. 5.22 FN with operand node N_{11} and temporary node N_{21+31}

$$[N_{11}]\ (x_{11} \mid y_{11})\ +\ [N_{21+31}]\ (x_{21}, x_{31} \mid y_{21}, y_{31}) \tag{5.65}$$

Further on, the vertical merging of the operand node N_{11} and the temporary node N_{21+31} from Fig.5.22 results into a product node $N_{11+(21+31)}$ in the form of a one-node FN. The latter can be described by the block scheme in Fig.5.23 and the topological expression in Eq.(5.66).

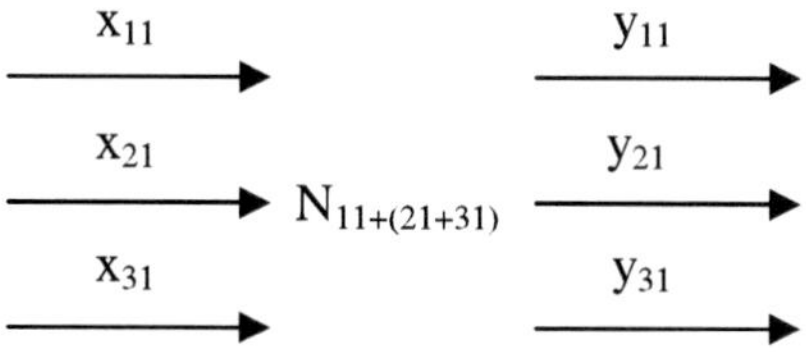

Fig. 5.23 FN with product node $N_{11+(21+31)}$

$$[N_{11+(21+31)}]\ (x_{11}, x_{21}, x_{31} \mid y_{11}, y_{21}, y_{31}) \tag{5.66}$$

In this case, the product node $N_{11+(21+31)}$ is also described by the Boolean matrix in Eq.(5.61) and the binary relation in Eq.(5.62). Therefore, $N_{11+(21+31)}$ is identical with $N_{(11+21)+31}$ and this identity is defined by Eq.(5.67) in accordance with Eq.(5.49) from Proof 5.3.

$$N_{(11+21)+31} = N_{11+(21+31)} = N_{11+21+31} \tag{5.67}$$

Example 5.6

This example considers a FN with three parallel operand nodes. The first and the second node N_{12} and N_{22} are taken from the four-node FN in Fig.3.1. These two nodes are described there by the Boolean matrices and the binary relations in Eqs.(3.14), (3.16), (3.18) and (3.20) whereas the third node N_{32} is described by the Boolean matrix in Eq.(5.68) and the binary relation in Eq.(5.69).

N_{32}: (5.68)

x_{32} \ y_{32}	1	2	3
1	0	0	1
2	0	1	0
3	1	0	0

$$N_{32}: \ \{(1, 3), (2, 2)\ (3, 1)\} \tag{5.69}$$

The nodes N_{12}, N_{22} and N_{32} represent a three-node FN that is an extended sub-network of the four-node FN from Fig.3.1. This three-node FN can be described by the block-scheme in Fig.5.24 and the topological expression in Eq.(5.70).

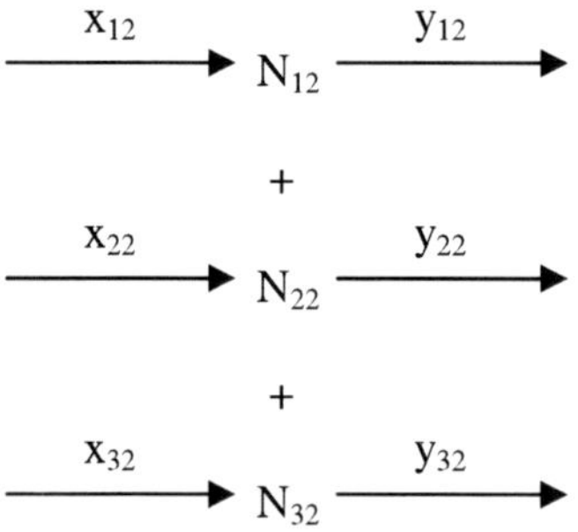

Fig. 5.24 FN with operand nodes N_{12}, N_{22} and N_{32}

$$[N_{12}]\ (x_{12} \mid y_{12}) + [N_{22}]\ (x_{22} \mid y_{22}) + [N_{32}]\ (x_{32} \mid y_{32}) \tag{5.70}$$

The vertical merging of the first operand node N_{12} and the second operand node N_{22} from Fig.5.24 results into a temporary node N_{12+22} that can be described by the Boolean matrix in Eq.(4.29) and the binary relation in Eq.(4.30). This temporary node is connected at the bottom with the third operand node N_{32} in the form of a two-node FN. The latter can be described by the block scheme in Fig.5.25 and the topological expression in Eq.(5.71).

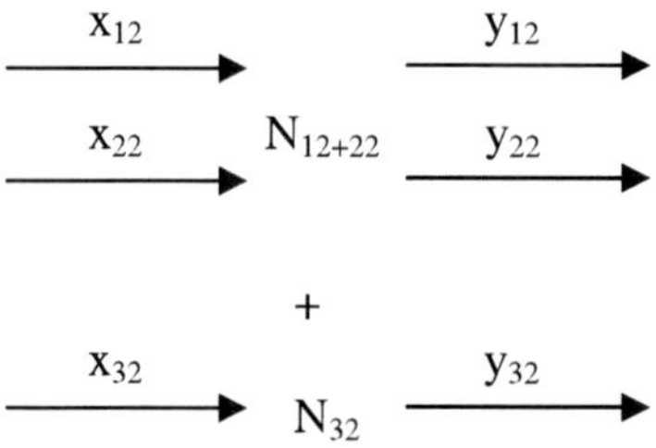

Fig. 5.25 FN with temporary node N_{12+22} and operand node N_{32}

$[N_{12+22}]\ (x_{12}, x_{22} \mid y_{12}, y_{22}) + [N_{32}]\ (x_{32} \mid y_{32})$ (5.71)

Further on, the vertical merging of the temporary node N_{12+22} and the operand node N_{32} from Fig.5.25 results into a product node $N_{(12+22)+32}$ in the form of a one-node FN. The latter can be described by the block scheme in Fig.5.26 and the topological expression in Eq.(5.72).

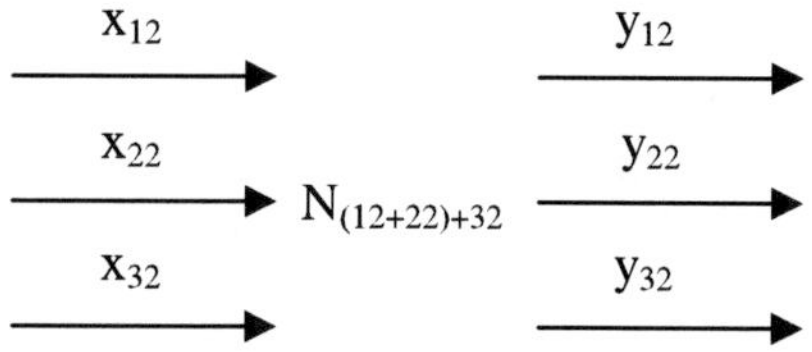

Fig. 5.26 FN with product node $N_{(12+22)+32}$

$[N_{(12+22)+32}]\ (x_{12}, x_{22}, x_{32} \mid y_{12}, y_{22}, y_{32})$ (5.72)

As a result, the product node $N_{(12+22)+32}$ is described by the Boolean matrix in Eq.(5.73) and the binary relation in Eq.(5.74). In this case, the labels and the elements of the Boolean matrix are represented by a compact notation. In particular, each of the capital letters A, B, C, D, E, F, G, H, I stands for three sequential rows and columns as indicated in brackets, 1_3 denotes the square Boolean matrix from Eq.(5.68) and 0_3 denotes a zero Boolean matrix of dimension 3X3.

$N_{(12+22)+32}$: (5.73)

y_{12}, y_{22}, y_{32} / x_{12}, x_{22}, x_{32}	A	B	C	D	E	F	G	H	I
A (111-113)	0_3	0_3	0_3	0_3	0_3	1_3	0_3	0_3	0_3
B (121-123)	0_3	0_3	0_3	1_3	0_3	0_3	0_3	0_3	0_3
C (131-133)	0_3	0_3	0_3	0_3	1_3	0_3	0_3	0_3	0_3
D (211-213)	0_3	0_3	0_3	0_3	0_3	0_3	0_3	0_3	1_3
E (221-223)	0_3	0_3	0_3	0_3	0_3	0_3	1_3	0_3	0_3
F (231-133)	0_3	0_3	0_3	0_3	0_3	0_3	0_3	1_3	0_3
G (311-313)	0_3	0_3	1_3	0_3	0_3	0_3	0_3	0_3	0_3
H (321-323)	1_3	0_3	0_3	0_3	0_3	0_3	0_3	0_3	0_3
I (331-333)	0_3	1_3	0_3	0_3	0_3	0_3	0_3	0_3	0_3

$N_{(12+22)+32}$: {(111, 233), (112, 232), (113, 231), (5.74)
(121, 213), (122, 212), (123, 211),
(131, 223), (132, 222), (133, 221),
(211, 333), (212, 332), (213, 331),
(221, 313), (222, 312), (223, 311),
(231, 323), (232, 322), (233, 321),
(311, 133), (312, 132), (313, 131),
(321, 113), (322, 112), (323, 111),
(331, 123), (332, 122), (333, 121)}

On the other hand, the vertical merging of the second operand node N_{22} and the third operand node N_{32} from Fig.5.24 results into a temporary node N_{22+32} that can be described by the Boolean matrix in Eq.(5.75) and the binary relation in Eq.(5.76). This temporary node is connected at the top with the first operand node N_{12} in the form of a two-node FN. The latter can be described by the block scheme in Fig.5.27 and the topological expression in Eq.(5.77).

N_{22+32} : (5.75)

x_{22}, x_{32} \ y_{22}, y_{32}	11	12	13	21	22	23	31	32	33
11	0	0	0	0	0	0	0	0	1
12	0	0	0	0	0	0	0	1	0
13	0	0	0	0	0	0	1	0	0
21	0	0	1	0	0	0	0	0	0
22	0	1	0	0	0	0	0	0	0
23	1	0	0	0	0	0	0	0	0
31	0	0	0	0	0	1	0	0	0
32	0	0	0	0	1	0	0	0	0
33	0	0	0	1	0	0	0	0	0

$$N_{22+32}: \{(11, 33), (12, 32), (13, 31), (21, 13), (22, 12), (23, 11), (31, 23), (32, 22), (33, 21)\} \quad (5.76)$$

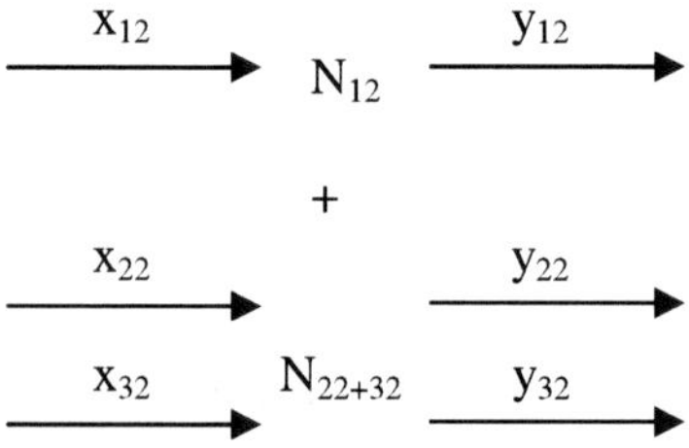

Fig. 5.27 FN with operand node N_{12} and temporary node N_{22+32}

$$[N_{12}]\,(x_{12} \mid y_{12}) + [N_{22+32}]\,(x_{22}, x_{32} \mid y_{22}, y_{32}) \quad (5.77)$$

Further on, the vertical merging of the operand node N_{12} and the temporary node N_{22+32} from Fig.5.27 results into a product node $N_{12+(22+32)}$ in the form of a one-node FN. The latter can be described by the block scheme in Fig.5.28 and the topological expression in Eq.(5.78).

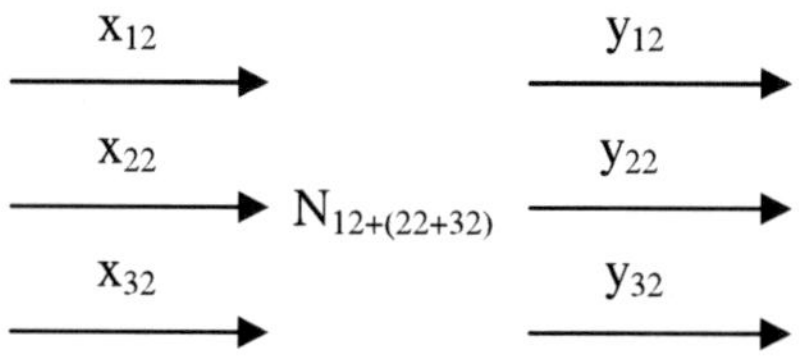

Fig. 5.28 FN with product node $N_{12+(22+32)}$

$$[N_{12+(22+32)}]\ (x_{12}, x_{22}, x_{32} \mid y_{12}, y_{22}, y_{32}) \tag{5.78}$$

In this case, the product node $N_{12+(22+32)}$ is also described by the Boolean matrix in Eq.(5.73) and the binary relation in Eq.(5.74). Therefore, $N_{12+(22+32)}$ is identical with $N_{(12+22)+32}$ and this identity is defined by Eq.(5.79) in accordance with Eq.(5.49) from Proof 5.3.

$$N_{(12+22)+32} = N_{12+(22+32)} = N_{12+22+32} \tag{5.79}$$

5.5 Variability of Vertical Splitting

Variability is a property related to the operation of vertical splitting when the latter is applied to a single node with the purpose of splitting it into three nodes. In particular, this property allows the splitting of an operand node A-B-C into three product nodes A, B and C to take place as a sequence of two unary splitting operations that can be applied either from top to bottom or from bottom to top. The property always holds when the underlying operation of vertical splitting can be applied in the manner specified above. In this case, the union of the inputs to the three product nodes A, B and C is the same as the input set to the operand node A-B-C whereas the union of the outputs from the three product nodes is the same as the output set from the operand node.

Proof 5.4

It has to be proved here that the operation of vertical splitting is variable in accordance with Eq.(5.80). In this case, the vertical splitting of any single operand node A-B-C from top to bottom should be equivalent to its vertical splitting from bottom to top.

$$A, (B\text{-}C) = (A\text{-}B), C = A, B, C \tag{5.80}$$

The proof is based on the use of a binary relation as a formal model for the operand node A-B-C, as shown in Eq.(5.81). In this case, the first and the second element of each pair in A-B-C are denoted by the triplet 'abc'.

$$
\begin{aligned}
A\text{-}B\text{-}C = \{&(a_1^{\,1} b_1^{\,1} c_1^{\,1}, a_2^{\,1} b_2^{\,1} c_2^{\,1}),\ldots,(a_1^{\,1} b_1^{\,1} c_1^{\,r}, a_2^{\,1} b_2^{\,1} c_2^{\,r}),\ldots, \\
&(a_1^{\,1} b_1^{\,q} c_1^{\,1}, a_2^{\,1} b_2^{\,q} c_2^{\,1}),\ldots,(a_1^{\,1} b_1^{\,q} c_1^{\,r}, a_2^{\,1} b_2^{\,q} c_2^{\,r}),\ldots, \\
&(a_1^{\,p} b_1^{\,1} c_1^{\,1}, a_2^{\,p} b_2^{\,1} c_2^{\,1}),\ldots,(a_1^{\,p} b_1^{\,1} c_1^{\,r}, a_2^{\,p} b_2^{\,q} c_2^{\,r}),\ldots, \\
&(a_1^{\,p} b_1^{\,q} c_1^{\,1}, a_2^{\,p} b_2^{\,q} c_2^{\,1}),\ldots,(a_1^{\,p} b_1^{\,q} c_1^{\,r}, a_2^{\,p} b_2^{\,1} c_2^{\,r})
\end{aligned}
\tag{5.81}
$$

The three individual elements in the first and the second triplet of any relational pair in A-B-C are denoted by the subscripts '1' and '2', respectively. In this case, the superscripts for these individual elements denote their target pairs in the product relations A, B and C whereby the operand relation A-B-C has 'p.q.r' pairs.

The vertical decomposition of the operand relation A-B-C from top to bottom gives the product relation A in Eq.(5.82) and the temporary relation B-C in Eq.(5.83).

$$A = \{(a_1^{\,1}, a_2^{\,1}),\ldots,(a_1^{\,p}, a_2^{\,p})\} \tag{5.82}$$

$$
\begin{aligned}
B\text{-}C = \{&(b_1^{\,1} c_1^{\,1}, b_2^{\,1} c_2^{\,1}),\ldots,(b_1^{\,1} c_1^{\,r}, b_2^{\,1} c_2^{\,r}),\ldots, \\
&(b_1^{\,q} c_1^{\,1}, b_2^{\,q} c_2^{\,1}),\ldots,(b_1^{\,q} c_1^{\,r}, b_2^{\,q} c_2^{\,r})\}
\end{aligned}
\tag{5.83}
$$

Further on, the vertical decomposition of the temporary relation B-C gives the product relations B and C, as shown in Eqs.(5.84)-(5.85).

$$B = \{(b_1^{\,1}, b_2^{\,1}),\ldots,(b_1^{\,q}, b_2^{\,q})\} \tag{5.84}$$

$$C = \{(c_1^{\,1}, c_2^{\,1}),\ldots,(c_1^{\,r}, c_2^{\,r})\} \tag{5.85}$$

On the other hand, the vertical decomposition of the operand relation A-B-C from bottom to top gives the temporary relation A-B in Eq.(5.86) and the product relation C which is identical with the product relation from Eq.(5.85).

$$
\begin{aligned}
A\text{-}B = \{&(a_1^{\,1} b_1^{\,1}, a_2^{\,1} b_2^{\,1}),\ldots,(a_1^{\,1} b_1^{\,q}, a_2^{\,1} b_2^{\,q}),\ldots, \\
&(a_1^{\,p} b_1^{\,1}, a_2^{\,p} b_2^{\,1}),\ldots,(a_1^{\,p} b_1^{\,q}, a_2^{\,p} b_2^{\,q})\}
\end{aligned}
\tag{5.86}
$$

In this case, the vertical decomposition of the temporary relation A-B gives the product relations A and B. As the latter are also identical with the product relations from Eq.(5.82) and Eq.(5.84), this implies the validity of Eq.(5.80) and concludes the proof.

Example 5.7

This example considers a one-node FN located in the first layer of a larger FN. This one-node FN has a single operand node $N_{11\text{-}21\text{-}31}$ that is described by the Boolean matrix in Eq.(5.61) and the binary relation in Eq.(5.62). The one-node FN can be described by the block-scheme in Fig.5.29 and the topological expression in Eq.(5.87).

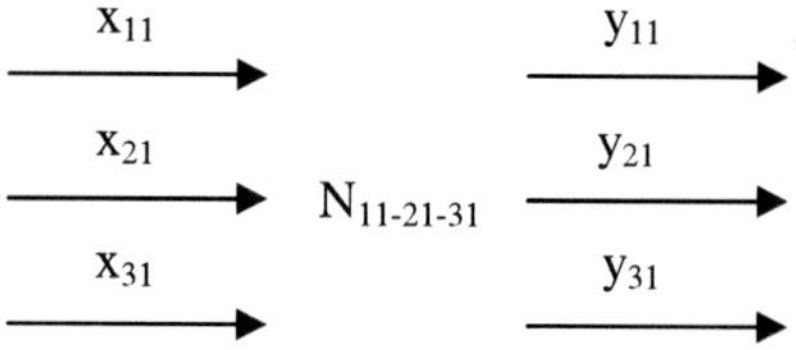

Fig. 5.29 FN with operand node $N_{11\text{-}21\text{-}31}$

$$[N_{11\text{-}21\text{-}31}]\ (x_{11}, x_{21}, x_{31} \mid y_{11}, y_{21}, y_{31}) \tag{5.87}$$

The vertical splitting of the operand node $N_{11\text{-}21\text{-}31}$ from top to bottom results into a product node N_{11} that is connected at the bottom with a temporary node $N_{21\text{-}31}$. In this case, the node N_{11} can be described by the Boolean matrix in Eq.(3.13) and the binary relation Eq.(3.17) whereas the node $N_{21\text{-}31}$ can be described by the Boolean matrix in Eq.(5.63) and the binary relation in Eq.(5.64).

The overall result of the above operation is a two-node FN that can be described by the block scheme in Fig.5.30 and the topological expression in Eq.(5.88).

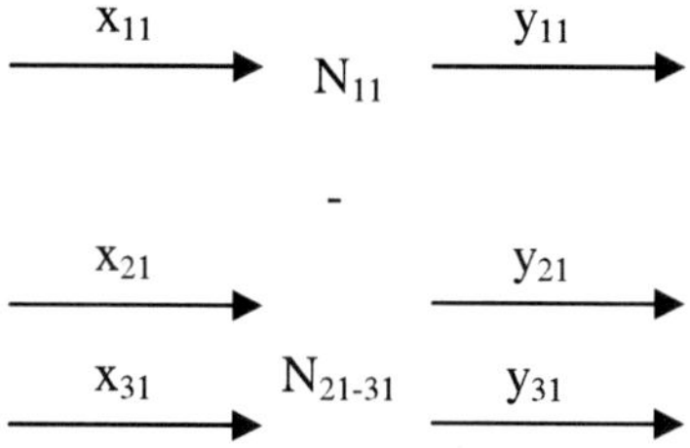

Fig. 5.30 FN with product node N_{11} and temporary node $N_{21\text{-}31}$

$$[N_{11}]\ (x_{11} \mid y_{11}) - [N_{21\text{-}31}]\ (x_{21}, x_{31} \mid y_{21}, y_{31}) \tag{5.88}$$

Further on, the vertical splitting of the temporary node $N_{21\text{-}31}$ results into two product nodes N_{21} and N_{31}. In this case, the node N_{21} can be described by the Boolean matrix in Eq.(3.15) and the binary relation Eq.(3.19) whereas the node N_{31} can be described by the Boolean matrix in Eq.(5.56) and the binary relation in Eq.(5.57).

The overall result of the above two operations is a three-node FN that can be described by the block scheme in Fig.5.31 and the topological expression in Eq.(5.89).

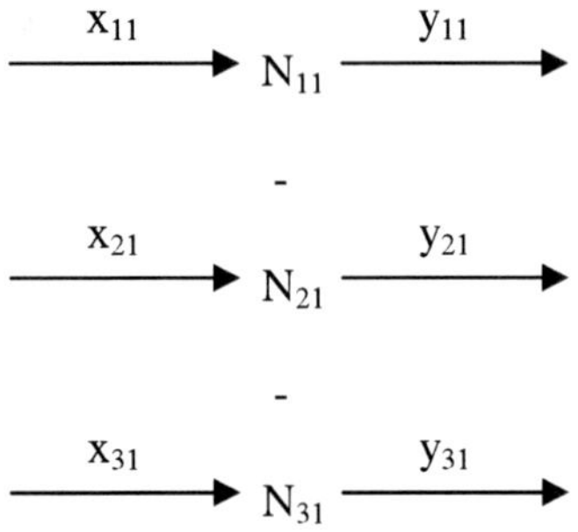

Fig. 5.31 FN with product nodes N_{11}, N_{21} and N_{31}

$$[N_{11}]\ (x_{11} \mid y_{11}) - [N_{21}]\ (x_{21} \mid y_{21}) - [N_{31}]\ (x_{31} \mid y_{31}) \tag{5.89}$$

On the other hand, the vertical splitting of the operand node $N_{11\text{-}21\text{-}31}$ from bottom to top results into a product node N_{31} that is connected at the top with a temporary node $N_{11\text{-}21}$. In this case, the node N_{31} can be described by the Boolean matrix in Eq.(5.56) and the binary relation in Eq.(5.57) whereas the node $N_{11\text{-}21}$ can be described by the Boolean matrix in Eq.(4.25) and the binary relation in Eq.(4.26).

The overall result of the above operation is a two-node FN that can be described by the block scheme in Fig.5.32 and the topological expression in Eq.(5.90).

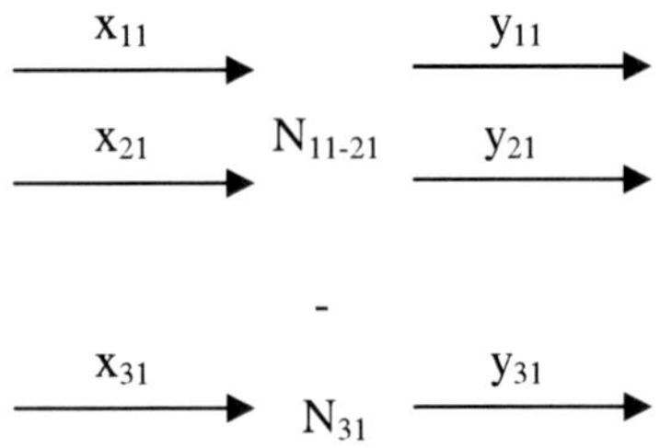

Fig. 5.32 FN with temporary node $N_{11\text{-}21}$ and product node N_{31}

$$[N_{11\text{-}21}]\ (x_{11}, x_{21} \mid y_{11}, y_{21}) - [N_{31}]\ (x_{31} \mid y_{31}) \tag{5.90}$$

Further on, the verical splitting of the temporary node $N_{11\text{-}21}$ results into two product nodes N_{11} and N_{21}. In this case, the node N_{11} can be described by the Boolean matrix in Eq.(3.13) and the binary relation in Eq.(3.17) whereas the node N_{21} can be described by the Boolean matrix in Eq.(3.15) and the binary relation in Eq.(3.19).

The overall result of the above two operations is a three-node FN that can be described by the block scheme in Fig.5.31 and the topological expression in Eq.(5.89).

Therefore, the vertical splitting of the operand node $N_{11\text{-}21\text{-}31}$ results into an identical set of product nodes $\{N_{11}, N_{21}, N_{31}\}$ for both cases of top-to-bottom and bottom-to-top splitting. This identity is defined by Eq.(5.91) in accordance with Eq.(5.80) from Proof 5.4.

$$N_{11}, N_{21\text{-}31} = N_{11\text{-}21}, N_{31} = N_{11}, N_{21}, N_{31} \tag{5.91}$$

Example 5.8

This example considers a one-node FN located in the second layer of a larger FN. This one-node FN has a single operand node $N_{12\text{-}22\text{-}32}$ that is described by the Boolean matrix in Eq.(5.73) and the binary relation in Eq.(5.74). The one-node FN can be described by the block-scheme in Fig.5.33 and the topological expression in Eq.(5.92).

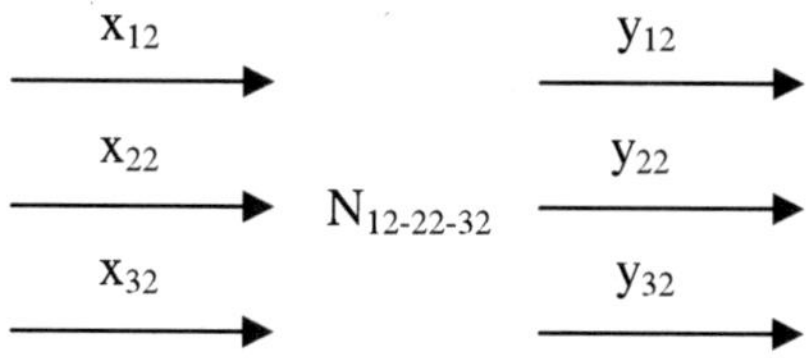

Fig. 5.33 FN with operand node $N_{12\text{-}22\text{-}32}$

$$[N_{12\text{-}22\text{-}32}]\ (x_{12}, x_{22}, x_{32} \mid y_{12}, y_{22}, y_{32}) \tag{5.92}$$

The vertical splitting of the operand node $N_{12\text{-}22\text{-}32}$ from top to bottom results into a product node N_{12} that is connected at the bottom with a temporary node $N_{22\text{-}32}$. In this case, the node N_{12} can be described by the Boolean matrix in Eq.(3.14) and the binary relation Eq.(3.18) whereas the node $N_{22\text{-}32}$ can be described by the Boolean matrix in Eq.(5.75) and the binary relation in Eq.(5.76).

The overall result of the above operation is a two-node FN that can be described by the block scheme in Fig.5.34 and the topological expression in Eq.(5.93).

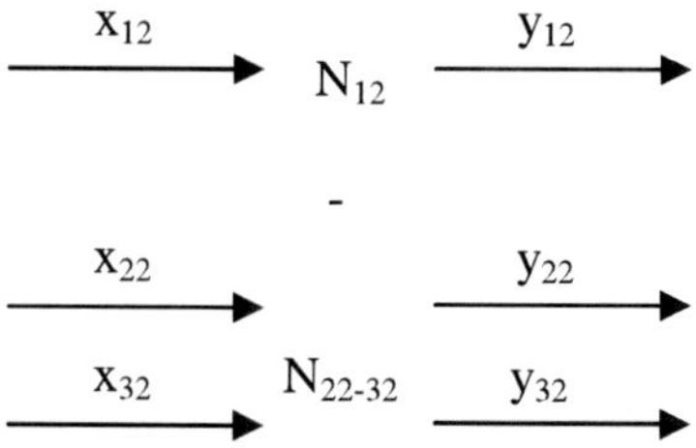

Fig. 5.34 Two-node FN with product node N_{11} and temporary node $N_{21\text{-}31}$

$[N_{12}]\ (x_{12} \mid y_{12})$ - $[N_{22\text{-}32}]\ (x_{22}, x_{32} \mid y_{22}, y_{32})$ (5.93)

Further on, the vertical splitting of the temporary node $N_{22\text{-}32}$ results into two product nodes N_{22} and N_{32}. In this case, the node N_{22} can be described by the Boolean matrix in Eq.(3.16) and the binary relation Eq.(3.20) whereas the node N_{32} can be described by the Boolean matrix in Eq.(5.68) and the binary relation in Eq.(5.69).

The overall result of the above two operations is a three-node FN that can be described by the block scheme in Fig.5.35 and the topological expression in Eq.(5.94).

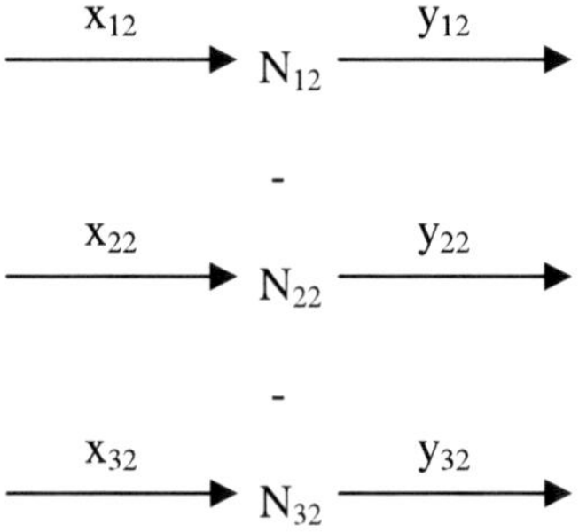

Fig. 5.35 FN with product nodes N_{12}, N_{22} and N_{32}

$[N_{12}]\ (x_{12} \mid y_{12})$ - $[N_{22}]\ (x_{22} \mid y_{22})$ - $[N_{32}]\ (x_{32} \mid y_{32})$ (5.94)

On the other hand, the vertical splitting of the operand node $N_{12\text{-}22\text{-}32}$ from bottom to top results into a product node N_{32} that is connected at the top with a temporary node $N_{12\text{-}22}$. In this case, the node N_{32} can be described by the Boolean matrix in Eq.(5.68) and the binary relation Eq.(5.69) whereas the node $N_{12\text{-}22}$ can be described by the Boolean matrix in Eq.(4.29) and the binary relation in Eq.(4.30).

The overall result of the above operation is a two-node FN that can be described by the block scheme in Fig.5.36 and the topological expression in Eq.(5.95).

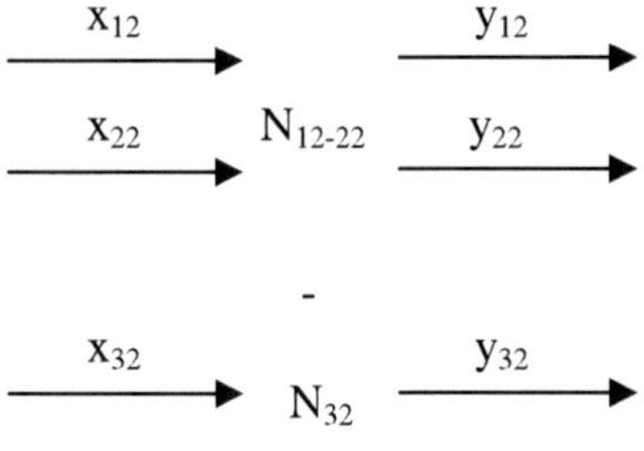

Fig. 5.36 FN with temporary node $N_{12\text{-}22}$ and product node N_{32}

$$[N_{12\text{-}22}]\ (x_{12}, x_{22} \mid y_{12}, y_{22}) - [N_{32}]\ (x_{32} \mid y_{32}) \quad (5.95)$$

Further on, the vertical splitting of the temporary node $N_{12\text{-}22}$ results into two product nodes N_{12} and N_{22}. In this case, the node N_{12} can be described by the Boolean matrix in Eq.(3.14) and the binary relation in Eq.(3.18) whereas the node N_{22} can be described by the Boolean matrix in Eq.(3.16) and the binary relation in Eq.(3.20).

The overall result of the above two operations is a three-node FN that can be described by the block scheme in Fig.5.35 and the topological expression in Eq.(5.94).

Therefore, the vertical splitting of the operand node $N_{12\text{-}22\text{-}32}$ results into an identical set of product nodes $\{N_{12}, N_{22}, N_{32}\}$ for both cases of top-to-bottom and bottom-to-top splitting. This identity is defined by Eq.(5.96) in accordance with Eq.(5.80) from Proof 5.4.

$$N_{12}, N_{22\text{-}32} = N_{12\text{-}22}, N_{32} = N_{12}, N_{22}, N_{32} \quad (5.96)$$

5.6 Associativity of Output Merging

Associativity is a property related to the operation of output merging when the latter is applied to three parallel nodes with common inputs for the purpose ofmerging them into a single node. In particular, this property allows the merging of three operand nodes A, B and C into a product node A;B;C to take place as a sequence of two binary merging operations that can be applied either from top to bottom or from bottom to top. The property can be applied when the outputs from the nodes A, B and C are self-standing. In this case, the input to the product node A;B;C is the same as the input to each of the operand nodes A, B and C whereas the output set from the product node is the union of the outputs from the operand nodes.

Proof 5.5

It has to be proved here that the operation of output merging is associative in accordance with Eq.(5.97). In this case, the output merging of any three operand nodes A, B and C from top to bottom should be equivalent to their vertical merging from bottom to top.

$$(A;B);C = A;(B;C) = A;B;C \quad (5.97)$$

The proof is based on the use of binary relations as formal models for the operand nodes A, B and C, as shown in Eqs. (5.98)-(5.100). In this case, the first elements of the relational pairs in A, B and C are denoted by the letter d whereas the second elements are denoted by the letters a, b and c, respectively.

$$A = \{(d_1^{\,1}, a_2^{\,1}),\ldots,(d_1^{\,1}, a_2^{\,p1}),\ldots,(d_1^{\,s}, a_2^{\,1}),\ldots,(d_1^{\,s}, a_2^{\,ps})\} \quad (5.98)$$

$$B = \{(d_1{}^1, b_2{}^1),\ldots,(d_1{}^1, b_2{}^{q1}),\ldots,(d_1{}^s, b_2{}^1),\ldots,(d_1{}^s, b_2{}^{qs})\} \quad (5.99)$$

$$C = \{(d_1{}^1, c_2{}^1),\ldots,(d_1{}^1, c_2{}^{r1}),\ldots,(d_1{}^s, c_2{}^1),\ldots,(d_1{}^s, c_2{}^{rs})\} \quad (5.100)$$

The first and the second element of any relational pair in A, B and C are denoted by the subscripts ‘1’ and ‘2’, respectively. However, the superscripts for the first elements of the relational pairs in A, B and C vary from 1 to s whereby s is the number of identical first elements in these pairs that represents the number of linguistic terms for the common input to A, B and C. As far as the superscripts for the second elements of the relational pairs in A, B and C are concerned, they vary from 1 to ps, 1 to qs and 1 to rs, respectively. In this case, the relation A has ‘p1+...+ps’ pairs, the relation B has 'q1+...+qs' pairs and the relation C has 'r1+...+rs’ pairs whereby p, q and r are the number of the linguistic terms for the outputs from A, B and C, respectively.

The output composition of the operand relations A and B gives the temporary relation A;B, as shown in Eq.(5.101).

$$\begin{aligned} A;B = \{&(d_1{}^1, a_2{}^1 b_2{}^1),\ldots,(d_1{}^1, a_2{}^1 b_2{}^{q1}),\ldots,(d_1{}^1, a_2{}^{p1} b_2{}^1),\ldots,(d_1{}^1, a_2{}^{p1} b_2{}^{q1}),\ldots, \\ &(d_1{}^s, a_2{}^1 b_2{}^1),\ldots,(d_1{}^s, a_2{}^1 b_2{}^{qs}),\ldots,(d_1{}^s, a_2{}^{ps} b_2{}^1),\ldots,(d_1{}^s, a_2{}^{ps} b_2{}^{qs})\} \end{aligned} \quad (5.101)$$

Further on, the output composition of the temporary relation A;B and the operand relation C gives the product relation (A;B);C, as shown in Eq.(5.102).

$$\begin{aligned} (A;B);C = \{&(d_1{}^1, a_2{}^1 b_2{}^1 c_2{}^1),\ldots,(d_1{}^1, a_2{}^1 b_2{}^1 c_2{}^{r1}),\ldots, \\ &(d_1{}^1, a_2{}^1 b_2{}^{q1} c_2{}^1),\ldots,(d_1{}^1, a_2{}^1 b_2{}^{q1} c_2{}^{r1}),\ldots, \\ &(d_1{}^1, a_2{}^{p1} b_2{}^1 c_2{}^1),\ldots,(d_1{}^1, a_2{}^{p1} b_2{}^1 c_2{}^{r1}),\ldots, \\ &(d_1{}^1, a_2{}^{p1} b_2{}^{q1} c_2{}^1),\ldots,(d_1{}^1, a_2{}^{p1} b_2{}^{q1} c_2{}^{r1}),\ldots, \\ &(d_1{}^s, a_2{}^1 b_2{}^1 c_2{}^1),\ldots,(d_1{}^s, a_2{}^1 b_2{}^1 c_2{}^{rs}),\ldots, \\ &(d_1{}^s, a_2{}^1 b_2{}^{qs} c_2{}^1),\ldots,(d_1{}^s, a_2{}^1 b_2{}^{qs} c_2{}^{rs}),\ldots, \\ &(d_1{}^s, a_2{}^{ps} b_2{}^1 c_2{}^1),\ldots,(d_1{}^s, a_2{}^{ps} b_2{}^1 c_2{}^{rs}),\ldots, \\ &(d_1{}^s, a_2{}^{ps} b_2{}^{qs} c_2{}^1),\ldots,(d_1{}^s, a_2{}^{ps} b_2{}^{qs} c_2{}^{rs})\} \end{aligned} \quad (5.102)$$

On the other hand, the output composition of the operand relations B and C gives the temporary relation B;C, as shown in Eq.(5.103).

$$\begin{aligned} B;C = \{&(d_1{}^1, b_2{}^1 c_2{}^1),\ldots,(d_1{}^1, b_2{}^1 c_2{}^{r1}),\ldots,(d_1{}^1, b_2{}^{q1} c_2{}^1),\ldots,(d_1{}^1, b_2{}^{q1} c_2{}^{r1}),\ldots, \\ &(d_1{}^s, b_2{}^1 c_2{}^1),\ldots,(d_1{}^s, b_2{}^1 c_2{}^{rs}),\ldots,(d_1{}^s, b_2{}^{qs} c_2{}^1),\ldots,(d_1{}^s, b_2{}^{qs} c_2{}^{rs})\} \end{aligned} \quad (5.103)$$

In this case, the output composition of the operand relation A and the temporary relation B;C gives the product relation A;(B;C). As the latter is identical with the product relation (A;B);C from Eq.(5.102), this implies the validity of Eq.(5.97) and concludes the proof.

Example 5.9

This example considers a FN with three parallel operand nodes with a common input $x_{11,21,31}$. The first and the second node N_{11} and N_{21} are taken from the four-node FN in Fig.3.1. These two nodes are described there by the Boolean matrices and the binary relations in Eqs.(3.13), (3.15), (3.17) and (3.19) whereas the third node N_{31} is described by the Boolean matrix in Eq.(5.104) and the binary relation in Eq.(5.105).

N_{31} : (5.104)

x_{31} \ y_{31}	1	2	3
1	0	0	1
2	0	1	0
3	1	0	0

$$N_{31}: \ \{(1, 3), (2, 2)\ (3, 1)\} \quad (5.105)$$

The nodes N_{11}, N_{21} and N_{31} represent a three-node FN that is an extended sub-network of the four-node FN from Fig.3.1. This three-node FN can be described by the block-scheme in Fig.5.37 and the topological expression in Eq.(5.106).

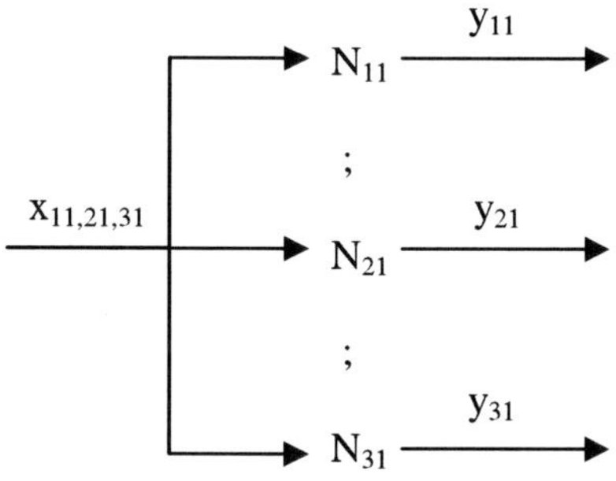

Fig. 5.37 FN with operand nodes N_{11}, N_{21}, N_{31} and common input

$$[N_{11}]\ (x_{11,21,31} \mid y_{11})\ ;\ [N_{21}]\ (x_{11,21,31} \mid y_{21})\ ;\ [N_{31}]\ (x_{11,21,31} \mid y_{31}) \quad (5.106)$$

The output merging of the first operand node N_{11} and the second operand node N_{21} from Fig.5.37 results into a temporary node $N_{11;21}$ that can be described by the Boolean matrix in Eq.(4.37) and the binary relation in Eq.(4.38). This temporary node is connected at the bottom with the third operand node N_{31} in the form of a two-node FN. The latter can be described by the block scheme in Fig.5.38 and the topological expression in Eq.(5.107).

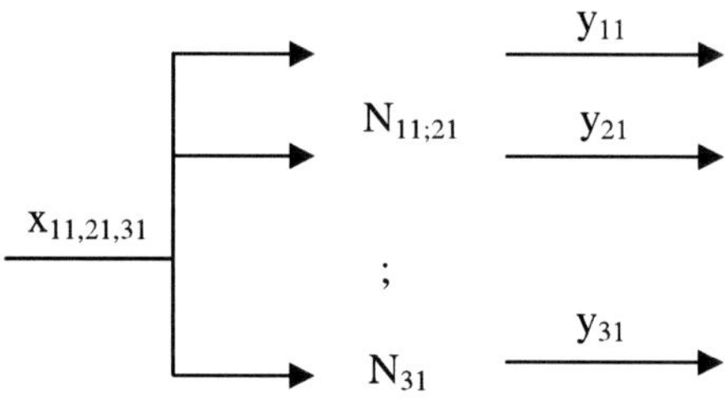

Fig. 5.38 FN with temporary node $N_{11;21}$ and operand node N_{31}

$$[N_{11;21}]\ (x_{11,21,31} \mid y_{11}, y_{21})\ ;\ [N_{31}]\ (x_{11,21,31} \mid y_{31}) \tag{5.107}$$

Further on, the output merging of the temporary node $N_{11;21}$ and the operand node N_{31} from Fig.5.38 results into a product node $N_{(11;21);31}$ in the form of a one-node FN. The latter can be described by the block scheme in Fig.5.39 and the topological expression in Eq.(5.108).

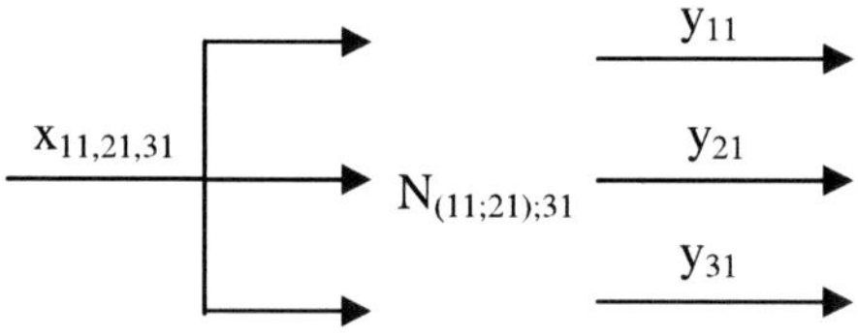

Fig. 5.39 FN with product node $N_{(11;21);31}$

$$[N_{(11;21);31}]\ (x_{11,21,31} \mid y_{11}, y_{21}, y_{31}) \tag{5.108}$$

As a result, the product node $N_{(11;21);31}$ is described by the Boolean matrix in Eq.(5.109) and the binary relation in Eq.(5.110). In this case, the labels and the elements of the Boolean matrix are represented by a compact notation. In particular, each of the capital letters A, B, C, D, E, F, G, H, I stands for three sequential columns in accordance with Eq.(5.61), 1_j , j=1,3 denotes the j-th Boolean row in the square Boolean matrix from Eq.(5.104) and 0_3 denotes a zero Boolean row of dimension 3.

$N_{(11;21);31}$: (5.109)

$x_{11,21,31}$ \ y_{11}, y_{21}, y_{31}	A	B	C	D	E	F	G	H	I
1	0_3	1_1	0_3	0_3	0_3	0_3	0_3	0_3	0_3
2	0_3	0_3	0_3	0_3	0_3	0_3	1_2	0_3	0_3
3	0_3	0_3	0_3	0_3	0_3	1_3	0_3	0_3	0_3

$N_{(11;21);31}$: {(1, 123), (2, 312) (3, 231)} (5.110)

On the other hand, the output merging of the second operand node N_{21} and the third operand node N_{31} from Fig.5.37 results into a temporary node $N_{21;31}$ that can be described by the Boolean matrix in Eq.(5.111) and the binary relation in Eq.(5.112). This temporary node is connected at the top with the first operand node N_{11} in the form of a two-node FN. The latter can be described by the block scheme in Fig.5.40 and the topological expression in Eq.(5.113).

$N_{21;31}$: (5.111)

$x_{11,21,31}$ \ y_{21}, y_{31}	11	12	13	21	22	23	31	32	33
1	0	0	0	0	0	1	0	0	0
2	0	1	0	0	0	0	0	0	0
3	0	0	0	0	0	0	1	0	0

$N_{21;31}$: {(1, 23), (1, 12), (1, 31) (5.112)

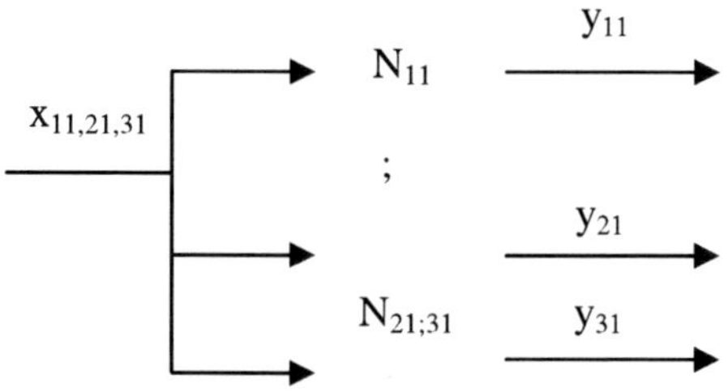

Fig. 5.40 FN with operand node N_{11} and temporary node $N_{21;31}$

$[N_{11}]\ (x_{11,21,31} \mid y_{11})\ ;\ [N_{21;31}]\ (x_{11,21,31} \mid y_{21}, y_{31})$ (5.113)

Further on, the output merging of the operand node N_{11} and the temporary node $N_{21;31}$ from Fig.5.40 results into a product node $N_{11;(21;31)}$ in the form of a one-node FN. The latter can be described by the block scheme in Fig.5.41 and the topological expression in Eq.(5.114).

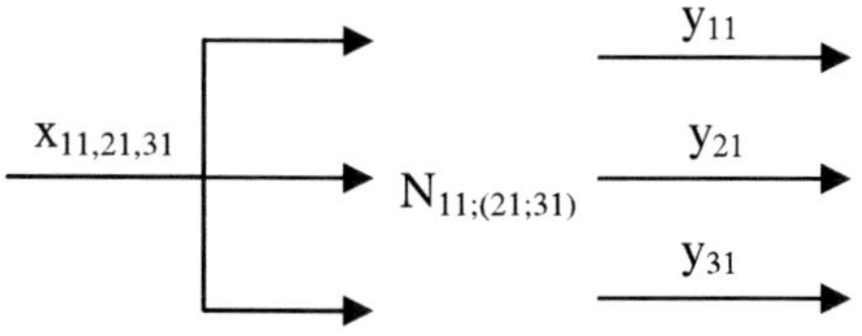

Fig. 5.41 FN with product node $N_{11;(21;31)}$

$$[N_{11;(21;31)}]\ (x_{11,21,31} \mid y_{11}, y_{21}, y_{31}) \tag{5.114}$$

In this case, the product node $N_{11;(21;31)}$ is also described by the Boolean matrix in Eq.(5.109) and the binary relation in Eq.(5.110). Therefore, $N_{11;(21;31)}$ is identical with $N_{(11;21);31}$ and this identity is defined by Eq.(5.115) in accordance with Eq.(5.97) from Proof 5.5.

$$N_{(11;21);31} = N_{11;(21;31)} = N_{11;21;31} \tag{5.115}$$

Example 5.10

This example considers a FN with three parallel operand nodes with a common input $x_{12,22,32}$. The first and the second node N_{12} and N_{22} are taken from the four-node FN in Fig.3.1. These two nodes are described there by the Boolean matrices and the binary relations in Eqs.(3.14), (3.16), (3.18) and (3.20) whereas the third node N_{32} is described by the Boolean matrix in Eq.(5.116) and the binary relation in Eq.(5.117).

N_{32} : (5.116)

y_{32} / x_{32}	1	2	3
1	0	0	1
2	0	1	0
3	1	0	0

$$N_{32}: \ \{(1, 3), (2, 2)\ (3, 1)\} \tag{5.117}$$

The nodes N_{12}, N_{22} and N_{32} represent a three-node FN that is an extended sub-network of the four-node FN from Fig.3.1. This three-node FN can be described by the block-scheme in Fig.5.42 and the topological expression in Eq.(5.118).

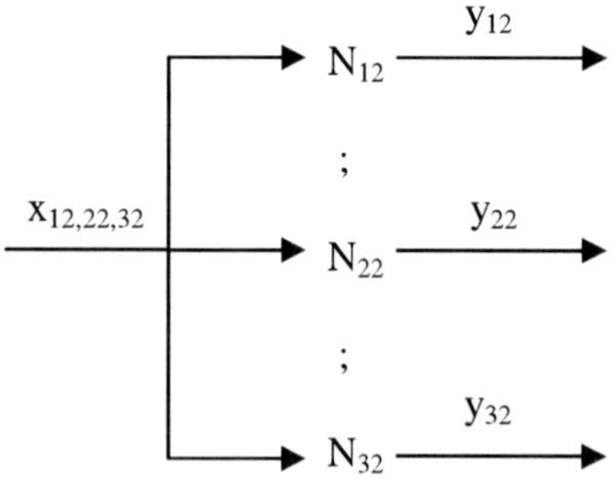

Fig. 5.42 FN with operand nodes N_{12}, N_{22}, N_{32} and common input

$$[N_{12}]\ (x_{12,22,32} \mid y_{12})\ ;\ [N_{22}]\ (x_{12,22,32} \mid y_{22})\ ;\ [N_{32}]\ (x_{12,22,32} \mid y_{32}) \tag{5.118}$$

The output merging of the first operand node N_{12} and the second operand node N_{22} from Fig.5.42 results into a temporary node $N_{12;22}$ that can be described by the Boolean matrix in Eq.(4.41) and the binary relation in Eq.(4.42). This temporary node is connected at the bottom with the third operand node N_{32} in the form of a two-node FN. The latter can be described by the block scheme in Fig.5.43 and the topological expression in Eq.(5.119).

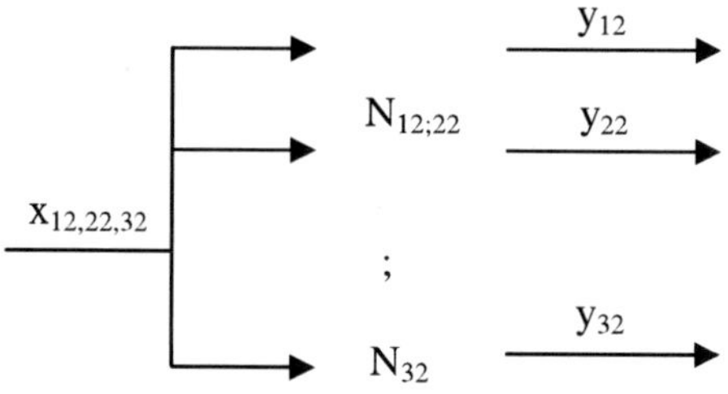

Fig. 5.43 FN with temporary node $N_{12;22}$ and operand node N_{32}

$$[N_{12;22}]\ (x_{12,22,32} \mid y_{12},\ y_{22})\ ;\ [N_{32}]\ (x_{12,22,32} \mid y_{32}) \tag{5.119}$$

Further on, the output merging of the temporary node $N_{12;22}$ and the operand node N_{32} from Fig.5.43 results into a product node $N_{(12;22);32}$ in the form of a one-node FN. The latter can be described by the block scheme in Fig.5.44 and the topological expression in Eq.(5.120).

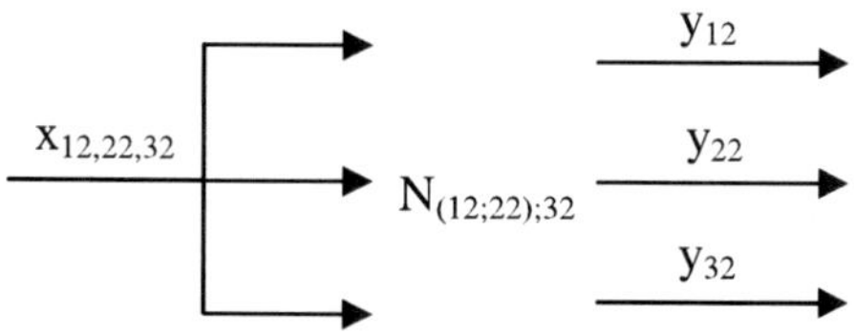

Fig. 5.44 FN with product node $N_{(12;22);32}$

$$[N_{(12;22);32}]\ (x_{12,22,32} \mid y_{12},\ y_{22},\ y_{32}) \tag{5.120}$$

As a result, the product node $N_{(12;22);32}$ is described by the Boolean matrix in Eq.(5.121) and the binary relation in Eq.(5.122). In this case, the labels and the elements of the Boolean matrix are represented by a compact notation. In particular, each of the capital letters A, B, C, D, E, F, G, H, I stands for three sequential columns in accordance with Eq.(5.61), 1_j , j=1,3 denotes the j-th Boolean row in the square Boolean matrix from Eq.(5.116) and 0_3 denotes a zero Boolean row of dimension 3.

$N_{(12;22);32}$: (5.121)

$x_{12,22,32}$ \ y_{12}, y_{22}, y_{32}	A	B	C	D	E	F	G	H	I
1	0_3	0_3	0_3	0_3	0_3	1_1	0_3	0_3	0_3
2	0_3	0_3	0_3	0_3	0_3	0_3	1_2	0_3	0_3
3	0_3	1_3	0_3	0_3	0_3	0_3	0_3	0_3	0_3

$N_{(12;22);32}$: {(1, 233), (2, 312) (3, 121)} (5.122)

On the other hand, the output merging of the second operand node N_{22} and the third operand node N_{32} from Fig.5.42 results into a temporary node $N_{22;32}$ that can be described by the Boolean matrix in Eq.(5.123) and the binary relation in Eq.(5.124). This temporary node is connected at the top with the first operand node N_{12} in the form of a two-node FN. The latter can be described by the block scheme in Fig.5.45 and the topological expression in Eq.(5.125).

$N_{22;32}$: (5.123)

$x_{12,22,32}$ \ y_{22}, y_{32}	11	12	13	21	22	23	31	32	33
1	0	0	0	0	0	0	0	0	1
2	0	1	0	0	0	0	0	0	0
3	0	0	0	1	0	0	0	0	0

$N_{22;32}$: {(1, 33), (2, 12), (3, 21) (5.124)

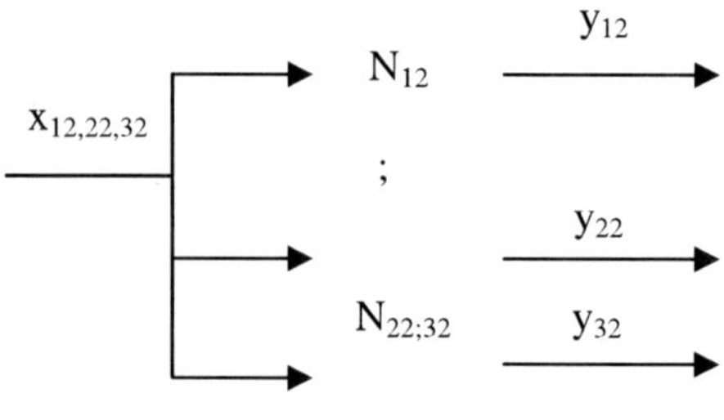

Fig. 5.45 FN with operand node N_{12} and temporary node $N_{22;32}$

$$[N_{12}]\ (x_{12,22,32} \mid y_{12})\ ;\ [N_{22;32}]\ (x_{12,22,32} \mid y_{22}, y_{32}) \quad (5.125)$$

Further on, the output merging of the operand node N_{12} and the temporary node $N_{22;32}$ from Fig.5.45 results into a product node $N_{12;(22;32)}$ in the form of a one-node FN. The latter can be described by the block scheme in Fig.5.46 and the topological expression in Eq.(5.126).

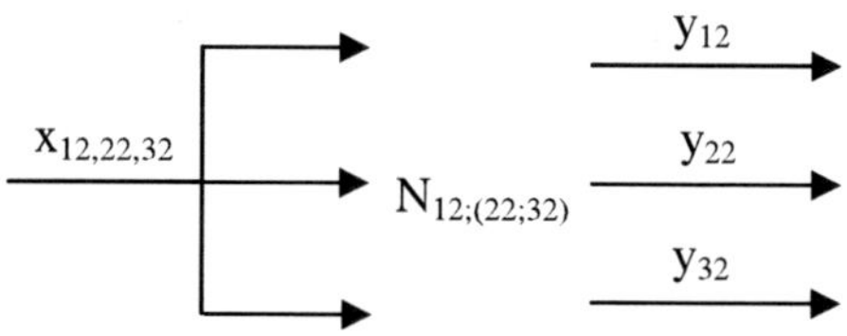

Fig. 5.46 FN with product node $N_{12;(22;32)}$

$$[N_{12;(22;32)}] (x_{12,22,32} \mid y_{12}, y_{22}, y_{32}) \quad (5.126)$$

In this case, the product node $N_{12;(22;32)}$ is also described by the Boolean matrix in Eq.(5.121) and the binary relation in Eq.(5.122). Therefore, $N_{12;(22;32)}$ is identical with $N_{(12;22);32}$ and this identity is defined by Eq.(5.127) in accordance with Eq.(5.97) from Proof 5.5.

$$N_{(12;22);32} = N_{12;(22;32)} = N_{12;22;32} \quad (5.127)$$

5.7 Variability of Output Splitting

Variability is a property related to the operation of output splitting when the latter is applied to a single node with the purpose of splitting it into three nodes with common inputs. In particular, this property allows the splitting of an operand node A:B:C into three product nodes A, B and C to take place as a sequence of two unary splitting operations that can be applied either from top to bottom or from bottom to top. The property always holds when the underlying operation of output splitting can be applied in the manner specified above. In this case, the common inputs to the three product nodes A, B and C are the same as the inputs to the operand node A:B:C whereas the union of the outputs from the three product nodes is the same as the output set from the operand node.

Proof 5.6

It has to be proved here that the operation of output splitting is variable in accordance with Eq.(5.128). In this case, the output splitting of any single operand node A:B:C from top to bottom should be equivalent to its vertical splitting from bottom to top.

$$A, (B:C) = (A:B), C = A, B, C \quad (5.128)$$

The proof is based on the use of a binary relation as a formal model for the operand node A:B:C, as shown in Eq.(5.129). In this case, the first element of each pair in A:B:C is denoted by the letter d whereas the second element of each pair is denoted by the triplet 'abc'.

$$A:B:C = \{(d_1^{\,1}, a_2^{\,1}\,b_2^{\,1}\,c_2^{\,1}),\ldots,(d_1^{\,1}, a_2^{\,1}\,b_2^{\,1}\,c_2^{\,r1}),\ldots,$$
$$(d_1^{\,1}, a_2^{\,1}\,b_2^{\,q1}\,c_2^{\,1}),\ldots,(d_1^{\,1}, a_2^{\,1}\,b_2^{\,q1}\,c_2^{\,r1}),\ldots,$$
$$(d_1^{\,1}, a_2^{\,p1}\,b_2^{\,1}\,c_2^{\,1}),\ldots,(d_1^{\,1}, a_2^{\,p1}\,b_2^{\,1}\,c_2^{\,r1}),\ldots,$$
$$(d_1^{\,1}, a_2^{\,p1}\,b_2^{\,q1}\,c_2^{\,1}),\ldots,(d_1^{\,1}, a_2^{\,p1}\,b_2^{\,q1}\,c_2^{\,r1}),\ldots,$$
$$(d_1^{\,s}, a_2^{\,1}\,b_2^{\,1}\,c_2^{\,1}),\ldots,(d_1^{\,s}, a_2^{\,1}\,b_2^{\,1}\,c_2^{\,rs}),\ldots,$$
$$(d_1^{\,s}, a_2^{\,1}\,b_2^{\,qs}\,c_2^{\,1}),\ldots,(d_1^{\,s}, a_2^{\,1}\,b_2^{\,qs}\,c_2^{\,rs}),\ldots,$$
$$(d_1^{\,s}, a_2^{\,ps}\,b_2^{\,1}\,c_2^{\,1}),\ldots,(d_1^{\,s}, a_2^{\,ps}\,b_2^{\,1}\,c_2^{\,rs}),\ldots,$$
$$(d_1^{\,s}, a_2^{\,ps}\,b_2^{\,qs}\,c_2^{\,1}),\ldots,(d_1^{\,s}, a_2^{\,ps}\,b_2^{\,qs}\,c_2^{\,rs})\} \quad (5.129)$$

The first element and the three individual elements in the triplet of any relational pair in A:B:C are denoted by the subscripts '1' and '2', respectively. In this case, the superscripts for all these elements denote their target pairs in the product relations A, B and C whereby the operand relation A:B:C has 'p1.q1.r1+...+ps.qs.rs' pairs.

The output decomposition of the operand relation A:B:C from top to bottom gives the product relation A in Eq.(5.130) and the temporary relation B:C in Eq.(5.131).

$$A = \{(d_1^{\,1}, a_2^{\,1}),\ldots,(d_1^{\,1}, a_2^{\,p1}),\ldots,(d_1^{\,s}, a_2^{\,1}),\ldots,(d_1^{\,s}, a_2^{\,ps})\} \quad (5.130)$$

$$B:C = \{(d_1^{\,1}, b_2^{\,1}\,c_2^{\,1}),\ldots,(d_1^{\,1}, b_2^{\,1}\,c_2^{\,r1}),\ldots,(d_1^{\,1}, b_2^{\,q1}\,c_2^{\,1}),\ldots,(d_1^{\,1}, b_2^{\,q1}\,c_2^{\,r1}),\ldots,$$
$$(d_1^{\,s}, b_2^{\,1}\,c_2^{\,1}),\ldots,(d_1^{\,s}, b_2^{\,1}\,c_2^{\,rs}),\ldots,(d_1^{\,s}, b_2^{\,qs}\,c_2^{\,1}),\ldots,(d_1^{\,s}, b_2^{\,qs}\,c_2^{\,rs})\} \quad (5.131)$$

Further on, the output decomposition of the temporary relation B:C gives the product relations B and C, as shown in Eqs.(5.132)-(5.133).

$$B = \{(d_1^{\,1}, b_2^{\,1}),\ldots,(d_1^{\,1}, b_2^{\,q1}),\ldots,(d_1^{\,s}, b_2^{\,1}),\ldots,(d_1^{\,s}, b_2^{\,qs})\} \quad (5.132)$$

$$C = \{(d_1^{\,1}, c_2^{\,1}),\ldots,(d_1^{\,1}, c_2^{\,r1}),\ldots,(d_1^{\,s}, c_2^{\,1}),\ldots,(d_1^{\,s}, c_2^{\,rs})\} \quad (5.133)$$

On the other hand, the output decomposition of the operand relation A:B:C from bottom to top gives the temporary relation A:B in Eq.(5.134) and the product relation C which is identical with the product relation from Eq.(5.133).

$$A:B = \{(d_1^{\,1}, a_2^{\,1}\,b_2^{\,1}),\ldots,(d_1^{\,1}, a_2^{\,1}\,b_2^{\,q1}),\ldots,(d_1^{\,1}, a_2^{\,p1}\,b_2^{\,1}),\ldots,(d_1^{\,1}, a_2^{\,p1}\,b_2^{\,q1}),\ldots,$$
$$(d_1^{\,s}, a_2^{\,1}\,b_2^{\,1}),\ldots,(d_1^{\,s}, a_2^{\,1}\,b_2^{\,qs}),\ldots,(d_1^{\,s}, a_2^{\,ps}\,b_2^{\,1}),\ldots,(d_1^{\,s}, a_2^{\,ps}\,b_2^{\,qs})\} \quad (5.134)$$

In this case, the output decomposition of the temporary relation A:B gives the product relations A and B. As the latter are also identical with the product relations from Eq.(5.130) and Eq.(5.132), this implies the validity of Eq.(5.128) and concludes the proof.

Example 5.11

This example considers a one-node FN located in the first layer of a larger FN. This one-node FN has a single operand node $N_{11:21:31}$ that is described by the Boolean matrix in Eq.(5.109) and the binary relation in Eq.(5.110). The one-node FN can be described by the block-scheme in Fig.5.47 and the topological expression in Eq.(5.135).

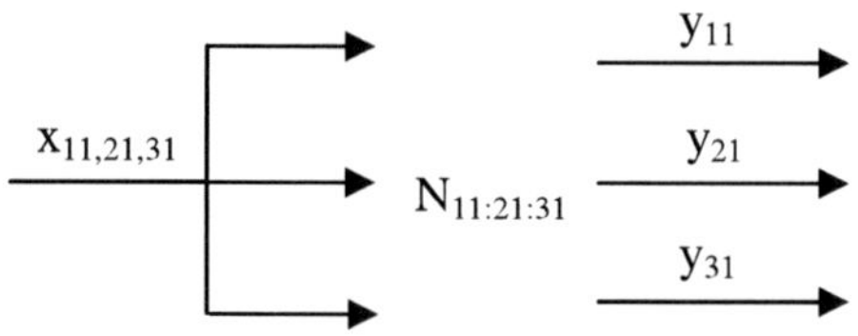

Fig. 5.47 FN with product node $N_{11:21:31}$

$$[N_{11:21:31}]\ (x_{11,21,31} \mid y_{11}, y_{21}, y_{31}) \qquad (5.135)$$

The output splitting of the operand node $N_{11:21:31}$ from top to bottom results into a product node N_{11} that is connected at the bottom with a temporary node $N_{21:31}$. In this case, the node N_{11} can be described by the Boolean matrix in Eq.(3.13) and the binary relation Eq.(3.17) whereas the node $N_{21:31}$ can be described by the Boolean matrix in Eq.(5.111) and the binary relation in Eq.(5.112).

The overall result of the above operation is a two-node FN that can be described by the block scheme in Fig.5.48 and the topological expression in Eq.(5.136).

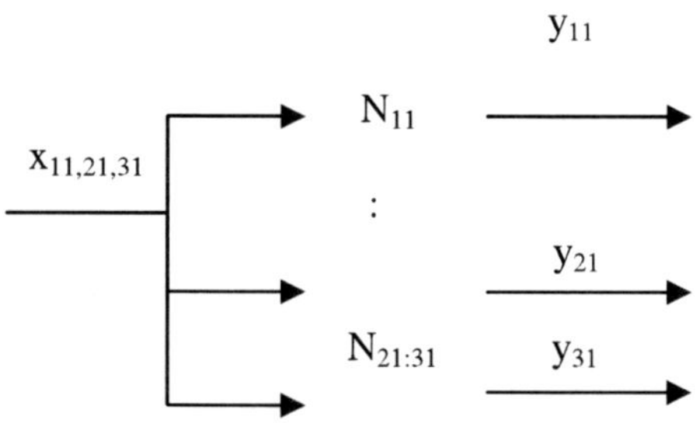

Fig. 5.48 FN with operand node N_{11} and temporary node $N_{21:31}$

$$[N_{11}]\ (x_{11,21,31} \mid y_{11}) : [N_{21:31}]\ (x_{11,21,31} \mid y_{21}, y_{31}) \qquad (5.136)$$

Further on, the output splitting of the temporary node $N_{21:31}$ results into two product nodes N_{21} and N_{31}. In this case, the node N_{21} can be described by the Boolean matrix in Eq.(3.15) and the binary relation Eq.(3.19) whereas the node N_{31} can be described by the Boolean matrix in Eq.(5.104) and the binary relation in Eq.(5.105).

The overall result of the above two operations is a three-node FN that can be described by the block scheme in Fig.5.49 and the topological expression in Eq.(5.137).

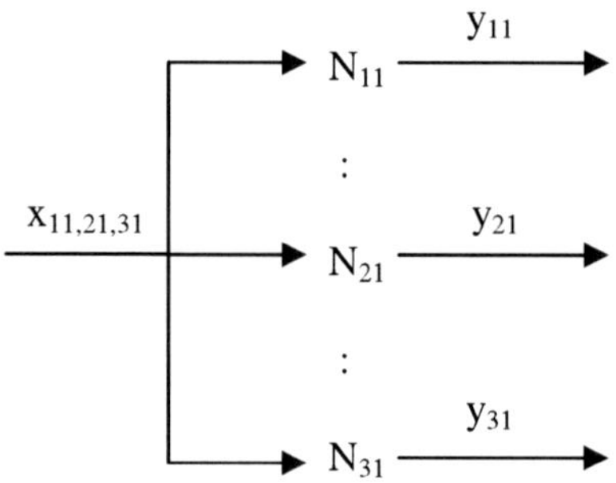

Fig. 5.49 FN with operand nodes N_{11}, N_{21}, N_{31} and common input

$$[N_{11}]\ (x_{11,21,31} \mid y_{11}) : [N_{21}]\ (x_{11,21,31} \mid y_{21}) : [N_{31}]\ (x_{11,21,31} \mid y_{31}) \tag{5.137}$$

On the other hand, the output splitting of the operand node $N_{11:21:31}$ from bottom to top results into a product node N_{31} that is connected at the top with a temporary node $N_{11:21}$. In this case, the node N_{31} can be described by the Boolean matrix in Eq.(5.104) and the binary relation Eq.(5.105) whereas the node $N_{11:21}$ can be described by the Boolean matrix in Eq.(4.37) and the binary relation in Eq.(4.38).

The overall result of the above operation is a two-node FN that can be described by the block scheme in Fig.5.50 and the topological expression in Eq.(5.138).

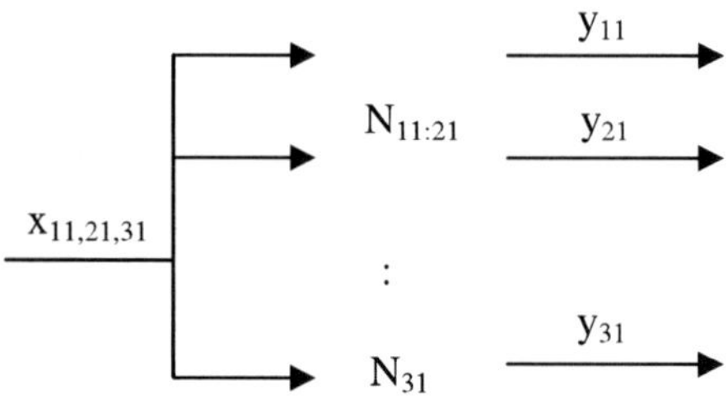

Fig. 5.50 FN with temporary node $N_{11:21}$ and operand node N_{31}

$$[N_{11:21}]\ (x_{11,21,31} |\ y_{11},\ y_{21}) : [N_{31}]\ (x_{11,21,31} |\ y_{31}) \tag{5.138}$$

Further on, the output splitting of the temporary node $N_{11:21}$ results into two product nodes N_{11} and N_{21}. In this case, the node N_{11} can be described by the Boolean matrix in Eq.(3.13) and the binary relation in Eq.(3.17) whereas the node N_{21} can be described by the Boolean matrix in Eq.(3.15) and the binary relation in Eq.(3.19).

The overall result of the above two operations is a three-node FN that can be described by the block scheme in Fig.5.49 and the topological expression in Eq.(5.137).

Therefore, the output splitting of the operand node $N_{11:21:31}$ results into an identical set of product nodes $\{N_{11},\ N_{21},\ N_{31}\}$ for both cases of top-to-bottom and bottom-to-top splitting. This identity is defined by Eq.(5.139) in accordance with Eq.(5.128) from Proof 5.6.

$$N_{11},\ N_{21:31} = N_{11:21},\ N_{31} = N_{11},\ N_{21},\ N_{31} \tag{5.139}$$

Example 5.12

This example considers a one-node FN located in the second layer of a larger FN. This one-node FN has a single operand node $N_{12:22:32}$ that is described by the Boolean matrix in Eq.(5.121) and the binary relation in Eq.(5.122). The one-node FN can be described by the block-scheme in Fig.5.51 and the topological expression in Eq.(5.140).

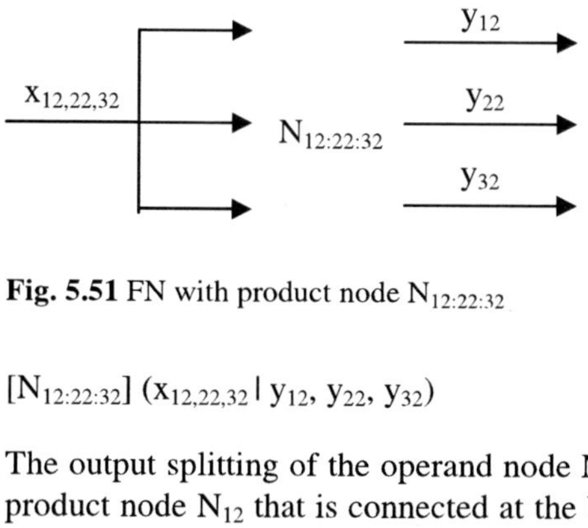

Fig. 5.51 FN with product node $N_{12:22:32}$

$$[N_{12:22:32}]\ (x_{12,22,32} \,|\, y_{12},\ y_{22},\ y_{32}) \tag{5.140}$$

The output splitting of the operand node $N_{12:22:32}$ from top to bottom results into a product node N_{12} that is connected at the bottom with a temporary node $N_{22:32}$. In this case, the node N_{12} can be described by the Boolean matrix in Eq.(3.14) and the binary relation Eq.(3.18) whereas the node $N_{22:32}$ can be described by the Boolean matrix in Eq.(5.123) and the binary relation in Eq.(5.124).

The overall result of the above operation is a two-node FN that can be described by the block scheme in Fig.5.52 and the topological expression in Eq.(5.141).

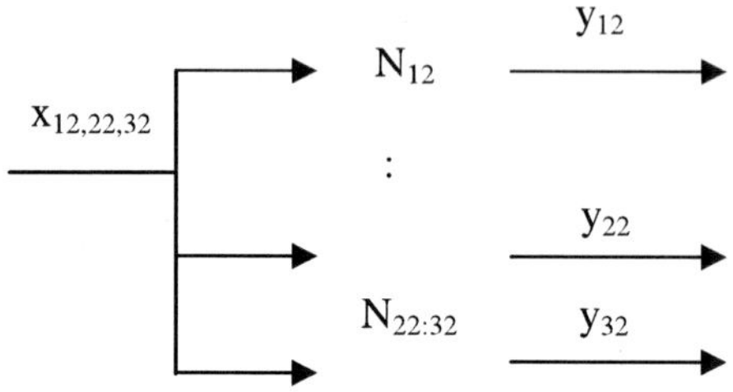

Fig. 5.52 FN with operand node N_{12} and temporary node $N_{22:32}$

$$[N_{12}]\ (x_{12,22,32} \mid y_{12}) : [N_{22:32}]\ (x_{12,22,32} \mid y_{22}, y_{32}) \tag{5.141}$$

Further on, the output splitting of the temporary node $N_{22:32}$ results into two product nodes N_{22} and N_{32}. In this case, the node N_{22} can be described by the Boolean matrix in Eq.(3.16) and the binary relation Eq.(3.20) whereas the node N_{32} can be described by the Boolean matrix in Eq.(5.116) and the binary relation in Eq.(5.117).

The overall result of the above two operations is a three-node FN that can be described by the block scheme in Fig.5.53 and the topological expression in Eq.(5.142).

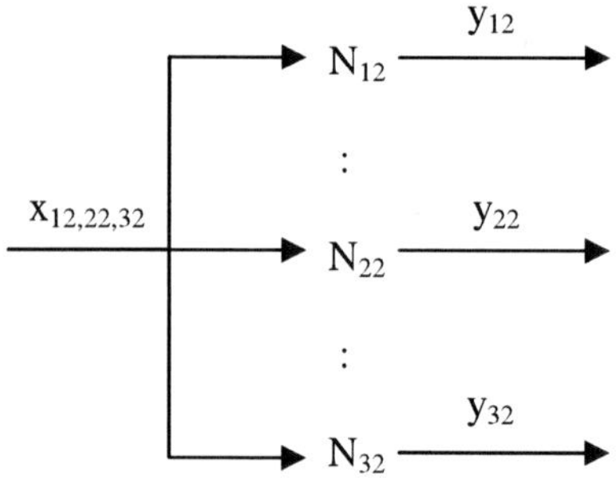

Fig. 5.53 FN with operand nodes N_{12}, N_{22}, N_{32} and common input

$$[N_{12}]\ (x_{12,22,32} \mid y_{12}) : [N_{22}]\ (x_{12,22,32} \mid y_{22}) : [N_{32}]\ (x_{12,22,32} \mid y_{32}) \tag{5.142}$$

On the other hand, the output splitting of the operand node $N_{12:22:32}$ from bottom to top results into a product node N_{32} that is connected at the top with a temporary node $N_{12:22}$. In this case, the node N_{32} can be described by the Boolean matrix in Eq.(5.116) and the binary relation Eq.(5.117) whereas the node $N_{12:22}$ can be described by the Boolean matrix in Eq.(4.41) and the binary relation in Eq.(4.42).

The overall result of the above operation is a two-node FN that can be described by the block scheme in Fig.5.54 and the topological expression in Eq.(5.143).

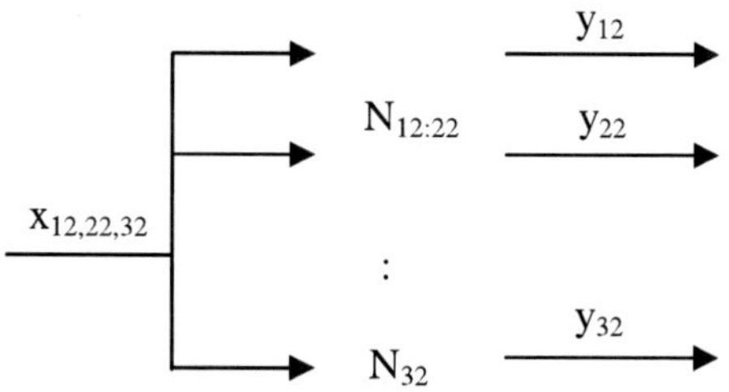

Fig. 5.54 FN with temporary node $N_{12:22}$ and operand node N_{32}

$$[N_{12:22}]\ (x_{12,22,32}|\ y_{12},\ y_{22}) : [N_{32}]\ (x_{12,22,32}|\ y_{32}) \tag{5.143}$$

Further on, the output splitting of the temporary node $N_{12:22}$ results into two product nodes N_{12} and N_{22}. In this case, the node N_{12} can be described by the Boolean matrix in Eq.(3.14) and the binary relation in Eq.(3.18) whereas the node N_{22} can be described by the Boolean matrix in Eq.(3.16) and the binary relation in Eq.(3.20).

The overall result of the above two operations is a three-node FN that can be described by the block scheme in Fig.5.53 and the topological expression in Eq.(5.142).

Therefore, the output splitting of the operand node $N_{12:22:32}$ results into an identical set of product nodes $\{N_{12}, N_{22}, N_{32}\}$ for both cases of top-to-bottom and bottom-to-top splitting. This identity is defined by Eq.(5.144) in accordance with Eq.(5.128) from Proof 5.6.

$$N_{12},\ N_{22:32} = N_{12:22},\ N_{32} = N_{12},\ N_{22},\ N_{32} \tag{5.144}$$

5.8 Mixed Properties of Operations

The structural properties of basic operations introduced in the preceding sections of this chapter are all homogenous, i.e. the properties are defined in the context of one type of operation. However, there are other properties that can be defined in a mixed context of more than one type of operation. For this purpose, there are usually brackets specifying the order of the operations involved.

This section introduces two groups of mixed properties. Each of these groups contains two properties - one in a merging and one in a splitting context. Each property is first proved and then illustrated by an example.

The first group shows the equivalence of two combinations of horizontal and vertical operations for a FN with at least two levels and two layers. The second group shows the equivalence of two combinations of horizontal, vertical and output operations for a FN with at least two levels and two layers whereby the nodes in the first layer are with common inputs.

The mixed properties are assumed to include only three operations. However, the extension of these properties to more complex mixed properties including more than three operations for a FN with more than two levels or two layers is straightforward.

Similarly to homogenous properties, mixed properties can be described by block schemes and topological expressions at network level as well as by Boolean matrices and binary relations at node level. For simplicity, each mixed property is described here mainly by block schemes and topological expressions whereby the associated Boolean matrices and binary relations are assumed to be embedded implicitly in the description.

For consistency, all mixed properties are presented by three stages. The first stage describes the initial state of the mixed property with the operand nodes before the application of any operations. The second stage describes an intermediate state of the property with some temporary nodes after the application of some operations. The third stage describes the final state of the property with the product nodes after the application of all operations.

Proof 5.7

It has to be proved here that the horizontal-horizontal-vertical merging of any four operand nodes A, B, C and D in a two-level-two-layer FN as the one in Fig.3.1 is equivalent to their vertical-vertical-horizontal merging in accordance with Eq.(5.145). In this case, the four nodes are assumed to be listed alphabetically from left to right and from top to bottom within the interconnection structure of this FN.

$$(A*B)+(C*D) = (A+C)*(B+D) \tag{5.145}$$

The proof is based on the use of binary relations as formal models for the operand nodes A, B, C and D, as shown in Eqs. (5.146)-(5.149). In this case, both elements of the relational pairs in A and C are denoted by the letters a and c, respectively. Also, the second elements of the relational pairs in B and D are denoted by the letters b and d whereas the first elements of the relational pairs in B and D are denoted by the letters a and c, respectively.

$$A = \{(a_1^1, a_2^1),\ldots,(a_1^p, a_2^p)\} \tag{5.146}$$

$$B = \{(a_2^1, b_2^1),\ldots,(a_2^1, b_2^{q1}),\ldots,(a_2^p, b_2^1),\ldots,(a_2^p, b_2^{qp})\} \tag{5.147}$$

$$C = \{(c_1^1, c_2^1),\ldots,(c_1^r, c_2^r)\} \tag{5.148}$$

$$D = \{(c_2^1, d_2^1),\ldots,(c_2^1, d_2^{s1}),\ldots,(c_2^r, d_2^1),\ldots,(c_2^r, d_2^{sr})\} \tag{5.149}$$

The first and the second element of any relational pair in A and C are denoted by the subscripts ‘1’ and ‘2’, respectively. However, both elements of any relational pair in B and D are denoted by the subscript ‘2’. This is due to the requirement for left composability of B and D, i.e. the first element of each pair in B must be identical with a second element of a pair in A whereas the first element of each pair in D must be identical with a second element of a pair in C.

The superscripts for both elements of any relational pair in A and C are identical as they indicate the corresponding number for each pair. In this case, the relation A has ‘p’ pairs and the relation C has ‘r’ pairs. The superscripts for the first elements of the relational pairs in B and D vary from 1 to p and 1 to r, respectively. As far as the superscripts for the second elements of the relational pairs in B and D are concerned, they vary from q1 to qp and s1 to sr, respectively. In this case, the relation B has ‘q1+...+qp’ pairs and the relation D has 's1+...+sr' pairs.

The horizontal composition of the operand relations A and B gives the temporary relation A*B, as shown in Eq.(5.150).

$$A*B = \{(a_1^1, b_2^1),\ldots,(a_1^1, b_2^{q1}),\ldots,(a_1^p, b_2^1),\ldots,(a_1^p, b_2^{qp})\} \quad (5.150)$$

Also, the horizontal composition of the operand relations C and D gives the temporary relation C*D, as shown in Eq.(5.151).

$$C*D = \{(c_1^1, d_2^1),\ldots,(c_1^1, d_2^{s1}),\ldots,(c_1^r, d_2^1),\ldots,(c_1^r, d_2^{sr})\} \quad (5.151)$$

Further on, the vertical composition of the temporary relation A*B and C*D gives the product relation (A*B)+(C*D), as shown in Eq.(5.152).

$$(A*B)+(C*D) = \quad (5.152)$$

$$\{(a_1^1 c_1^1, b_2^1 d_2^1),\ldots,(a_1^1 c_1^1, b_2^1 d_2^{s1}),\ldots,(a_1^1 c_1^r, b_2^1 d_2^1),\ldots,(a_1^1 c_1^r, b_2^1 d_2^{sr}),\ldots,$$

$$(a_1^1 c_1^1, b_2^{q1} d_2^1),\ldots,(a_1^1 c_1^1, b_2^{q1} d_2^{s1}),\ldots,(a_1^1 c_1^r, b_2^{q1} d_2^1),\ldots,(a_1^1 c_1^r, b_2^{q1} d_2^{sr}),\ldots,$$

$$(a_1^p c_1^1, b_2^1 d_2^1),\ldots,(a_1^p c_1^1, b_2^1 d_2^{s1}),\ldots,(a_1^p c_1^r, b_2^1 d_2^1),\ldots,(a_1^p c_1^r, b_2^1 d_2^{sr}),\ldots,$$

$$(a_1^p c_1^1, b_2^{qp} d_2^1),\ldots,(a_1^p c_1^1, b_2^{qp} d_2^{s1}),\ldots,(a_1^p c_1^r, b_2^{qp} d_2^1),\ldots,(a_1^p c_1^r, b_2^{qp} d_2^{sr})\}$$

On the other hand, the vertical composition of the operand relations A and C gives the temporary relation A+C, as shown in Eq.(5.153).

$$A+C = \{(a_1^1 c_1^1, a_2^1 c_2^1),\ldots,(a_1^1 c_1^r, a_2^1 c_2^r),\ldots, \quad (5.153)$$

$$(a_1^p c_1^1, a_2^p c_2^1),\ldots,(a_1^p c_1^r, a_2^p c_2^r)\}$$

Also, the vertical composition of the operand relations B and D gives the temporary relation B+D, as shown in Eq.(5.154).

$$B+D = \tag{5.154}$$

$$\{(a_2^1\, c_2^1,\, b_2^1\, d_2^1),\ldots,(a_2^1\, c_2^1,\, b_2^1\, d_2^{s1}),\ldots,(a_2^1\, c_2^r,\, b_2^1\, d_2^1),\ldots,(a_2^1\, c_2^r,\, b_2^1\, d_2^{sr}),\ldots,$$

$$(a_2^1\, c_2^1,\, b_2^{q1}\, d_2^1),\ldots,(a_2^1\, c_2^1,\, b_2^{q1}\, d_2^{s1}),\ldots,(a_2^1\, c_2^r,\, b_2^{q1}\, d_2^1),\ldots,(a_2^1\, c_2^r,\, b_2^{q1}\, d_2^{sr}),\ldots,$$

$$(a_2^p\, c_2^1,\, b_2^1\, d_2^1),\ldots,(a_2^p\, c_2^1,\, b_2^1\, d_2^{s1}),\ldots,(a_2^p\, c_2^r,\, b_2^1\, d_2^1),\ldots,(a_2^p\, c_2^r,\, b_2^1\, d_2^{sr}),\ldots,$$

$$(a_2^p\, c_2^1,\, b_2^{qp}\, d_2^1),\ldots,(a_2^p\, c_2^1,\, b_2^{qp}\, d_2^{s1}),\ldots,(a_2^p\, c_2^r,\, b_2^{qp}\, d_2^1),\ldots,(a_2^p\, c_2^r,\, b_2^{qp}\, d_2^{sr})\}$$

In this case, the horizontal composition of the temporary relations A+C and B+D gives the product relation (A+C)*(B+D). As the latter is identical with the product relation (A*B)+(C*D) from Eq.(5.152), this shows the validity of Eq.(5.145) and concludes the proof.

Example 5.13

This example illustrates the equivalence of horizontal-horizontal-vertical merging and vertical-vertical-horizontal merging for four operand nodes A, B, C and D in the context of Proof 5.7. First, node A is horizontally merged with node B into a temporary node A*B and node C is horizontally merged with node D into a temporary node C*D. Then, node A*B is vertically merged with node C*D into a product node (A*B)+(C*D). Alternatively, node A is vertically merged with node C into a temporary node A+C and node B is vertically merged with node D into a temporary node B+D. Then, node A+C is horizontally merged with node B+D into a product node (A+C)*(B+D). All relevant states of this mixed property are described by the block schemes and the topological expressions in Figs.5.55-5.60 and Eqs.(5.155)-(5.160).

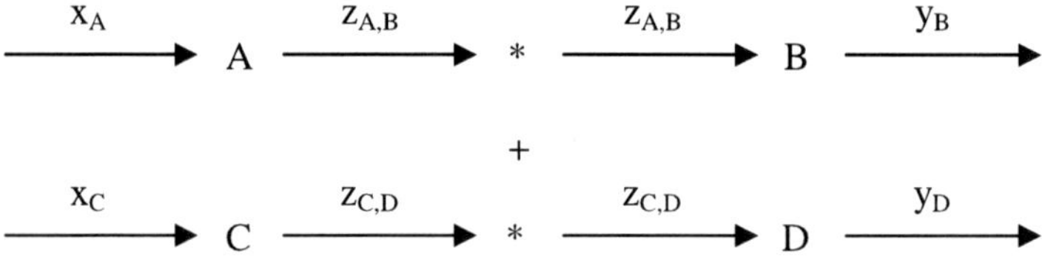

Fig. 5.55 Initial state for horizontal-horizontal-vertical merging

$$\{[A]\,(x_A\,|\,z_{A,B}) * [B]\,(z_{A,B}\,|\,y_B)\} + \{[C]\,(x_C\,|\,z_{C,D}) * [D]\,(z_{C,D}\,|\,y_D)\} \tag{5.155}$$

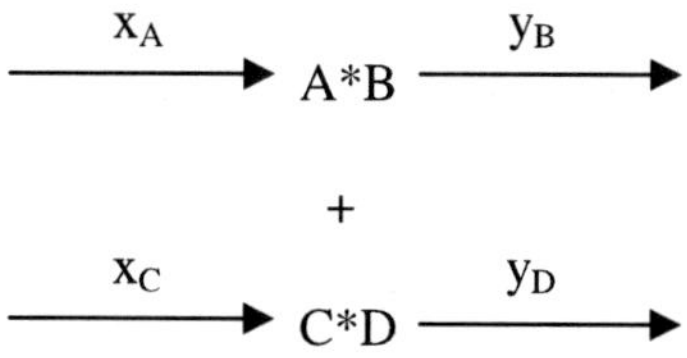

Fig. 5.56 Intermediate state for horizontal-horizontal-vertical merging

$$[A*B]\ (x_A \mid y_B) + [C*D]\ (x_C \mid y_D) \tag{5.156}$$

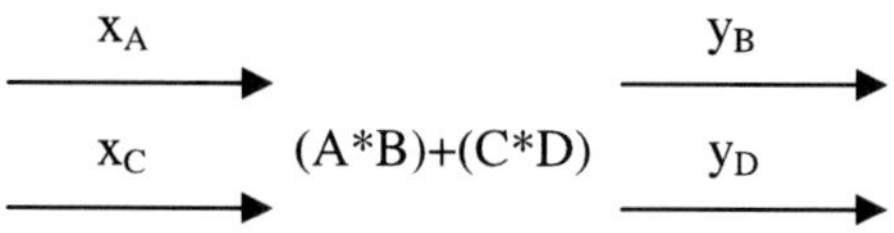

Fig. 5.57 Final state for horizontal-horizontal-vertical merging

$$[(A*B)+(C*D)]\ (x_A, x_C \mid y_B, y_D\ ) \tag{5.157}$$

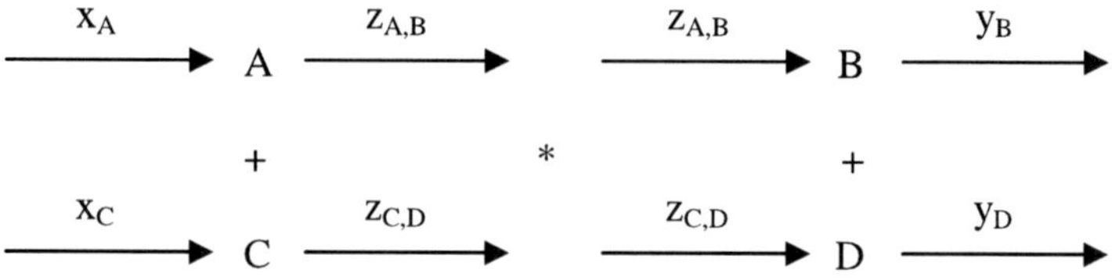

Fig. 5.58 Initial state for vertical-vertical-horizontal merging

$$\{[A]\ (x_A \mid z_{A,B}) + [C]\ (x_C \mid z_{C,D})\} * \{[B]\ (z_{A,B} \mid y_B) * [D]\ (z_{C,D} \mid y_D)\} \tag{5.158}$$

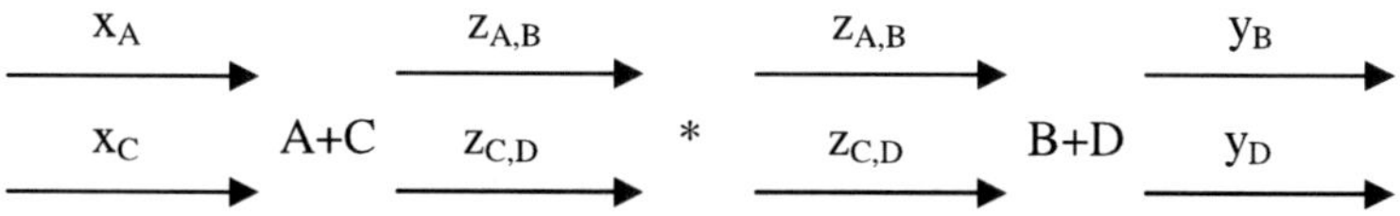

Fig. 5.59 Intermediate state for vertical-vertical-horizontal merging

$$[A+C]\ (x_A, x_C \mid z_{A,B}, z_{C,D}) * [B+D]\ (z_{A,B}, z_{C,D} \mid y_B, y_D) \tag{5.159}$$

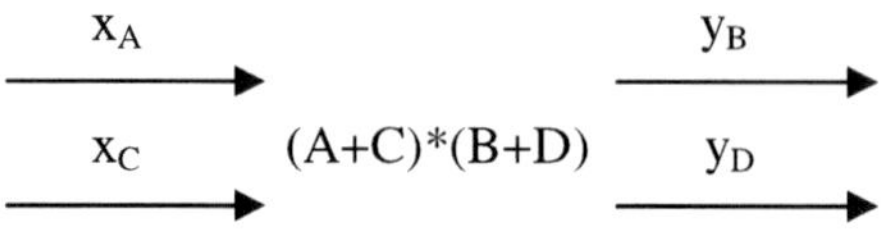

Fig. 5.60 Final state for vertical-vertical-horizontal merging

$$[(A*B)+(C*D)]\ (x_A, x_C \mid y_B, y_D\ ) \tag{5.160}$$

Proof 5.8

It has to be proved here that the vertical-horizontal-horizontal splitting of any single operand node (A/B)-(C/D) is equivalent the horizontal-vertical-vertical splitting of the same node (A-C)/(B-D) in accordance with Eq.(5.161). In this case, the four product nodes A, B, C and D are assumed to be listed alphabetically from left to right and from top to bottom within the interconnection structure of a two-level-two-layer FN as the one in Fig.3.1.

$$(A/B)\text{-}(C/D) = (A\text{-}C)/(B\text{-}D) \tag{5.161}$$

The proof is based on the use of a binary relation as a formal model for the two versions of the operand node (A/B)-(C/D) and (A-C)/(B-D), as shown in Eq. (5.162). In this case, the first elements of the relational pairs in (A/B)-(C/D) and (A-C)/(B-D) are duplets of type 'ac' whereas the second elements of these relational pairs are duplets of type 'bd'.

$$(A/B)\text{-}(C/D) = (A\text{-}C)/(B\text{-}D) = \tag{5.162}$$

$$\{(a_1^{1} c_1^{1}, b_2^{1} d_2^{1}),\ldots,(a_1^{1} c_1^{1}, b_2^{1} d_2^{s1}),\ldots,(a_1^{1} c_1^{r}, b_2^{1} d_2^{1}),\ldots,(a_1^{1} c_1^{r}, b_2^{1} d_2^{sr}),\ldots,$$

$$(a_1^{1} c_1^{1}, b_2^{q1} d_2^{1}),\ldots,(a_1^{1} c_1^{1}, b_2^{q1} d_2^{s1}),\ldots,(a_1^{1} c_1^{r}, b_2^{q1} d_2^{1}),\ldots,(a_1^{1} c_1^{r}, b_2^{q1} d_2^{sr}),\ldots,$$

$$(a_1^{p} c_1^{1}, b_2^{1} d_2^{1}),\ldots,(a_1^{p} c_1^{1}, b_2^{1} d_2^{s1}),\ldots,(a_1^{p} c_1^{r}, b_2^{1} d_2^{1}),\ldots,(a_1^{p} c_1^{r}, b_2^{1} d_2^{sr}),\ldots,$$

$$(a_1^{p} c_1^{1}, b_2^{qp} d_2^{1}),\ldots,(a_1^{p} c_1^{1}, b_2^{qp} d_2^{s1}),\ldots,(a_1^{p} c_1^{r}, b_2^{qp} d_2^{1}),\ldots,(a_1^{p} c_1^{r}, b_2^{qp} d_2^{sr})\}$$

The first and the second duplets of any relational pair in (A/B)-(C/D) and (A-C)/(B-D) are denoted by the subscripts ‘1’ and ‘2’, respectively. As far as the associated superscripts are concerned, they denote the target pairs in the product relations A, B, C and D.

The vertical decomposition of the operand relation (A/B)-(C/D) gives the temporary relations A/B and C/D, as shown in Eqs.(5.163)-(5.164).

$$A/B = \{(a_1^{1}, b_2^{1}),\ldots,(a_1^{1}, b_2^{q1}),\ldots,(a_1^{p}, b_2^{1}),\ldots,(a_1^{p}, b_2^{qp})\} \tag{5.163}$$

$$C/D = \{(c_1^{1}, d_2^{1}),\ldots,(c_1^{1}, d_2^{s1}),\ldots,(c_1^{r}, d_2^{1}),\ldots,(c_1^{r}, d_2^{sr})\} \tag{5.164}$$

Further on, the horizontal decomposition of the temporary relations A/B and C/D gives the product relations A, B and C, D, respectively, as shown in Eqs. (5.165)-(5.168).

$$A = \{(a_1^1, a_2^1),\ldots,(a_1^p, a_2^p)\} \quad (5.165)$$

$$B = \{(a_2^1, b_2^1),\ldots,(a_2^1, b_2^{q1}),\ldots,(a_2^p, b_2^1),\ldots,(a_2^p, b_2^{qp})\} \quad (5.166)$$

$$C = \{(c_1^1, c_2^1),\ldots,(c_1^r, c_2^r)\} \quad (5.167)$$

$$D = \{(c_2^1, d_2^1),\ldots,(c_2^1, d_2^{s1}),\ldots,(c_2^r, d_2^1),\ldots,(c_2^r, d_2^{sr})\} \quad (5.168)$$

Alternatively, the horizontal decomposition of the operand relation (A-C)/(B-D) gives the temporary relations A-C and B-D, as shown in Eqs.(5.169)-(5.170).

$$A\text{-}C = \{(a_1^1 c_1^1, a_2^1 c_2^1),\ldots,(a_1^1 c_1^r, a_2^1 c_2^r),\ldots, \quad (5.169)$$
$$(a_1^p c_1^1, a_2^p c_2^1),\ldots,(a_1^p c_1^r, a_2^p c_2^r)\}$$

$$B\text{-}D = \quad (5.170)$$

$$\{(a_2^1 c_2^1, b_2^1 d_2^1),\ldots,(a_2^1 c_2^1, b_2^1 d_2^{s1}),\ldots,(a_2^1 c_2^r, b_2^1 d_2^1),\ldots,(a_2^1 c_2^r, b_2^1 d_2^{sr}),\ldots,$$
$$(a_2^1 c_2^1, b_2^{q1} d_2^1),\ldots,(a_2^1 c_2^1, b_2^{q1} d_2^{s1}),\ldots,(a_2^1 c_2^r, b_2^{q1} d_2^1),\ldots,(a_2^1 c_2^r, b_2^{q1} d_2^{sr}),\ldots,$$
$$(a_2^p c_2^1, b_2^1 d_2^1),\ldots,(a_2^p c_2^1, b_2^1 d_2^{s1}),\ldots,(a_2^p c_2^r, b_2^1 d_2^1),\ldots,(a_2^p c_2^r, b_2^1 d_2^{sr}),\ldots,$$
$$(a_2^p c_2^1, b_2^{qp} d_2^1),\ldots,(a_2^p c_2^1, b_2^{qp} d_2^{s1}),\ldots,(a_2^p c_2^r, b_2^{qp} d_2^1),\ldots,(a_2^p c_2^r, b_2^{qp} d_2^{sr})\}$$

Further on, the vertical decomposition of the temporary relations A-B and C-D gives the product relations A, B and C, D, respectively. As the latter are identical with the product relations from Eqs.(5.165)-(5.168), this shows the validity of Eq.(5.161) and concludes the proof.

Example 5.14

This example illustrates the equivalence of vertical-horizontal-horizontal splitting and horizontal-vertical-vertical splitting for an operand node with the two versions (A/B)-(C/D) and (A-C)/(B-D) in the context of Proof 5.8. First, node (A/B)-(C/D) is vertically split into temporary nodes A/B and C/D. Then, these two temporary nodes are horizontally split into product nodes A, B and C, D, respectively. Alternatively, node (A-C)/(B-D) is horizontally split into temporary nodes A-C and B-D. Then, these two temporary nodes are vertically split into product nodes A,C and B, D, respectively. All relevant states of this mixed property are described by the block schemes and the topological expressions in Figs.5.61-5.66 and Eqs.(5.171)-(5.176).

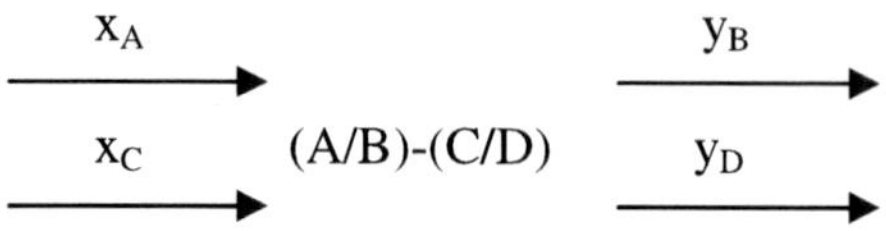

Fig. 5.61 Initial state for vertical-horizontal-horizontal splitting

$$[(A/B)\text{-}(C/D)]\ (x_A, x_C \mid y_B, y_D) \tag{5.171}$$

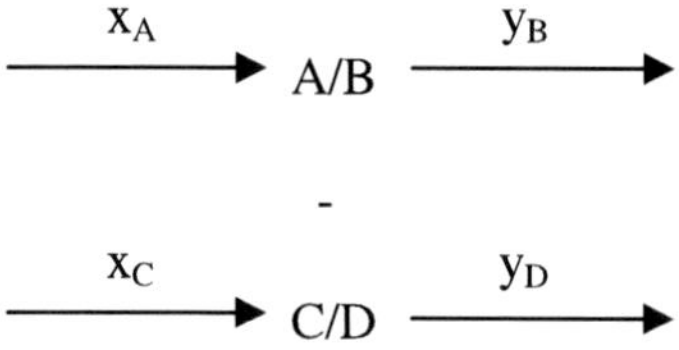

Fig. 5.62 Intermediate state for vertical-horizontal-horizontal splitting

$$[A/B]\ (x_A \mid y_B)\ \text{-}\ [C/D]\ (x_C \mid y_D) \tag{5.172}$$

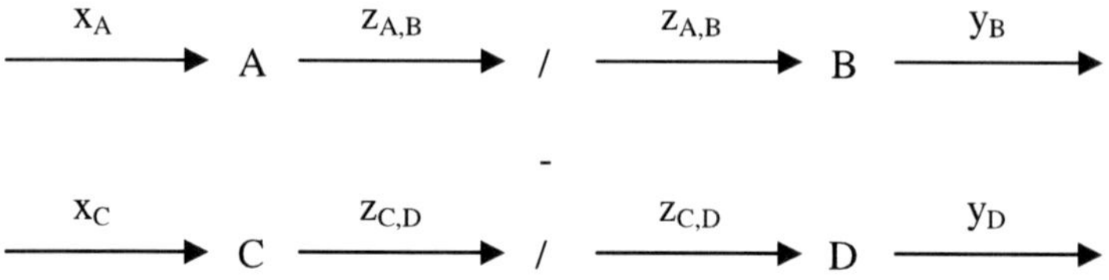

Fig. 5.63 Final state for vertical-horizontal-horizontal splitting

$$\{[A]\ (x_A \mid z_{A,B})\ /\ [B]\ (z_{A,B} \mid y_B)\}\ \text{-}\ \{[C]\ (x_C \mid z_{C,D})\ /\ [D]\ (z_{C,D} \mid y_D)\} \tag{5.173}$$

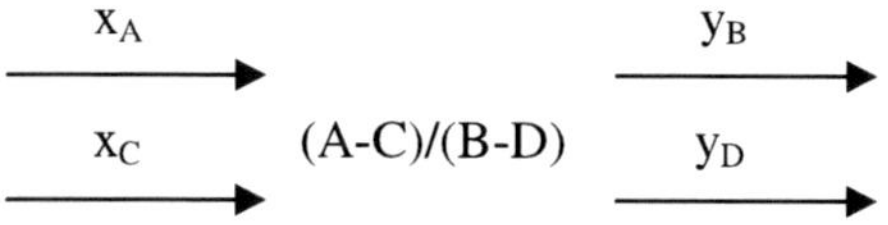

Fig. 5.64 Initial state for horizontal-vertical-vertical splitting

$$[(A\text{-}C)/(B\text{-}D)]\ (x_A, x_C \mid y_B, y_D) \tag{5.174}$$

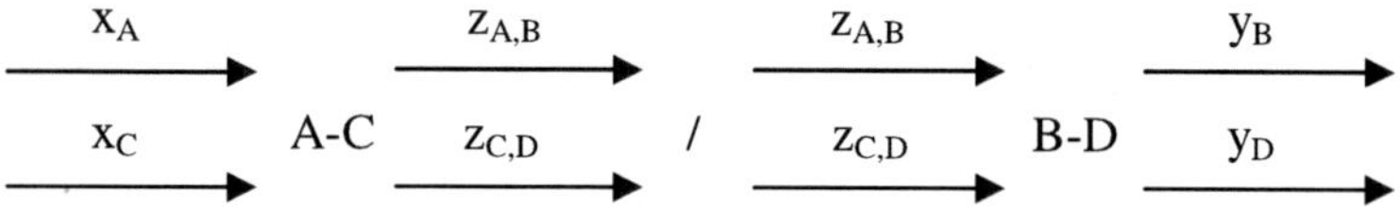

Fig. 5.65 Intermediate state for horizontal-vertical-vertical splitting

$$[A\text{-}C]\ (x_A, x_C \mid z_{A,B}, z_{C,D})\ /\ [B\text{-}D]\ (z_{A,B}, z_{C,D} \mid y_B, y_D) \tag{5.175}$$

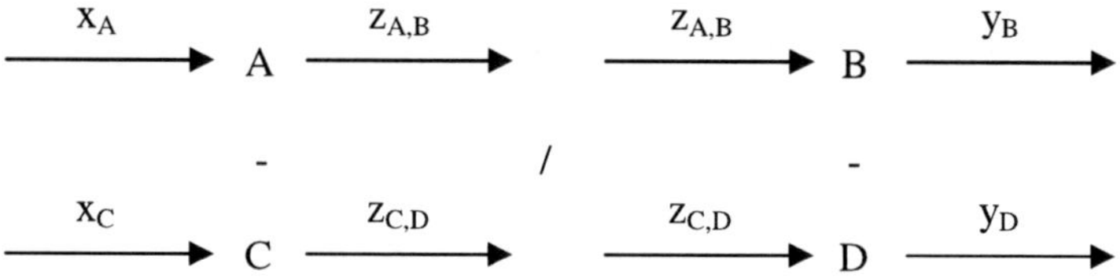

Fig. 5.66 Final state for horizontal-vertical-vertical splitting

$$\{[A]\ (x_A \mid z_{A,B})\ -\ [C]\ (x_C \mid z_{C,D})\}\ /\ \{[B]\ (z_{A,B} \mid y_B)\ -\ [D]\ (z_{C,D} \mid y_D)\} \tag{5.176}$$

Proof 5.9

It has to be proved here that the horizontal-horizontal-output merging of any four operand nodes A, B, C and D in a two-level-two-layer FN as the one in Fig.3.1 is equivalent to their output-vertical-horizontal merging in accordance with Eq.(5.177). In this case, the four nodes are assumed to be listed alphabetically from left to right and from top to bottom within the interconnection structure of this FN whereby the nodes in the first layer A and C are with common inputs.

$$(A*B);(C*D) = (A;C)*(B+D) \tag{5.177}$$

The proof is based on the use of binary relations as formal models for the operand nodes A, B, C and D, as shown in Eqs. (5.178)-(5.181). In this case, the first elements of the relational pairs in A and C are denoted by the letter e whereas the second elements of these relational pairs are denoted by the letter a and c, respectively. Also, the second elements of the relational pairs in B and D are denoted by the letters b and d whereas the first elements of these relational pairs are denoted by the letters a and c, respectively.

$$A = \{(e_1^{\,1}, a_2^{\,1}),\ldots,(e_1^{\,1}, a_2^{\,p1}),\ldots,(e_1^{\,r}, a_2^{\,1}),\ldots,(e_1^{\,r}, a_2^{\,pr})\} \tag{5.178}$$

$$B = \{(a_2^{\,1}, b_2^{\,1}),\ldots,(a_2^{\,1}, b_2^{\,s1}),\ldots,(a_2^{\,p1}, b_2^{\,1}),\ldots,$$
$$(a_2^{\,p1}, b_2^{\,s,p1}),\ldots,(a_2^{\,pr}, b_2^{\,1}),\ldots,(a_2^{\,pr}, b_2^{\,s,pr})\} \tag{5.179}$$

$$C = \{(e_1^{\,1}, c_2^{\,1}),\ldots,(e_1^{\,1}, c_2^{\,q1}),\ldots,(e_1^{\,r}, c_2^{\,1}),\ldots,(e_1^{\,r}, c_2^{\,qr})\} \tag{5.180}$$

$$D = \{(c_2^{1}, d_2^{1}),\ldots,(c_2^{1}, d_2^{t1}),\ldots,(c_2^{q1}, d_2^{1}),\ldots,$$
$$(c_2^{q1}, d_2^{t,q1}),\ldots,(c_2^{qr}, d_2^{1}),\ldots,(c_2^{qr}, d_2^{t,qr})\} \quad (5.181)$$

The first and the second element of any relational pair in A and C are denoted by the subscripts ‘1’ and ‘2’, respectively. However, both elements of any relational pair in B and D are denoted by the subscript ‘2’. This is due to the requirement for left composability of B and D, i.e. the first element of each pair in B must be identical with a second element of a pair in A whereas the first element of each pair in D must be identical with a second element of a pair in C.

The superscripts for the first elements of any relational pair in A and C vary from 1 to r whereas the superscripts for the second elements of these pairs vary from 1 to pr and 1 to qr, respectively. In this case, the relation A has ‘p1+...+pr’ pairs and the relation C has ‘q1+...+qr’ pairs. The superscripts for the first elements of the relational pairs in B and D vary from 1 to pr via p1 and from 1 to qr via q1, respectively. As far as the superscripts for the second elements of the relational pairs in B and D are concerned, they vary from 1 to s1, 1 to p1, 1 to pr for B and from 1 to t1, 1 to q1, 1 to qr for D. In this case, the relation B has ‘s1+...+p1+...+pr’ pairs and the relation D has 't1+...+q1+...+qr' pairs.

The horizontal composition of the operand relations A and B gives the temporary relation A*B, as shown in Eq.(5.182).

$$A*B = \{(e_1^{1}, b_2^{1}),\ldots,(e_1^{1}, b_2^{s1}),\ldots,(e_1^{1}, b_2^{s,p1}),\ldots,$$
$$(e_1^{r}, b_2^{1}),\ldots,(e_1^{r}, b_2^{s1}),\ldots,(e_1^{r}, b_2^{s,pr})\} \quad (5.182)$$

Also, the horizontal composition of the operand relations C and D gives the temporary relation C*D, as shown in Eq.(5.183).

$$C*D = \{(e_1^{1}, d_2^{1}),\ldots,(e_1^{1}, d_2^{t1}),\ldots,(e_1^{1}, d_2^{t,q1}),\ldots,$$
$$(e_1^{r}, d_2^{1}),\ldots,(e_1^{r}, d_2^{t1}),\ldots,(e_1^{r}, d_2^{t,qr})\} \quad (5.183)$$

Further on, the output composition of the temporary relation A*B and C*D gives the product relation (A*B);(C*D), as shown in Eq.(5.184).

$$(A*B);(C*D) = \{(e_1^{1}, b_2^{1} d_2^{1}),\ldots,(e_1^{1}, b_2^{1} d_2^{t1}),\ldots,(e_1^{1}, b_2^{1} d_2^{t,q1}),\ldots,$$
$$(e_1^{1}, b_2^{s1} d_2^{1}),\ldots,(e_1^{1}, b_2^{s1} d_2^{t1}),\ldots,(e_1^{1}, b_2^{s1} d_2^{t,q1}),\ldots,$$
$$(e_1^{1}, b_2^{s,p1} d_2^{1}),\ldots,(e_1^{1}, b_2^{s,p1} d_2^{t1}),\ldots,(e_1^{1}, b_2^{s,p1} d_2^{t,q1}),\ldots,$$
$$(e_1^{r}, b_2^{1} d_2^{1}),\ldots,(e_1^{r}, b_2^{1} d_2^{t1}),\ldots,(e_1^{r}, b_2^{1} d_2^{t,qr}),\ldots,$$
$$(e_1^{r}, b_2^{s1} d_2^{1}),\ldots,(e_1^{r}, b_2^{s1} d_2^{t1}),\ldots,(e_1^{r}, b_2^{s1} d_2^{t,qr}),\ldots,$$
$$(e_1^{r}, b_2^{s,pr} d_2^{1}),\ldots,(e_1^{r}, b_2^{s,pr} d_2^{t1}),\ldots,(e_1^{r}, b_2^{s,pr} d_2^{t,qr})\} \quad (5.184)$$

On the other hand, the output composition of the operand relations A and C gives the temporary relation A;C, as shown in Eq.(5.185).

$$A;C = \{(e_1^{1}, a_2^{1} c_2^{1}),\ldots,(e_1^{1}, a_2^{1} c_2^{q1}),\ldots,(e_1^{1}, a_2^{p1} c_2^{1}),\ldots,(e_1^{1}, a_2^{p1} c_2^{q1}),\ldots, (e_1^{r}, a_2^{1} c_2^{1}),\ldots,(e_1^{r}, a_2^{1} c_2^{qr}),\ldots,(e_1^{r}, a_2^{pr} c_2^{1}),\ldots,(e_1^{r}, a_2^{pr} c_2^{qr})\} \quad (5.185)$$

Also, the vertical composition of the operand relations B and D gives the temporary relation B+D, as shown in Eq.(5.186).

$$\begin{aligned} B+D = \{&(a_2^{1} c_2^{1}, b_2^{1} d_2^{1}),\ldots,(a_2^{1} c_2^{1}, b_2^{1} d_2^{t1}),\ldots,(a_2^{1} c_2^{q1}, b_2^{1} d_2^{1}),\ldots, \\ &(a_2^{1} c_2^{q1}, b_2^{1} d_2^{t,q1}),\ldots,(a_2^{1} c_2^{qr}, b_2^{1} d_2^{1}),\ldots,(a_2^{1} c_2^{qr}, b_2^{1} d_2^{t,qr}),\ldots, \\ &(a_2^{1} c_2^{1}, b_2^{s1} d_2^{1}),\ldots,(a_2^{1} c_2^{1}, b_2^{s1} d_2^{t1}),\ldots,(a_2^{1} c_2^{q1}, b_2^{s1} d_2^{1}),\ldots, \\ &(a_2^{1} c_2^{q1}, b_2^{s1} d_2^{t,q1}),\ldots,(a_2^{1} c_2^{qr}, b_2^{s1} d_2^{1}),\ldots,(a_2^{1} c_2^{qr}, b_2^{s1} d_2^{t,qr}),\ldots, \\ &(a_2^{p1} c_2^{1}, b_2^{1} d_2^{1}),\ldots,(a_2^{p1} c_2^{1}, b_2^{1} d_2^{t1}),\ldots,(a_2^{p1} c_2^{q1}, b_2^{1} d_2^{1}),\ldots, \\ &(a_2^{p1} c_2^{q1}, b_2^{1} d_2^{t,q1}),\ldots,(a_2^{p1} c_2^{qr}, b_2^{1} d_2^{1}),\ldots,(a_2^{p1} c_2^{qr}, b_2^{1} d_2^{t,qr}),\ldots, \\ &(a_2^{p1} c_2^{1}, b_2^{s,p1} d_2^{1}),\ldots,(a_2^{p1} c_2^{1}, b_2^{s,p1} d_2^{t1}),\ldots,(a_2^{p1} c_2^{q1}, b_2^{s,p1} d_2^{1}),\ldots, \\ &(a_2^{p1} c_2^{q1}, b_2^{s,p1} d_2^{t,q1}),\ldots,(a_2^{p1} c_2^{qr}, b_2^{s,p1} d_2^{1}),\ldots,(a_2^{p1} c_2^{qr}, b_2^{s,p1} d_2^{t,qr}),\ldots, \\ &(a_2^{pr} c_2^{1}, b_2^{1} d_2^{1}),\ldots,(a_2^{pr} c_2^{1}, b_2^{1} d_2^{t1}),\ldots,(a_2^{pr} c_2^{q1}, b_2^{1} d_2^{1}),\ldots, \\ &(a_2^{pr} c_2^{q1}, b_2^{1} d_2^{t,q1}),\ldots,(a_2^{pr} c_2^{qr}, b_2^{1} d_2^{1}),\ldots,(a_2^{pr} c_2^{qr}, b_2^{1} d_2^{t,qr}),\ldots, \\ &(a_2^{pr} c_2^{1}, b_2^{s,pr} d_2^{1}),\ldots,(a_2^{pr} c_2^{1}, b_2^{s,pr} d_2^{t1}),\ldots,(a_2^{pr} c_2^{q1}, b_2^{s,pr} d_2^{1}),\ldots, \\ &(a_2^{pr} c_2^{q1}, b_2^{s,pr} d_2^{t,q1}),\ldots,(a_2^{pr} c_2^{qr}, b_2^{s,pr} d_2^{1}),\ldots,(a_2^{pr} c_2^{qr}, b_2^{s,pr} d_2^{t,qr})\} \end{aligned} \quad (5.186)$$

In this case, the horizontal composition of the temporary relations A;C and B+D gives the product relation (A;C)*(B+D). As the latter is identical with the product relation (A*B);(C*D) from Eq.(5.184), this shows the validity of Eq.(5.177) and concludes the proof.

Example 5.15

This example illustrates the equivalence of horizontal-horizontal-output merging and output-vertical-horizontal merging for four operand nodes A, B, C and D with common inputs in the context of Proof 5.9. First, node A is horizontally merged with node B into a temporary node A*B and node C is horizontally merged with node D into a temporary node C*D. Then, node A*B is output merged with node C*D into a product node (A*B);(C*D). Alternatively, node A is output merged with node C into a temporary node A;C and node B is vertically merged with node

D into a temporary node B+D. Then, node A;C is horizontally merged with node B+D into a product node (A;C)*(B+D). All relevant states of this mixed property are described by the block schemes and the topological expressions in Figs.5.67-5.72 and Eqs.(5.187)-(5.192).

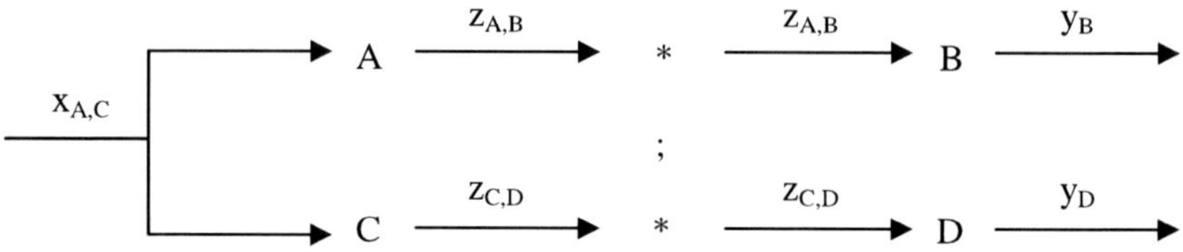

Fig. 5.67 Initial state for horizontal-horizontal-output merging

$$\{[A]\,(x_{A,C} \,|\, z_{A,B}) * [B]\,(z_{A,B} \,|\, y_B)\} \;;\; \{[C]\,(x_{A,C} \,|\, z_{C,D}) * [D]\,(z_{C,D} \,|\, y_D)\} \qquad (5.187)$$

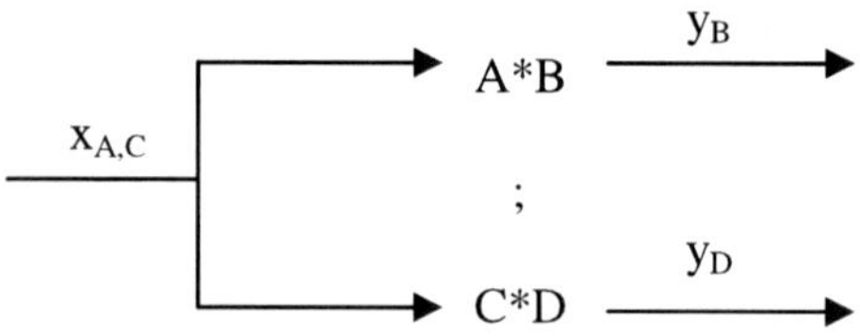

Fig. 5.68 Intermediate state for horizontal-horizontal-output merging

$$[A*B]\,(x_{A,C} \,|\, y_B) \;;\; [C*D]\,(x_{A,C} \,|\, y_D) \qquad (5.188)$$

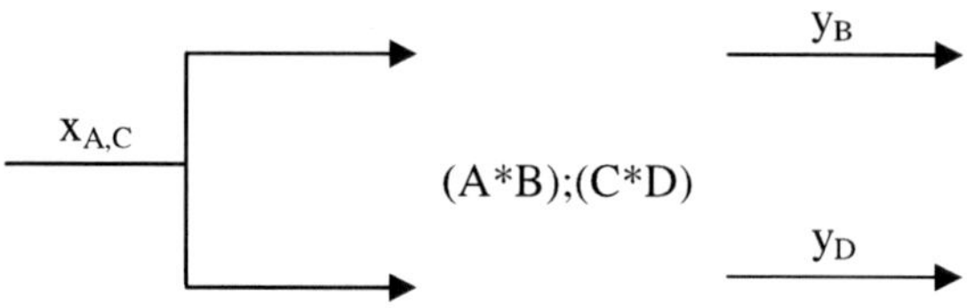

Fig. 5.69 Final state for horizontal-horizontal-output merging

$$[(A*B);(C*D)]\,(x_{A,C} \,|\, y_B,\, y_D\,) \qquad (5.189)$$

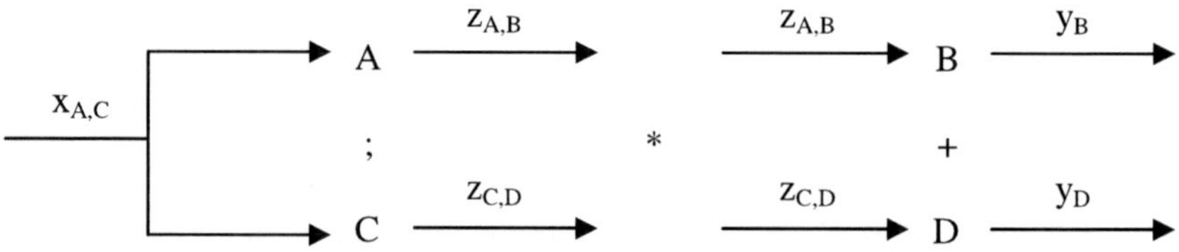

Fig. 5.70 Initial state for output-vertical-horizontal merging

$$\{[A]\ (x_{A,C} \mid z_{A,B})\ ;\ [C]\ (x_{A,C} \mid z_{C,D})\}\ *\ \{[B]\ (z_{A,B} \mid y_B) + [D]\ (z_{C,D} \mid y_D)\} \qquad (5.190)$$

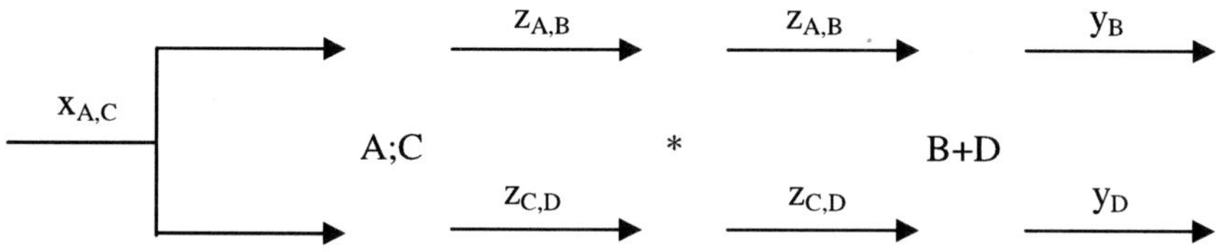

Fig. 5.71 Intermediate state for output-vertical-horizontal merging

$$[A;C]\ (x_{A,C} \mid z_{A,B},\ z_{C,D})\ *\ [B+D]\ (z_{A,B},\ z_{C,D} \mid y_B,\ y_D) \qquad (5.191)$$

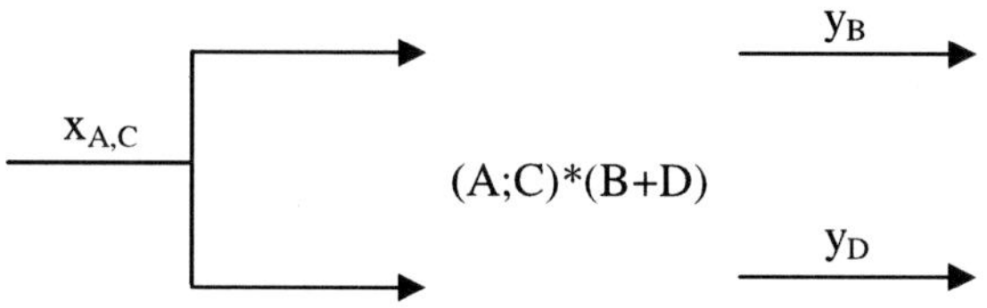

Fig. 5.72 Final state for output-vertical-horizontal merging

$$[(A;C)*(B+D)]\ (x_{A,C} \mid y_B,\ y_D\ ) \qquad (5.192)$$

Proof 5.10

It has to be proved here that the output-horizontal-horizontal splitting of any single operand node (A/B):(C/D) is equivalent the horizontal-output-vertical splitting of the same node (A:C)/(B-D) in accordance with Eq.(5.193). In this case, the four product nodes A, B, C and D are assumed to be listed alphabetically from left to right and from top to bottom within the interconnection structure of a two-level-two-layer FN as the one in Fig.3.1.

(A/B):(C/D) = (A:C)/(B-D) (5.193)

The proof is based on the use of a binary relation as a formal model for the two versions of the operand node (A/B):(C/D) and (A:C)/(B-D), as shown in Eq. (5.194). In this case, the first element of the relational pairs in (A/B)-(C/D) and (A-C)/(B-D) is the letter e whereas the second elements of these relational pairs are duplets of type 'bd'.

$$(A/B):(C/D) = (A:C)/(B\text{-}D) = \tag{5.194}$$

$$\{(e_1{}^1, b_2{}^1 d_2{}^1),\ldots,(e_1{}^1, b_2{}^1 d_2{}^{t1}),\ldots,(e_1{}^1, b_2{}^1 d_2{}^{t,q1}),\ldots,$$
$$(e_1{}^1, b_2{}^{s1} d_2{}^1),\ldots,(e_1{}^1, b_2{}^{s1} d_2{}^{t1}),\ldots,(e_1{}^1, b_2{}^{s1} d_2{}^{t,q1}),\ldots,$$
$$(e_1{}^1, b_2{}^{s,p1} d_2{}^1),\ldots,(e_1{}^1, b_2{}^{s,p1} d_2{}^{t1}),\ldots,(e_1{}^1, b_2{}^{s,p1} d_2{}^{t,q1}),\ldots,$$
$$(e_1{}^r, b_2{}^1 d_2{}^1),\ldots,(e_1{}^r, b_2{}^1 d_2{}^{t1}),\ldots,(e_1{}^r, b_2{}^1 d_2{}^{t,qr}),\ldots,$$
$$(e_1{}^r, b_2{}^{s1} d_2{}^1),\ldots,(e_1{}^r, b_2{}^{s1} d_2{}^{t1}),\ldots,(e_1{}^r, b_2{}^{s1} d_2{}^{t,qr}),\ldots,$$
$$(e_1{}^r, b_2{}^{s,pr} d_2{}^1),\ldots,(e_1{}^r, b_2{}^{s,pr} d_2{}^{t1}),\ldots,(e_1{}^r, b_2{}^{s,pr} d_2{}^{t,qr})\}$$

The first and the second elements of any relational pair in (A/B):(C/D) and (A:C)/(B-D) are denoted by the subscripts '1' and '2', respectively. As far as the associated superscripts are concerned, they denote the target pairs in the product relations A, B, C and D.

The output decomposition of the operand relation (A/B):(C/D) gives the temporary relations A/B and C/D, as shown in Eqs.(5.195)-(5.196).

$$A/B = \{(e_1{}^1, b_2{}^1),\ldots,(e_1{}^1, b_2{}^{s1}),\ldots,(e_1{}^1, b_2{}^{s,p1}),\ldots, \tag{5.195}$$
$$(e_1{}^r, b_2{}^1),\ldots,(e_1{}^r, b_2{}^{s1}),\ldots,(e_1{}^r, b_2{}^{s,pr})\}$$

$$C/D = \{(e_1{}^1, d_2{}^1),\ldots,(e_1{}^1, d_2{}^{t1}),\ldots,(e_1{}^1, d_2{}^{t,q1}),\ldots, \tag{5.196}$$
$$(e_1{}^r, d_2{}^1),\ldots,(e_1{}^r, d_2{}^{t1}),\ldots,(e_1{}^r, d_2{}^{t,qr})\}$$

Further on, the horizontal decomposition of the temporary relations A/B and C/D gives the product relations A, B and C, D, respectively, as shown in Eqs. (5.197)-(5.200).

$$A = \{(e_1{}^1, a_2{}^1),\ldots,(e_1{}^1, a_2{}^{p1}),\ldots,(e_1{}^r, a_2{}^1),\ldots,(e_1{}^r, a_2{}^{pr})\} \tag{5.197}$$

$$B = \{(a_2{}^1, b_2{}^1),\ldots,(a_2{}^1, b_2{}^{s1}),\ldots,(a_2{}^{p1}, b_2{}^1),\ldots, \tag{5.198}$$
$$(a_2{}^{p1}, b_2{}^{s,p1}),\ldots,(a_2{}^{pr}, b_2{}^1),\ldots,(a_2{}^{pr}, b_2{}^{s,pr})\}$$

$$C = \{(e_1{}^1, c_2{}^1),\ldots,(e_1{}^1, c_2{}^{q1}),\ldots,(e_1{}^r, c_2{}^1),\ldots,(e_1{}^r, c_2{}^{qr})\} \tag{5.199}$$

$$D = \{(c_2{}^1, d_2{}^1),\ldots,(c_2{}^1, d_2{}^{t1}),\ldots,(c_2{}^{q1}, d_2{}^1),\ldots, \tag{5.200}$$
$$(c_2{}^{q1}, d_2{}^{t,q1}),\ldots,(c_2{}^{qr}, d_2{}^1),\ldots,(c_2{}^{qr}, d_2{}^{t,qr})\}$$

Alternatively, the horizontal decomposition of the operand relation (A:C)/(B-D) gives the temporary relations A:C and B-D, as shown in Eqs.(5.201)-(5.202).

$$A{:}C = \{(e_1{}^1, a_2{}^1 c_2{}^1),\ldots,(e_1{}^1, a_2{}^1 c_2{}^{q1}),\ldots,(e_1{}^1, a_2{}^{p1} c_2{}^1),\ldots,(e_1{}^1, a_2{}^{p1} c_2{}^{q1}),\ldots, \tag{5.201}$$
$$(e_1{}^r, a_2{}^1 c_2{}^1),\ldots,(e_1{}^r, a_2{}^1 c_2{}^{qr}),\ldots,(e_1{}^r, a_2{}^{pr} c_2{}^1),\ldots,(e_1{}^r, a_2{}^{pr} c_2{}^{qr})\}$$

$$B\text{-}D = \{(a_2{}^1 c_2{}^1, b_2{}^1 d_2{}^1),\ldots,(a_2{}^1 c_2{}^1, b_2{}^1 d_2{}^{t1}),\ldots,(a_2{}^1 c_2{}^{q1}, b_2{}^1 d_2{}^1),\ldots, \tag{5.202}$$
$$(a_2{}^1 c_2{}^{q1}, b_2{}^1 d_2{}^{t,q1}),\ldots,(a_2{}^1 c_2{}^{qr}, b_2{}^1 d_2{}^1),\ldots,(a_2{}^1 c_2{}^{qr}, b_2{}^1 d_2{}^{t,qr}),\ldots,$$
$$(a_2{}^1 c_2{}^1, b_2{}^{s1} d_2{}^1),\ldots,(a_2{}^1 c_2{}^1, b_2{}^{s1} d_2{}^{t1}),\ldots,(a_2{}^1 c_2{}^{q1}, b_2{}^{s1} d_2{}^1),\ldots,$$
$$(a_2{}^1 c_2{}^{q1}, b_2{}^{s1} d_2{}^{t,q1}),\ldots,(a_2{}^1 c_2{}^{qr}, b_2{}^{s1} d_2{}^1),\ldots,(a_2{}^1 c_2{}^{qr}, b_2{}^{s1} d_2{}^{t,qr}),\ldots,$$
$$(a_2{}^{p1} c_2{}^1, b_2{}^1 d_2{}^1),\ldots,(a_2{}^{p1} c_2{}^1, b_2{}^1 d_2{}^{t1}),\ldots,(a_2{}^{p1} c_2{}^{q1}, b_2{}^1 d_2{}^1),\ldots,$$
$$(a_2{}^{p1} c_2{}^{q1}, b_2{}^1 d_2{}^{t,q1}),\ldots,(a_2{}^{p1} c_2{}^{qr}, b_2{}^1 d_2{}^1),\ldots,(a_2{}^{p1} c_2{}^{qr}, b_2{}^1 d_2{}^{t,qr}),\ldots,$$
$$(a_2{}^{p1} c_2{}^1, b_2{}^{s,p1} d_2{}^1),\ldots,(a_2{}^{p1} c_2{}^1, b_2{}^{s,p1} d_2{}^{t1}),\ldots,(a_2{}^{p1} c_2{}^{q1}, b_2{}^{s,p1} d_2{}^1),\ldots,$$
$$(a_2{}^{p1} c_2{}^{q1}, b_2{}^{s,p1} d_2{}^{t,q1}),\ldots,(a_2{}^{p1} c_2{}^{qr}, b_2{}^{s,p1} d_2{}^1),\ldots,(a_2{}^{p1} c_2{}^{qr}, b_2{}^{s,p1} d_2{}^{t,qr}),\ldots,$$
$$(a_2{}^{pr} c_2{}^1, b_2{}^1 d_2{}^1),\ldots,(a_2{}^{pr} c_2{}^1, b_2{}^1 d_2{}^{t1}),\ldots,(a_2{}^{pr} c_2{}^{q1}, b_2{}^1 d_2{}^1),\ldots,$$
$$(a_2{}^{pr} c_2{}^{q1}, b_2{}^1 d_2{}^{t,q1}),\ldots,(a_2{}^{pr} c_2{}^{qr}, b_2{}^1 d_2{}^1),\ldots,(a_2{}^{pr} c_2{}^{qr}, b_2{}^1 d_2{}^{t,qr}),\ldots,$$
$$(a_2{}^{pr} c_2{}^1, b_2{}^{s,pr} d_2{}^1),\ldots,(a_2{}^{pr} c_2{}^1, b_2{}^{s,pr} d_2{}^{t1}),\ldots,(a_2{}^{pr} c_2{}^{q1}, b_2{}^{s,pr} d_2{}^1),\ldots,$$
$$(a_2{}^{pr} c_2{}^{q1}, b_2{}^{s,pr} d_2{}^{t,q1}),\ldots,(a_2{}^{pr} c_2{}^{qr}, b_2{}^{s,pr} d_2{}^1),\ldots,(a_2{}^{pr} c_2{}^{qr}, b_2{}^{s,pr} d_2{}^{t,qr})\}$$

Further on, the output decomposition of the temporary relation A:C and the vertical decomposition of the temporary relation B-D gives the product relations A, B and C, D, respectively. As the latter are identical with the product relations from Eqs.(5.197)-(5.200), this shows the validity of Eq.(5.193) and concludes the proof.

Example 5.16

This example illustrates the equivalence of output-horizontal-horizontal splitting and horizontal-output-vertical splitting for an operand node with the two versions (A/B):(C/D) and (A:C)/(B-D) in the context of Proof 5.10. First, node (A/B):(C/D) is output split into temporary nodes A/B and C/D. Then, these two temporary nodes are horizontally split into product nodes A, B and C, D, respectively. Alternatively, node (A:C)/(B-D) is horizontally split into temporary nodes A:C and B-D. Then, these two temporary nodes are output and vertically split into product nodes A,C and B, D, respectively. All relevant states of this mixed property are described by the block schemes and the topological expressions in Figs.5.73-5.78 and Eqs.(5.203)-(5.208).

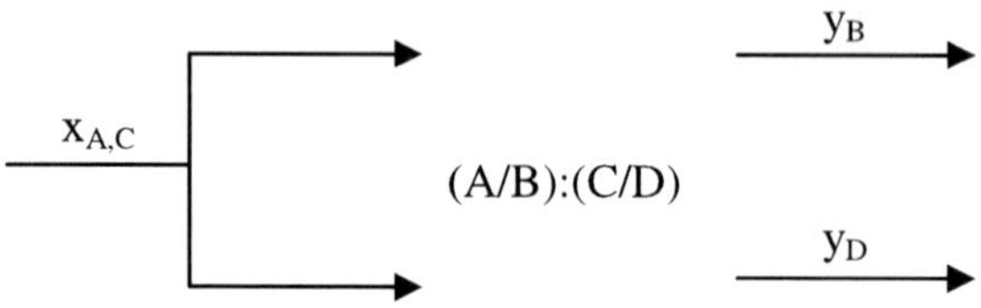

Fig. 5.73 Initial state for output horizontal-horizontal splitting

$$[(A/B):(C/D)]\ (x_{A,C} \mid y_B,\ y_D\ ) \tag{5.203}$$

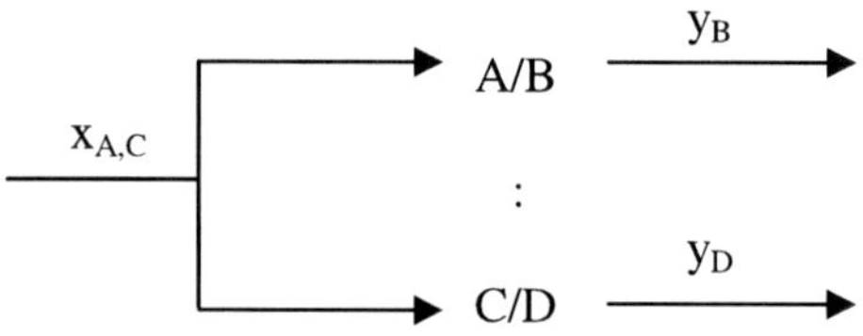

Fig. 5.74 Intermediate state for output-horizontal-horizontal splitting

$$[A/B]\ (x_{A,C} \mid y_B)\ :\ [C/D]\ (x_{A,C} \mid y_D) \tag{5.204}$$

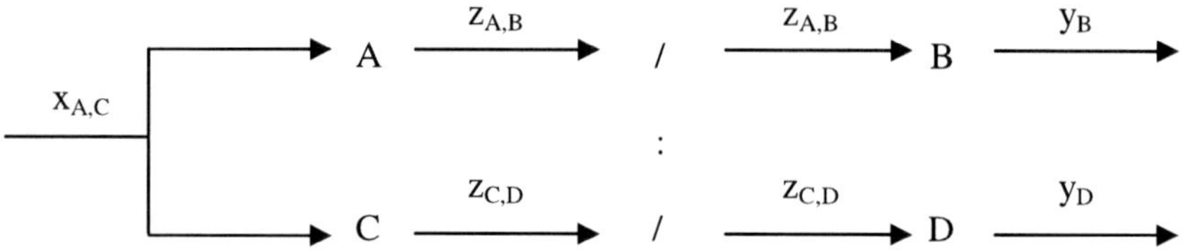

Fig. 5.75 Final state for output horizontal-horizontal splitting

$$\{[A]\ (x_{A,C} \mid z_{A,B})\ /\ [B]\ (z_{A,B} \mid y_B)\}\ :\ \{[C]\ (x_{A,C} \mid z_{C,D})\ /\ [D]\ (z_{C,D} \mid y_D)\} \tag{5.205}$$

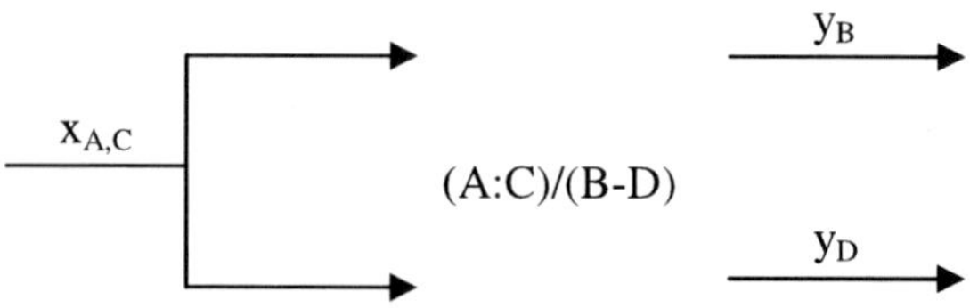

Fig. 5.76 Initial state for horizontal-output-vertical spitting

$$[(A:C)/(B-D)]\ (x_{A,C} \mid y_B,\ y_D\ ) \tag{5.206}$$

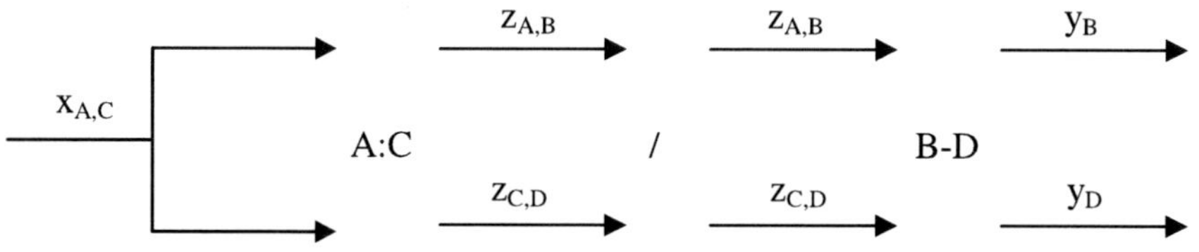

Fig. 5.77 Intermediate state for horizontal-output-vertical splitting

$$[A:C]\ (x_{A,C} \mid z_{A,B},\ z_{C,D})\ /\ [B\text{-}D]\ (z_{A,B},\ z_{C,D} \mid y_B,\ y_D) \tag{5.207}$$

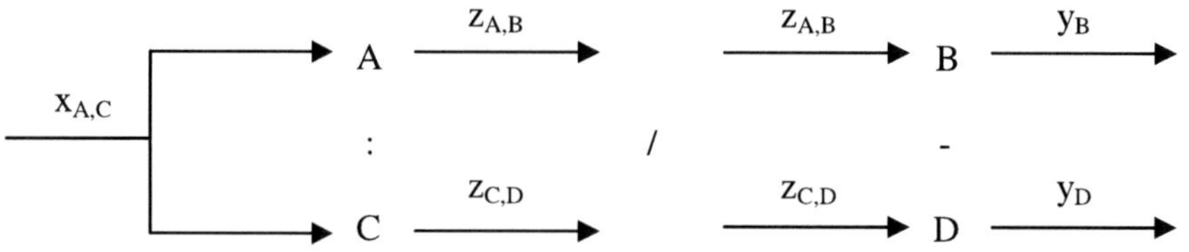

Fig. 5.78 Final state for horizontal-output-vertical splitting

$$\{[A]\ (x_{A,C} \mid z_{A,B}) : [C]\ (x_{A,C} \mid z_{C,D})\}\ /\ \{[B]\ (x_{A,B} \mid y_B)\ \text{-}\ [D]\ (z_{C,D} \mid y_D)\} \tag{5.208}$$

5.9 Comparison of Structural Properties

The structural properties of basic operations introduced in this chapter are central to the linguistic composition approach used in the book. This applies particularly to properties of merging operations which are used for composing the networked rule bases within a FN into a linguistically equivalent single rule base for a fuzzy system. On the contrary, properties of splitting operations are used for decomposing a single rule base for a fuzzy system into linguistically equivalent networked rule bases within a FN. However, in some cases properties of splitting operations may complement properties of merging operations in the context of the linguistic composition approach and this is shown by some examples further in this book.

The application of structural properties is governed by the location of brackets in the corresponding topological expressions. In particular, operations in brackets are carried out first in the case of associativity and last in the case of variability.

The characteristics of different types of structural properties of basic operations are summarised in Table 5.1.

Table 5.1 Characteristics of structural properties of basic operations

Structural properties	Composition	Brackets
Associativity of horizontal merging	Yes	First
Variability of horizontal splitting	No	Last
Associativity of vertical merging	Yes	First
Variability of vertical splitting	No	Last
Associativity of output merging	Yes	First
Variability of output splitting	No	Last

The next chapter introduces more advanced concepts from the theoretical framework for FNs. In particular, it discusses several types of advanced operations in FNs.

Chapter 6
Advanced Operations in Fuzzy Networks

6.1 Introduction to Advanced Operations

The structural properties of basic operations introduced in Chapter 5 can be applied to FNs with a large number of nodes. These nodes may be sequential, parallel or with common inputs. However, the structure of a FN may include more complex connections among the nodes which would require some preliminary manipulation before the properties of the basic operations can be applied. Therefore, it is necessary to define a number of additional advanced operations that can be used in this context.

Advanced operations make possible the manipulation of nodes within a FN with an arbitrarily complex structure. In this respect, there are two types of advanced operations in FNs – operations based on node transformation and operations based on node identification. Node transformation based operations are used for analysis whereby all nodes in a FN are known and the task is to find an equivalent fuzzy system in accordance with the linguistic composition approach. As opposed to this, node identification based operations are used for design whereby some nodes in a FN are unknown and the task is to find these nodes in a way that would guarantee a prespecified performance of the equivalent fuzzy system. In either case, all connections among nodes are assumed to be known.

Some of the advanced operations introduced further in this chapter are similar to operations used in applied mathematics and control theory. However, these advanced operations are also novel in that they are applied to a FN which is a novel concept.

All advanced operations are illustrated with examples of nodes with a fairly small number of inputs, outputs and intermediate variables but the extension of these examples to cases of higher dimension is straightforward. Besides this, the advanced operations related to design are first formulated at a more abstract level as generic problems and their solutions are given in a general form. The problems and the examples are based on the use of Boolean matrices or binary relations as formal models for FNs at node level as these formal models lend themselves easily to advanced manipulation in the context of the linguistic composition approach. Therefore, the advanced operations can be viewed as sophisticated building blocks for the simplification of an arbitrarily complex FN to a fuzzy system.

A. Gegov: Fuzzy Networks for Complex Systems, STUDFUZZ 259, pp. 109–160.
springerlink.com

6.2 Node Transformation for Input Augmentation

Node transformation is usually applied for input augmentation when two or more nodes in a particular layer of a FN have some common inputs as well as inputs that are not common to all these nodes. In this case, it is necessary to augment the nodes with the missing common inputs such that all nodes have only common inputs. The purpose of this augmentation is to allow the output merging operation and its property of associativity to be applied to all nodes in the first layer of the FN. As a result, the nodes with the augmented inputs have to be transformed appropriately to reflect the presence of these inputs. In this context, input augmentation can always be applied due to the possibility of extending the set of inputs to a node with an arbitrary number of additional inputs.

When a Boolean matrix is used as a formal model for a node during input augmentation, the transformation of this node represents an expansion of the associated matrix along its rows. In particular, the product matrix is obtained by replicating each row from the operand matrix as many times as the number of permutations of linguistic terms for the augmented inputs minus one. The location of the replicated rows in the product matrix depends on the place of the augmented inputs in the extended set of inputs.

Node transformation can also be applied for input augmentation in the context of binary relations when such a relation is used as a formal model for the operand node. In this case, the transformation of the operand node represents a special type of relational extension that is applied only to the first elements from the pairs of the operand relation whereas the second elements remain unchanged. In particular, the first elements in the pairs of the operand relation and the product relation represent all possible permutations of linguistic terms of inputs from the original and the extended set of inputs, respectively. During this process, the unchanged second elements in the pairs of the operand relation are replicated in the pairs of the product relation as many times as the number of permutations of linguistic terms for the augmented inputs minus one.

Example 6.1

This example considers an operand node N with input x and output y whose input is augmented with an input x^{AI} to the top, i.e. x^{AI} is located before x in the extended set of inputs. This node can be described by the Boolean matrix in Eq.(6.1) and the binary relation in Eq.(6.2). In this context, node N represents a one-node FN that can be described by the block-scheme in Fig.6.1 and the topological expression in Eq.(6.3).

N: (6.1)

x \ y	1	2	3
1	0	1	0
2	0	0	1
3	1	0	0

$$N: \{(1, 2), (2, 3), (3, 1)\} \tag{6.2}$$

Fig. 6.1 One-node FN with one input before augmentation

$$[N]\ (x \mid y) \tag{6.3}$$

As a result of this input augmentation, the operand node N is transformed into a product node N^{AI} with input set $\{x^{AI}, x\}$ and output y. This node can be described by the Boolean matrix in Eq.(6.4) and the binary relation in Eq.(6.5). In this context, node N^{AI} represents a one-node FN that can be described by the block-scheme in Fig.6.2 and the topological expression in Eq.(6.6).

N^{AI}: (6.4)

x^{AI}, x \ y	1	2	3
11	0	1	0
12	0	0	1
13	1	0	0
21	0	1	0
22	0	0	1
23	1	0	0
31	0	1	0
32	0	0	1
33	1	0	0

$$N^{AI}: \{(11, 2), (12, 3)\ (13, 1), (21, 2), (22, 3)\ (23, 1), (31, 2), (32, 3)\ (33, 1)\} \tag{6.5}$$

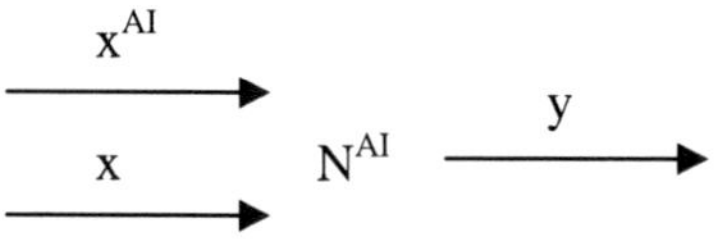

Fig. 6.2 One-node FN with one input after augmentation to the top

$$[N^{AI}]\ (x^{AI}, x \mid y) \tag{6.6}$$

Example 6.2

This example considers an operand node N with input x and output y whose input is augmented with an input x^{AI} to the bottom, i.e. x^{AI} is located after x in the extended set of inputs. This node can be described by the Boolean matrix in

Eq.(6.7) and the binary relation in Eq.(6.8). In this context, node N represents a one-node FN that can be described by the block-scheme in Fig.6.1 and the topological expression in Eq.(6.3).

N: (6.7)

x \ y	1	2	3
1	0	1	0
2	1	0	0
3	0	0	1

$$N: \{(1, 2), (2, 1), (3, 3)\} \tag{6.8}$$

As a result of this input augmentation, the operand node N is transformed into a product node N^{AI} with input set $\{x, x^{AI}\}$ and output y. This node can be described by the Boolean matrix in Eq.(6.9) and the binary relation in Eq.(6.10). In this context, node N^{AI} represents a one-node FN that can be described by the block-scheme in Fig.6.3 and the topological expression in Eq.(6.11).

N^{AI}: (6.9)

x, x^{AI} \ y	1	2	3
11	0	1	0
12	0	1	0
13	0	1	0
21	1	0	0
22	1	0	0
23	1	0	0
31	0	0	1
32	0	0	1
33	0	0	1

$$\begin{aligned} N^{AI}: \{&(11, 2), (12, 2)\ (13, 2), \\ &(21, 1), (22, 1)\ (23, 1), \\ &(31, 3), (32, 3)\ (33, 3)\} \end{aligned} \tag{6.10}$$

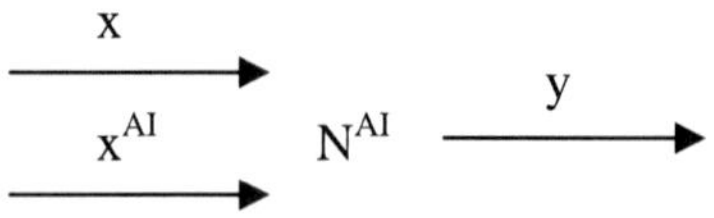

Fig. 6.3 One-node FN with one input after augmentation to the bottom

$$[N^{AI}]\ (x, x^{AI} \mid y) \tag{6.11}$$

Example 6.3

This example considers an operand node N with input set $\{x_1, x_2\}$ and output y whose input set is augmented with an input x^{AI} to the top, i.e. x^{AI} is located before x_1 and x_2 in the extended set of inputs. This node can be described by the Boolean matrix in Eq.(6.12) and the binary relation in Eq.(6.13). In this context, node N represents a one-node FN that can be described by the block-scheme in Fig.6.4 and the topological expression in Eq.(6.14).

N: (6.12)

x_1, x_2 \ y	1	2	3
11	0	0	1
12	0	0	1
21	0	1	0
22	1	0	0

$$N: \{(11, 3), (12, 3), (21, 2), (22,1)\} \quad (6.13)$$

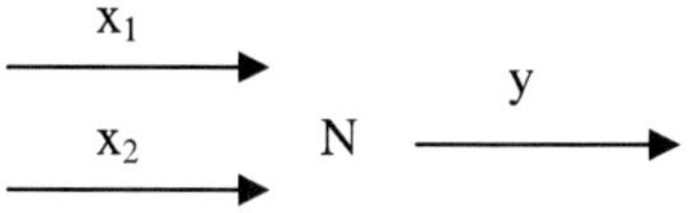

Fig. 6.4 One-node FN with two inputs before augmentation

$$[N]\ (x_1, x_2 \mid y) \quad (6.14)$$

As a result of this input augmentation, the operand node N is transformed into a product node N^{AI} with an input set $\{x^{AI}, x_1, x_2\}$ and output y. This node can be described by the Boolean matrix in Eq.(6.15) and the binary relation in Eq.(6.16). In this context, node N^{AI} represents a one-node FN that can be described by the block-scheme in Fig.6.5 and the topological expression in Eq.(6.17).

N^{AI}: (6.15)

x^{AI}, x_1, x_2 \ y	1	2	3
111	0	0	1
112	0	0	1
121	0	1	0
122	1	0	0
211	0	0	1
212	0	0	1
221	0	1	0
222	1	0	0

$$N^{AI}: \{(111, 3), (112, 3), (121, 2), (122, 1), (211, 3), (212, 3), (221, 2), (222, 1)\} \quad (6.16)$$

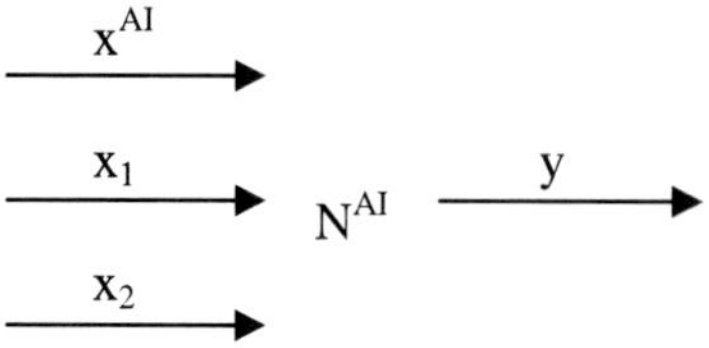

Fig. 6.5 One-node FN with two inputs after augmentation to the top

$$[N^{AI}]\ (x^{AI}, x_1, x_2 \mid y) \tag{6.17}$$

Example 6.4

This example considers an operand node N with input set $\{x_1, x_2\}$ and output y whose input set is augmented with an input x^{AI} in the middle, i.e. x^{AI} is located between x_1 and x_2 in the extended set of inputs. This node can be described by the Boolean matrix in Eq.(6.18) and the binary relation in Eq.(6.19). In this context, node N represents a one-node FN that can be described by the block-scheme in Fig.6.4 and the topological expression in Eq.(6.14).

N: (6.18)

x_1, x_2 \ y	1	2	3
11	0	0	1
12	0	1	0
21	0	1	0
22	1	0	0

$$N: \{(11, 3), (12, 2), (21, 2), (22,1)\} \tag{6.19}$$

As a result of this input augmentation, the operand node N is transformed into a product node N^{AI} with an input set $\{x_1, x^{AI}, x_2\}$ and output y. This node can be described by the Boolean matrix in Eq.(6.20) and the binary relation in Eq.(6.21). In this context, node N^{AI} represents a one-node FN that can be described by the block-scheme in Fig.6.6 and the topological expression in Eq.(6.22).

N^{AI}: (6.20)

x_1, x^{AI}, x_2 \ y	1	2	3
111	0	0	1
112	0	1	0
121	0	0	1
122	0	1	0
211	0	1	0
212	1	0	0
221	0	1	0
222	1	0	0

N^{AI}: {(111, 3), (112, 2), (121, 3), (122, 2),
(211, 2), (212, 1), (221, 2), (222, 1)} (6.21)

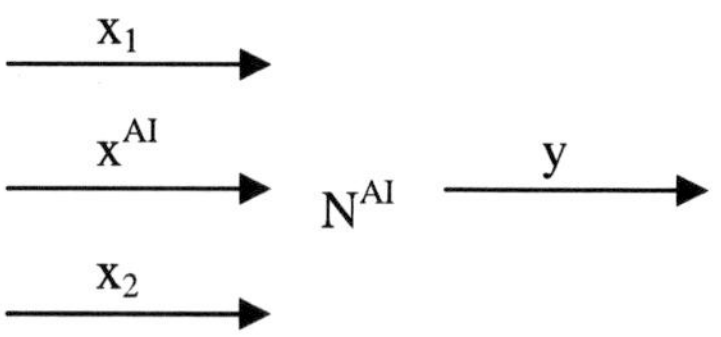

Fig. 6.6 One-node FN with two inputs after augmentation in the middle

$[N^{AI}]$ $(x_1, x^{AI}, x_2 \mid y)$ (6.22)

Example 6.5

This example considers an operand node N with input set $\{x_1, x_2\}$ and output y whose input set is augmented with an input x^{AI} to the bottom, i.e. x^{AI} is located after x_1 and x_2 in the extended set of inputs. This node can be described by the Boolean matrix in Eq.(6.23) and the binary relation in Eq.(6.24). In this context, node N represents a one-node FN that can be described by the block-scheme in Fig.6.4 and the topological expression in Eq.(6.14).

N: (6.23)

x_1, x_2 \ y	1	2	3
11	0	0	1
12	0	1	0
21	1	0	0
22	1	0	0

N: {(11, 3), (12, 2), (21, 1), (22,1)} (6.24)

As a result of this input augmentation, the operand node N is transformed into a product node N^{AI} with an input set $\{x_1, x_2, x^{AI}\}$ and output y. This node can be described by the Boolean matrix in Eq.(6.25) and the binary relation in Eq.(6.26). In this context, node N^{AI} represents a one-node FN that can be described by the block-scheme in Fig.6.7 and the topological expression in Eq.(6.27).

N^{AI}: (6.25)

y	1	2	3
x_1, x_2, x^{AI}			
111	0	0	1
112	0	0	1
121	0	1	0
122	0	1	0
211	1	0	0
212	1	0	0
221	1	0	0
222	1	0	0

$$N^{AI}: \{(111, 3), (112, 3), (121, 2), (122, 2), (211, 1), (212, 1), (221, 1), (222, 1)\} \tag{6.26}$$

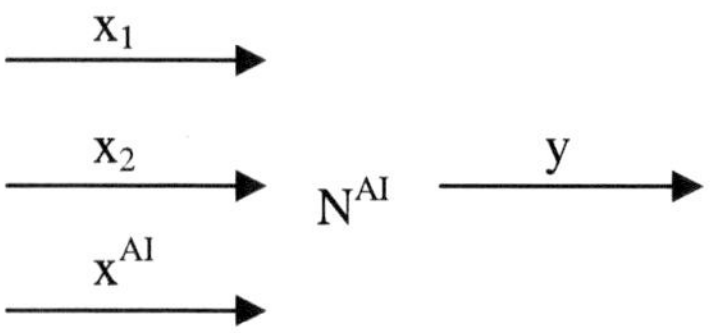

Fig. 6.7 One-node FN with two inputs after augmentation to the bottom

$$[N^{AI}]\ (x_1, x_2, x^{AI} \mid y) \tag{6.27}$$

6.3 Node Transformation for Output Permutation

Node transformation is usually applied for output permutation when two or more adjacent nodes in the same level of a FN have some connections with crossing paths. In this case, it is necessary to permute the output points of these connections such that the corresponding paths become parallel. The purpose of this permutation is to allow the horizontal merging operation and its property of associativity to be applied to all nodes in this level of the FN. As a result, the nodes with the permuted outputs have to be transformed appropriately to reflect the changed ordering of these outputs. In this context, output permutation can always be applied due to the possibility of rearranging the output set of a node for an arbitrary number of outputs.

When a Boolean matrix is used as a formal model for a node during output permutation, the transformation of this node is based on relocation of the non-zero columns of the associated matrix. In particular, the product matrix is obtained by moving each non-zero column from the operand matrix under a column label with linguistic terms permuted in accordance with the associated permuted outputs. The space vacated by a relocated non-zero column in the product matrix is filled with a zero column unless another non-zero column is moved there as part of the overall node transformation process.

Node transformation can also be applied for output permutation in the context of binary relations when such a relation is used as a formal model for the operand node. In this case, the transformation of the operand node represents a special type of relational manipulation that is applied only to the second elements from the pairs of the operand relation whereas the first elements remain unchanged. In particular, the first elements in the pairs of the operand relation and the product relation represent all possible permutations of linguistic terms for the inputs, respectively. During this process, the second elements in the pairs of the operand relation are obtained by permuting the corresponding linguistic terms for the outputs in accordance with the output permutation.

Example 6.6

This example considers an operand node N with input x and output set $\{y_1, y_2\}$ whose outputs are permuted, i.e. y_2 comes first and y_1 comes second in the rearranged set of outputs. Before the permutation, this node can be described by the Boolean matrix in Eq.(6.28) and the binary relation in Eq.(6.29). In this context, node N represents a one-node FN that can be described by the block-scheme in Fig.6.8 and the topological expression in Eq.(6.30).

N: (6.28)

x \ y_1, y_2	11	12	13	21	22	23	31	32	33
1	0	0	0	0	0	1	0	0	0
2	0	0	0	0	0	0	1	0	0
3	0	1	0	0	0	0	0	0	0

$$N: \{(1, 23), (2, 31), (3, 12)\} \quad (6.29)$$

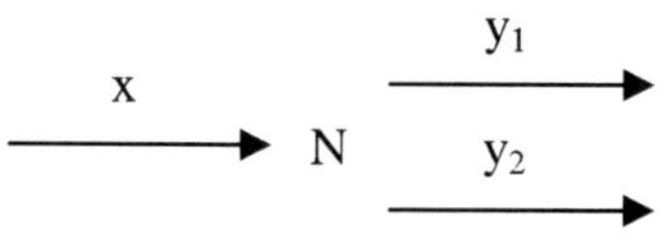

Fig. 6.8 One-node FN with two outputs before permutation

$$[N]\ (x \mid y_1, y_2) \quad (6.30)$$

As a result of this output permutation, the operand node N is transformed into a product node N^{PO} with input x and output set $\{y_2, y_1\}$. This node can be described by the Boolean matrix in Eq.(6.31) and the binary relation in Eq.(6.32). In this context, node N^{PO} represents a one-node FN that can be described by the block-scheme in Fig.6.9 and the topological expression in Eq.(6.33).

N^{PO}: (6.31)

x \ y_2, y_1	11	12	13	21	22	23	31	32	33
1	0	0	0	0	0	0	0	1	0
2	0	0	1	0	0	0	0	0	0
3	0	0	0	1	0	0	0	0	0

$$N^{PO} : \{(1, 32), (2, 13), (3, 21)\} \quad (6.32)$$

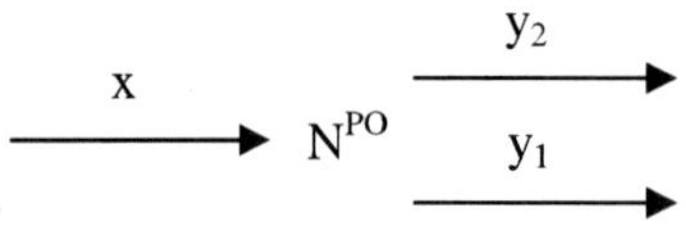

Fig. 6.9 One-node FN with two outputs after permutation

$$[N^{PO}]\ (x \mid y_2, y_1) \quad (6.33)$$

Example 6.7

This example considers an operand node N with input x and output set $\{y_1, y_2, y_3\}$ whose outputs are permuted in a middle-top-bottom manner, i.e. y_2 comes first, y_1 comes second and y_3 stays third in the rearranged set of outputs. Before the permutation, this node can be described by the Boolean matrix in Eq.(6.34) and the binary relation in Eq.(6.35). In this context, node N represents a one-node FN that can be described by the block-scheme in Fig.6.10 and the topological expression in Eq.(6.36).

N: (6.34)

x \ y_1, y_2, y_3	111	112	121	122	211	212	221	222
1	0	0	1	0	0	0	0	0
2	0	0	0	0	1	0	0	0
3	0	0	0	1	0	0	0	0

$$N : \{(1, 121), (2, 211), (3, 122)\} \quad (6.35)$$

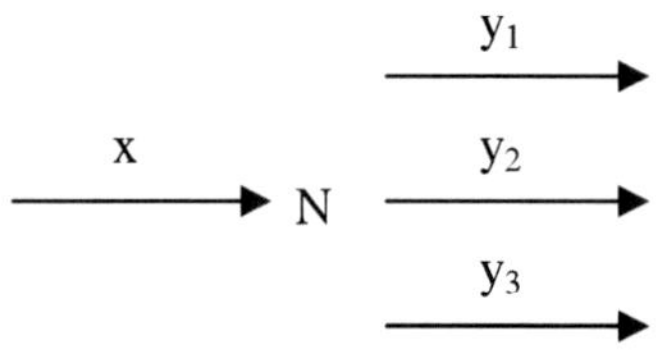

Fig. 6.10 One-node FN with three outputs before permutation

$[N]\ (x \mid y_1, y_2, y_3)$ (6.36)

As a result of this output permutation, the operand node N is transformed into a product node N^{PO} with input x and output set $\{y_2, y_1, y_3\}$. This node can be described by the Boolean matrix in Eq.(6.37) and the binary relation in Eq.(6.38). In this context, node N^{PO} represents a one-node FN that can be described by the block-scheme in Fig.6.11 and the topological expression in Eq.(6.39).

N^{PO}: (6.37)

x \ y_2, y_1, y_3	111	112	121	122	211	212	221	222
1	0	0	0	0	1	0	0	0
2	0	0	1	0	0	0	0	0
3	0	0	0	0	0	1	0	0

N^{PO}: {(1, 211), (2, 121), (3, 212)} (6.38)

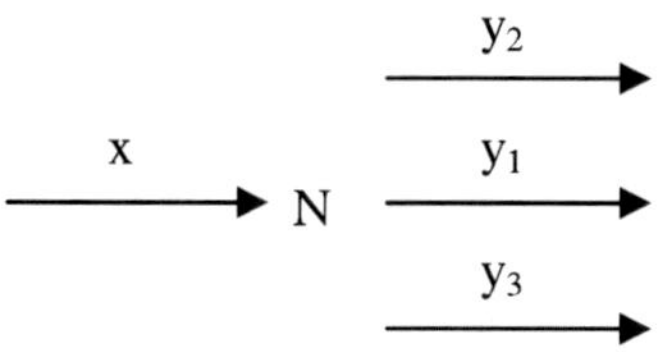

Fig. 6.11 One-node FN with three outputs after middle-top-bottom permutation

$[N]\ (x \mid y_2, y_1, y_3)$ (6.39)

Example 6.8

This example considers an operand node N with input x and output set $\{y_1, y_2, y_3\}$ whose outputs are permuted in a top-bottom-middle manner, i.e. y_1 stays first, y_3 comes second and y_2 comes third in the rearranged set of outputs. Before the permutation, this node can be described by the Boolean matrix in Eq.(6.40) and the binary relation in Eq.(6.41). In this context, node N represents a one-node FN that can be described by the block-scheme in Fig.6.10 and the topological expression in Eq.(6.36).

N: (6.40)

x \ y_1, y_2, y_3	111	112	121	122	211	212	221	222
1	0	1	0	0	0	0	0	0
2	0	0	1	0	0	0	0	0
3	0	0	0	0	0	1	0	0

N: {(1, 112), (2, 121), (3, 212)} (6.41)

As a result of this output permutation, the operand node N is transformed into a product node N^{PO} with input x and output set $\{y_1, y_3, y_2\}$. This node can be described by the Boolean matrix in Eq.(6.42) and the binary relation in Eq.(6.43). In this context, node N^{PO} represents a one-node FN that can be described by the block-scheme in Fig.6.12 and the topological expression in Eq.(6.44).

N^{PO}: (6.42)

x \ y_1, y_3, y_2	111	112	121	122	211	212	221	222
1	0	0	1	0	0	0	0	0
2	0	1	0	0	0	0	0	0
3	0	0	0	0	0	0	1	0

N^{PO}: {(1, 121), (2, 112), (3, 221)} (6.43)

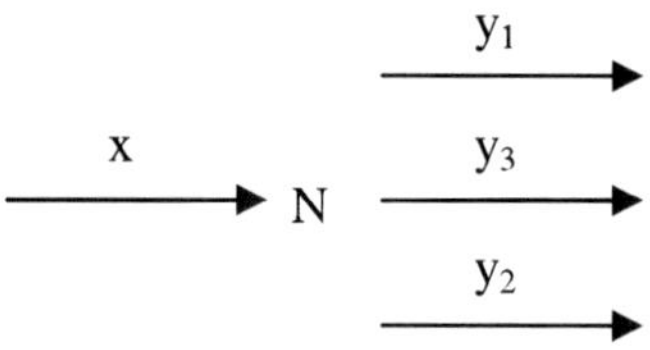

Fig. 6.12 One-node FN with three outputs after top-bottom-middle permutation

[N] (x | y_1, y_3, y_2) (6.44)

Example 6.9

This example considers an operand node N with input x and output set $\{y_1, y_2, y_3\}$ whose outputs are permuted in a bottom-middle-top manner, i.e. y_3 comes first, y_2 stays second and y_1 comes third in the rearranged set of outputs. Before the permutation, this node can be described by the Boolean matrix in Eq.(6.45) and the binary relation in Eq.(6.46). In this context, node N represents a one-node FN that can be described by the block-scheme in Fig.6.10 and the topological expression in Eq.(6.36).

N: (6.45)

x \ y_1, y_2, y_3	111	112	121	122	211	212	221	222
1	0	1	0	0	0	0	0	0
2	0	0	0	0	1	0	0	0
3	0	0	0	1	0	0	0	0

N: {(1, 112), (2, 211), (3, 122)} (6.46)

As a result of this output permutation, the operand node N is transformed into a product node N^{PO} with input x and output set $\{y_3, y_2, y_1\}$. This node can be described by the Boolean matrix in Eq.(6.47) and the binary relation in Eq.(6.48). In this context, node N^{PO} represents a one-node FN that can be described by the block-scheme in Fig.6.13 and the topological expression in Eq.(6.49).

N^{PO}: (6.47)

x \ y_3, y_2, y_1	111	112	121	122	211	212	221	222
1	0	0	0	0	1	0	0	0
2	0	1	0	0	0	0	0	0
3	0	0	0	0	0	0	1	0

$$N^{PO}: \{(1, 211), (2, 112), (3, 221)\} \quad (6.48)$$

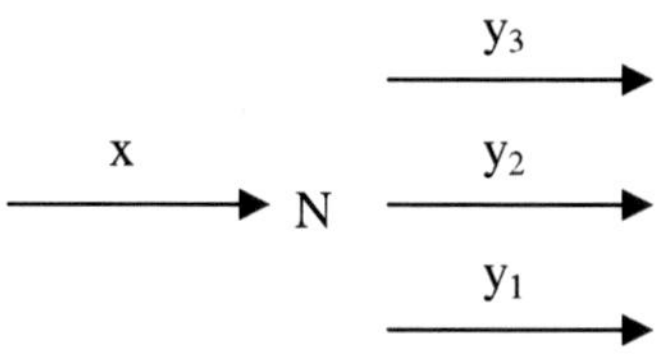

Fig. 6.13 One-node FN with three outputs after bottom-middle-top permutation

$$[N]\,(x \mid y_3, y_2, y_1) \quad (6.49)$$

Example 6.10

This example considers an operand node N with input x and output set $\{y_1, y_2, y_3\}$ whose outputs are permuted in a middle-bottom-top manner, i.e. y_2 comes first, y_3 comes second and y_1 comes third in the rearranged set of outputs. Before the permutation, this node can be described by the Boolean matrix in Eq.(6.50) and the binary relation in Eq.(6.51). In this context, node N represents a one-node FN that can be described by the block-scheme in Fig.6.10 and the topological expression in Eq.(6.36).

N: (6.50)

x \ y_1, y_2, y_3	111	112	121	122	211	212	221	222
1	0	1	0	0	0	0	0	0
2	0	0	0	1	0	0	0	0
3	0	0	1	0	0	0	0	0

$$N: \{(1, 112), (2, 122), (3, 121)\} \quad (6.51)$$

As a result of this output permutation, the operand node N is transformed into a product node N^{PO} with input x and output set $\{y_2, y_3, y_1\}$. This node can be described by the Boolean matrix in Eq.(6.52) and the binary relation in Eq.(6.53). In this context, node N^{PO} represents a one-node FN that can be described by the block-scheme in Fig.6.14 and the topological expression in Eq.(6.54).

N^{PO}: (6.52)

y_2, y_3, y_1	111	112	121	122	211	212	221	222
x								
1	0	0	1	0	0	0	0	0
2	0	0	0	0	0	0	1	0
3	0	0	0	0	1	0	0	0

N^{PO}: {(1, 121), (2, 221), (3, 211)} (6.53)

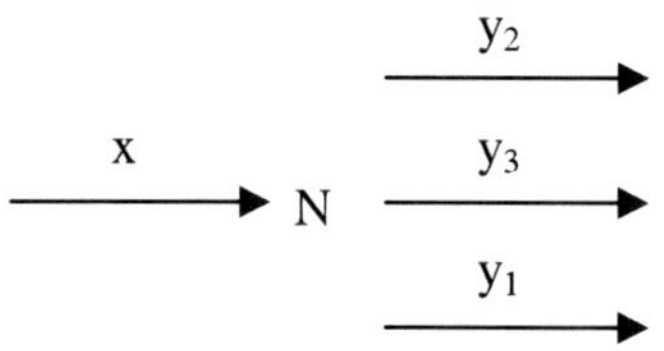

Fig. 6.14 One-node FN with three outputs after middle-bottom-top permutation

[N] (x | y_2, y_3, y_1) (6.54)

Example 6.11

This example considers an operand node N with input x and output set {y_1, y_2, y_3} whose outputs are permuted in a bottom-top-middle manner, i.e. y_3 comes first, y_1 comes second and y_2 comes third in the rearranged set of outputs. Before the permutation, this node can be described by the Boolean matrix in Eq.(6.55) and the binary relation in Eq.(6.56). In this context, node N represents a one-node FN that can be described by the block-scheme in Fig.6.10 and the topological expression in Eq.(6.36).

N: (6.55)

y_1, y_2, y_3	111	112	121	122	211	212	221	222
x								
1	0	0	0	0	1	0	0	0
2	0	0	0	0	0	0	1	0
3	0	0	0	0	0	1	0	0

N: {(1, 211), (2, 221), (3, 212)} (6.56)

As a result of this output permutation, the operand node N is transformed into a product node N^{PO} with input x and output set {y_3, y_1, y_2}. This node can be described by the Boolean matrix in Eq.(6.57) and the binary relation in Eq.(6.58). In this context, node N^{PO} represents a one-node FN that can be described by the block-scheme in Fig.6.15 and the topological expression in Eq.(6.59).

N^{PO}: (6.57)

x \ y_3, y_1, y_2	111	112	121	122	211	212	221	222
1	0	0	1	0	0	0	0	0
2	0	0	0	1	0	0	0	0
3	0	0	0	0	0	0	1	0

N^{PO}: {(1, 121), (2, 122), (3, 221)} (6.58)

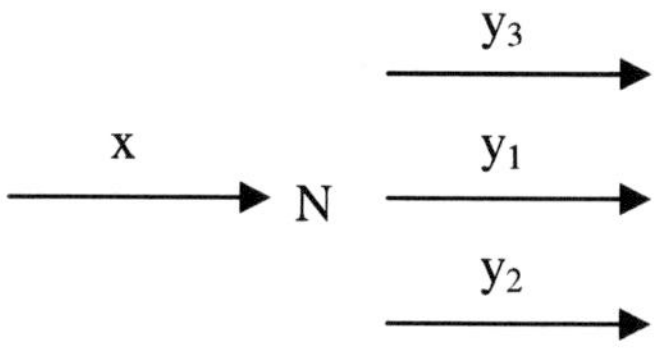

Fig. 6.15 One-node FN with three outputs after bottom-top-middle permutation

$[N]\ (x \mid y_3, y_1, y_2)$ (6.59)

6.4 Node Transformation for Feedback Equivalence

Node transformation is usually applied for feedback equivalence when some outputs from one or more nodes in a FN are fed back unchanged as inputs to the same or other nodes. In this case, it is necessary to reflect this identical feedback equivalently in the formal models for these nodes. The purpose of this equivalence is to allow the nodes with feedback to become operands in any merging operations as nodes without feedback. As a result, the nodes with feedback equivalence have to be transformed appropriately to reflect the presence of identical feedback. In this context, feedback equivalence can always be applied due to the possibility of representing the linguistic terms for any output from a node as the same linguistic terms for a corresponding input to this or other node.

When a Boolean matrix is used as a formal model for a node during feedback equivalence, the transformation of this node represents a modification of the associated operand matrix. In particular, the product matrix is obtained by making each element from the universal operand matrix that represents identical feedback equal to 1 and making all other elements equal to 0. The location of the non-zero elements depends on the ordering of the inputs and the outputs for the node as well as which outputs are fed back and as which inputs these outputs are fed back.

Node transformation can also be applied for feedback equivalence in the context of binary relations when such a relation is used as a formal model for the operand node. In this case, the transformation of the operand node represents a special type of relational modification whereby each pair of the universal operand relation that represents identical feedback is preserved and all other pairs are removed.

In the two paragraphs above, the operand matrix and the operand relation are referred to as 'universal' as they both may represent a universal type of node that must be changed to reflect an arbitrary type of identical feedback. Also, the assumption that all elements of the operand matrix and all pairs of the operand relation that represent identical feedback are made equal to 1 and preserved, respectively, refers to a node with a complete rule base, i.e. when all possible permutations of linguistic terms for inputs are present. Otherwise, when the rule base is incomplete, the associated elements in the product matrix are made equal to zero and the associated pairs in the product relation are removed.

Example 6.12

This example considers an operand node N with input set {z, x} and output set {z, y} whereby one of its outputs is mapped to one of its inputs by identical feedback in a top-top manner, i.e. the first output z is fed back unchanged as a first input. This node can be described by the universal Boolean matrix in Eq.(6.60) and the universal binary relation in Eq.(6.61). In this context, node N represents a one-node FN that can be described by the block-scheme in Fig.6.16 and the topological expression in Eq.(6.62).

N: (6.60)

z, x \ z, y	11	12	21	22
11	1	1	1	1
12	1	1	1	1
21	1	1	1	1
22	1	1	1	1

N: {(11, 11), (11, 12), (11, 21), (11, 22),
(12, 11), (12, 12), (12, 21), (12, 22),
(21, 11), (21, 12), (21, 21), (21, 22),
(22, 11), (22, 12), (22, 21), (22, 22)} (6.61)

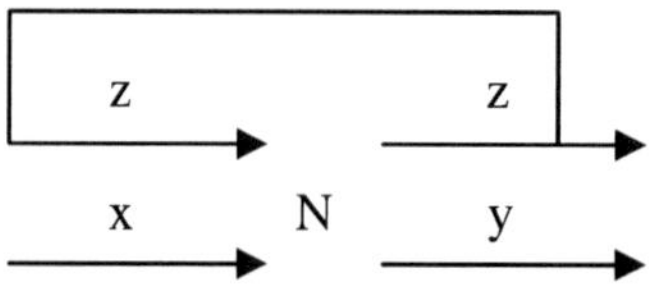

Fig. 6.16 One-node FN with single feedback before top-top equivalence

[N] (z, x | z, y) (6.62)

As a result of this feedback equivalence, the operand node N is transformed into a product node N^{EF} with input set {x^{EF}, x} and output set {y^{EF}, y}. This node can be described by the Boolean matrix in Eq.(6.63) and the binary relation in Eq.(6.64). In this context, node N^{EF} represents a one-node FN that can be described by the block-scheme in Fig.6.17 and the topological expression in Eq.(6.65).

N^{EF} : (6.63)

x^{EF}, x \ y^{EF}, y	11	12	21	22
11	1	1	0	0
12	1	1	0	0
21	0	0	1	1
22	0	0	1	1

$$N^{EF}: \{(11, 11), (11, 12), (12, 11), (12, 12), (21, 21), (21, 22), (22, 21), (22, 22)\} \quad (6.64)$$

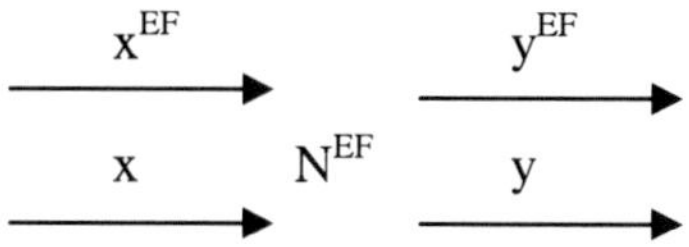

Fig. 6.17 One-node FN with single feedback after top-top equivalence

$$[N^{EF}] (x^{EF}, x \mid y^{EF}, y) \quad (6.65)$$

Example 6.13

This example considers an operand node N with input set {x, z} and output set {z, y} whereby one of its outputs is mapped to one of its inputs by identical feedback in a top-bottom manner, i.e. the first output z is fed back unchanged as a second input. This node can be described by the universal Boolean matrix in Eq.(6.60) and the universal binary relation in Eq.(6.61) which are based on the input set and the output set for the node. In this context, node N represents a one-node FN that can be described by the block-scheme in Fig.6.18 and the topological expression in Eq.(6.66).

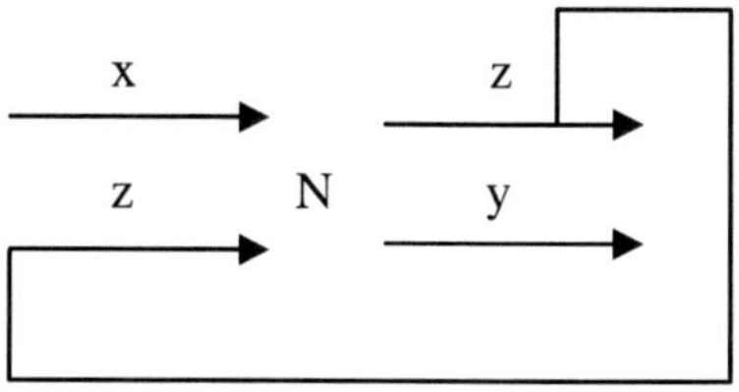

Fig. 6.18 One-node FN with single feedback before top-bottom equivalence

[N] (x, z | z, y) (6.66)

As a result of this feedback equivalence, the operand node N is transformed into a product node N^{EF} with input set {x, x^{EF}} and output set {y^{EF}, y}. This node can be described by the Boolean matrix in Eq.(6.67) and the binary relation in Eq.(6.68). In this context, node N^{EF} represents a one-node FN that can be described by the block-scheme in Fig.6.19 and the topological expression in Eq.(6.69).

N^{EF} : (6.67)

x, x^{EF} \ y^{EF}, y	11	12	21	22
11	1	1	0	0
12	0	0	1	1
21	1	1	0	0
22	0	0	1	1

N^{EF} : {(11, 11), (11, 12),
(12, 21), (12, 22),
(21, 11), (21, 12),
(22, 21), (22, 22)} (6.68)

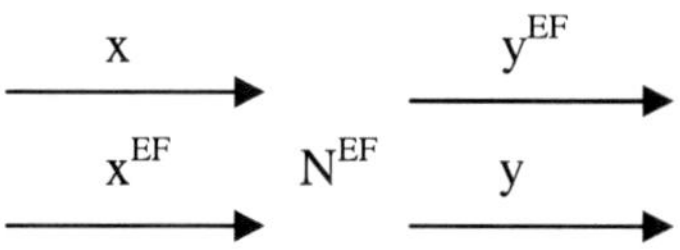

Fig. 6.19 One-node FN with single feedback after top-bottom equivalence

[N^{EF}] (x, x^{EF} | y^{EF}, y) (6.69)

Example 6.14

This example considers an operand node N with input set {z, x} and output set {y, z} whereby one of its outputs is mapped to one of its inputs by identical feedback in a bottom-top manner, i.e. the second output z is fed back unchanged as a first input. This node can be described by the universal Boolean matrix in Eq.(6.60) and the universal binary relation in Eq.(6.61) which are based on the input set and the output set for the node. In this context, node N represents a one-node FN that can be described by the block-scheme in Fig.6.20 and the topological expression in Eq.(6.70).

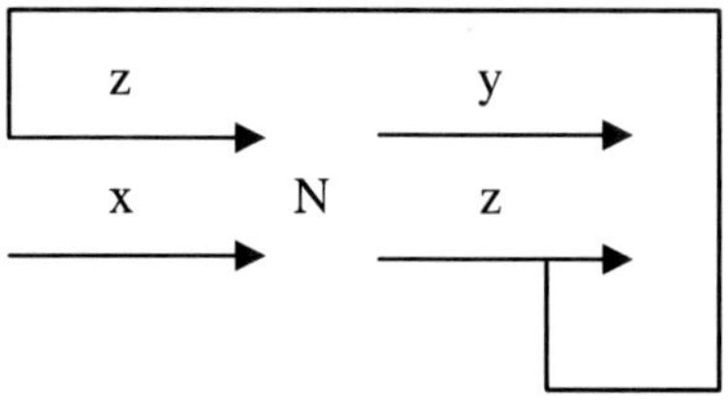

Fig. 6.20 One-node FN with single feedback before bottom-top equivalence

[N] (z, x | y, z) (6.70)

As a result of this feedback equivalence, the operand node N is transformed into a product node N^{EF} with input set $\{x^{EF}, x\}$ and output set $\{y, y^{EF}\}$. This node can be described by the Boolean matrix in Eq.(6.71) and the binary relation in Eq.(6.72). In this context, node N^{EF} represents a one-node FN that can be described by the block-scheme in Fig.6.19 and the topological expression in Eq.(6.73).

N^{EF}: (6.71)

x^{EF}, x \ y, y^{EF}	11	12	21	22
11	1	0	1	0
12	1	0	1	0
21	0	1	0	1
22	0	1	0	1

N^{EF}: {(11, 11), (11, 21),
(12, 11), (12, 21),
(21, 12), (21, 22),
(22, 12), (22, 22)} (6.72)

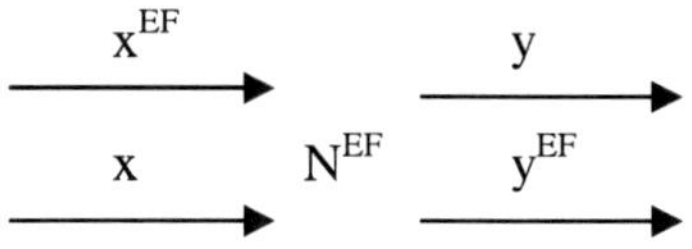

Fig. 6.21 One-node FN with single feedback after bottom-top equivalence

$[N^{EF}]$ (x^{EF}, x | y, y^{EF}) (6.73)

Example 6.15

This example considers an operand node N with input set {x, z} and output set {y, z} whereby one of its outputs is mapped to one of its inputs by identical feedback in a bottom-bottom manner, i.e. the second output z is fed back unchanged as a second input. This node can be described by the universal Boolean matrix in Eq.(6.60) and the universal binary relation in Eq.(6.61) which are based on the input set and the output set for the node. In this context, node N represents a one-node FN that can be described by the block-scheme in Fig.6.22 and the topological expression in Eq.(6.74).

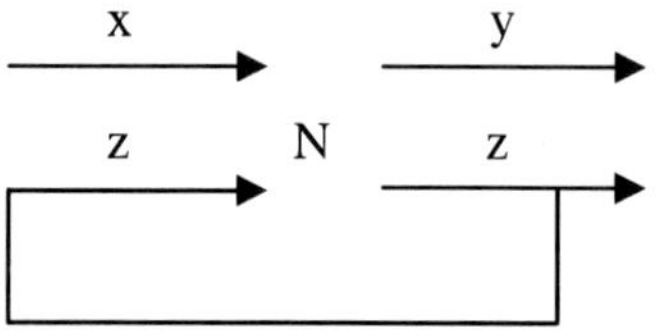

Fig. 6.22 One-node FN with single feedback before bottom-bottom equivalence

[N] (x, z | y, z) (6.74)

As a result of this feedback equivalence, the operand node N is transformed into a product node N^{EF} with input set {x, x^{EF}} and output set {y, y^{EF}}. This node can be described by the Boolean matrix in Eq.(6.75) and the binary relation in Eq.(6.76). In this context, node N^{EF} represents a one-node FN that can be described by the block-scheme in Fig.6.23 and the topological expression in Eq.(6.77).

N^{EF}: (6.75)

x, x^{EF} \ y, y^{EF}	11	12	21	22
11	1	0	1	0
12	0	1	0	1
21	1	0	1	0
22	0	1	0	1

N^{EF}: {(11, 11), (11, 21),
(12, 12), (12, 22),
(21, 11), (21, 21),
(22, 12), (22, 22)} (6.76)

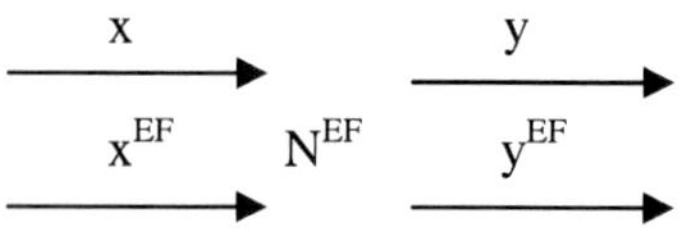

Fig. 6.23 One-node FN with single feedback after bottom-bottom equivalence

$[N^{EF}]$ (x, x^{EF} | y, y^{EF}) (6.77)

Figure 6.24 and the topological expression in Eq.(6.84). The known nodes A and C are described by the Boolean matrices and the binary relations in Eqs.(6.85)-(6.86) and Eqs.(6.87)-(6.88).

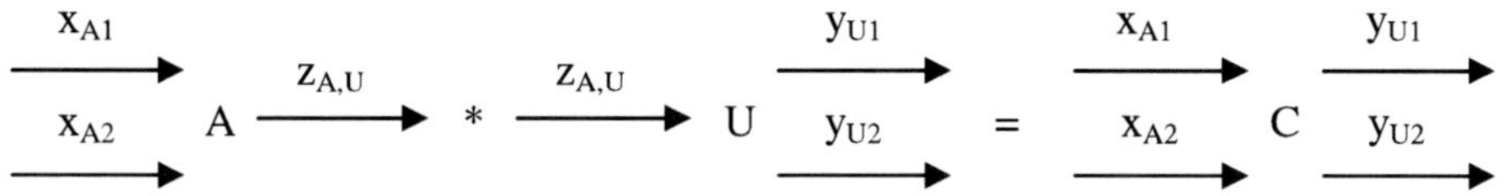

Fig. 6.24 FN with nodes A and U horizontally merged into node C

$$[A]\ (x_{A1}, x_{A2} \mid z_{A,U}) * [U]\ (z_{A,U} \mid y_{U1}, y_{U2}) = [C]\ (x_{A1}, x_{A2} \mid y_{U1}, y_{U2}) \tag{6.84}$$

A: (6.85)

x_{A1}, x_{A2} \ $z_{A,U}$	1	2	3
11	1	0	0
12	1	0	0
21	0	0	1
22	0	0	1

A: {(11, 1), (12, 1), (21, 3), (22, 3)} (6.86)

C: (6.87)

x_{A1}, x_{A2} \ y_{U1}, y_{U2}	11	12	21	22
11	0	0	0	1
12	0	0	0	1
21	1	0	0	0
22	1	0	0	0

C: {(11, 22), (12, 22), (21, 11), (22, 11)} (6.88)

The columns of the Boolean matrix for the unknown node U can be found from Eq.(6.82) which is presented in a detailed form for this example by Eqs.(6.89)-(6.92).

$$\begin{aligned}
&\max\ [\min(1, u_{11}), \min(0, u_{21}), \min(0, u_{31})] = 0\\
&\max\ [\min(1, u_{11}), \min(0, u_{21}), \min(0, u_{31})] = 0\\
&\max\ [\min(0, u_{11}), \min(0, u_{21}), \min(1, u_{31})] = 1\\
&\max\ [\min(0, u_{11}), \min(0, u_{21}), \min(1, u_{31})] = 1
\end{aligned} \tag{6.89}$$

$$\begin{aligned}
&\max\ [\min(1, u_{12}), \min(0, u_{22}), \min(0, u_{32})] = 0\\
&\max\ [\min(1, u_{12}), \min(0, u_{22}), \min(0, u_{32})] = 0\\
&\max\ [\min(0, u_{12}), \min(0, u_{22}), \min(1, u_{32})] = 0\\
&\max\ [\min(0, u_{12}), \min(0, u_{22}), \min(1, u_{32})] = 0
\end{aligned} \tag{6.90}$$

$$\begin{aligned} &\max\ [\min(1, u_{13}), \min(0, u_{23}), \min(0, u_{33})] = 0 \\ &\max\ [\min(1, u_{13}), \min(0, u_{23}), \min(0, u_{33})] = 0 \\ &\max\ [\min(0, u_{13}), \min(0, u_{23}), \min(1, u_{33})] = 0 \\ &\max\ [\min(0, u_{13}), \min(0, u_{23}), \min(1, u_{33})] = 0 \end{aligned} \tag{6.91}$$

$$\begin{aligned} &\max\ [\min(1, u_{14}), \min(0, u_{24}), \min(0, u_{34})] = 1 \\ &\max\ [\min(1, u_{14}), \min(0, u_{24}), \min(0, u_{34})] = 1 \\ &\max\ [\min(0, u_{14}), \min(0, u_{24}), \min(1, u_{34})] = 0 \\ &\max\ [\min(0, u_{14}), \min(0, u_{24}), \min(1, u_{34})] = 0 \end{aligned} \tag{6.92}$$

The variation in the number of solutions for the set of 4 systems of Boolean equations in Eqs.(6.89)-(6.92) with 4 equations and 3 unknowns each can be found from Eq.(6.83) which is presented in a specific form for this example by Eq.(6.93).

$$4^1 + 1 = 5 \tag{6.93}$$

The solution for node U is given by the Boolean matrices and the binary relations in Eqs.(6.94)-(6.103). In this case, each subscript i=1,5 for U represents an individual solution from the solution set. Each individual solution represents a consistent rule base, i.e. a rule base whose Boolean matrix has not more than one non-zero element in each row.

U_1: (6.94)

$z_{A,U}$ \ y_{U1}, y_{U2}	11	12	21	22
1	0	0	0	1
2	0	0	0	0
3	1	0	0	0

U_1: {(1, 22), (3, 11)} (6.95)

U_2: (6.96)

$z_{A,U}$ \ y_{U1}, y_{U2}	11	12	21	22
1	0	0	0	1
2	1	0	0	0
3	1	0	0	0

U_2: {(1, 22), (2, 11), (3, 11)} (6.97)

U_3: (6.98)

$z_{A,U}$ \ y_{U1}, y_{U2}	11	12	21	22
1	0	0	0	1
2	0	1	0	0
3	1	0	0	0

U_3: {(1, 22), (2, 12), (3, 11)} (6.99)

U_4: (6.100)

$z_{A,U}$ \ y_{U1}, y_{U2}	11	12	21	22
1	0	0	0	1
2	0	0	1	0
3	1	0	0	0

U_4: {(1, 22), (2, 21), (3, 11)} (6.101)

U_5: (6.102)

$z_{A,U}$ \ y_{U1}, y_{U2}	11	12	21	22
1	0	0	0	1
2	0	0	0	1
3	1	0	0	0

U_5: {(1, 22), (2, 22), (3, 11)} (6.103)

Problem 6.2

In this problem, node identification is considered in the context of horizontal merging of two nodes U and B into a node C whereby only B and C are given. Therefore, it is necessary to identify the unknown node U on the basis of the known nodes B and C. This problem can be described in a general form by the Boolean matrix equation in Eq.(6.104) where U, B and C represent the Boolean matrices for the above three nodes. The dimensions and the detailed descriptions of these Boolean matrices are given in Eqs.(6.105)-(6.107).

$$U * B = C \qquad (6.104)$$

$$U_p \times_q = \begin{matrix} u_{11} \ldots\ldots u_{1q} \\ \ldots\ldots\ldots\ldots\ldots \\ u_{p1} \ldots\ldots u_{pq} \end{matrix} \qquad (6.105)$$

$$B_q \times_r = \begin{matrix} b_{11} \ldots\ldots b_{1r} \\ \ldots\ldots\ldots\ldots\ldots \\ b_{q1} \ldots\ldots b_{qr} \end{matrix} \qquad (6.106)$$

$$C_p \times_r = \begin{matrix} c_{11} \ldots\ldots c_{1r} \\ \ldots\ldots\ldots\ldots\ldots \\ c_{p1} \ldots\ldots c_{pr} \end{matrix} \qquad (6.107)$$

The Boolean matrix equation in Eq.(6.104) can be represented as a set of 'p' systems of Boolean equations whereby each system consists of 'r' equations and 'q' unknowns. This set of systems of Boolean equations is given by Eq.(6.108).

$$\max\ [\min\ (u_{11}, b_{11}),\ldots, \min\ (u_{1q}, b_{q1})] = c_{11} \qquad (6.108)$$

$$\ldots\ldots\ldots\ldots\ldots\ldots\ldots\ldots\ldots\ldots\ldots\ldots\ldots\ldots\ldots$$

$$\max\ [\min\ (u_{11}, b_{1r}),\ldots, \min\ (u_{1q}, b_{qr})] = c_{1r}$$

$$\ldots\ldots\ldots\ldots\ldots\ldots\ldots\ldots\ldots\ldots\ldots\ldots\ldots\ldots\ldots$$

$$\max\ [\min\ (u_{p1}, b_{11}),\ldots, \min\ (u_{pq}, b_{q1})] = c_{p1}$$

$$\ldots\ldots\ldots\ldots\ldots\ldots\ldots\ldots\ldots\ldots\ldots\ldots\ldots\ldots\ldots$$

$$\max\ [\min\ (u_{p1}, b_{1r}),\ldots, \min\ (u_{pq}, b_{qr})] = c_{pr}$$

The solution for each system of Boolean equations in Eq.(6.108) represents a row in the unknown Boolean matrix U from the Boolean matrix equation in Eq.(6.104). The most trivial way of solving each system of Boolean equations is to generate all possible permutations of '0's and '1's for the unknowns. In this case, a multiple solution is very likely to exist, especially if some rows in the Boolean matrix B contain only zero elements. These zero elements will have an overriding effect on the elements in the corresponding columns of the Boolean matrix U, i.e. the latter can be taken as either '0's or '1's in the solution. In particular, the potential variation V in the number of solutions for the Boolean matrix equation in Eq.(6.104) is given by the general formula in Eq.(6.109) where 'p' is the number of rows in U and 's' is the number of zero rows in B.

$$V = (2^p)^s \qquad (6.109)$$

Example 6.17

This example considers a FN with two sequential nodes U and B whereby $\{x_{U1}, x_{U2}\}$ is the input set for U, $z_{U,B}$ is the intermediate variable and $\{y_{B1}, y_{B2}\}$ is the output set for B. These nodes are horizontally merged into node C that represents the equivalent fuzzy system for this FN, as shown by the block-scheme in Figure 6.25 and the topological expression in Eq.(6.110). The known nodes B and C are described by the Boolean matrices and the binary relations in Eqs.(6.111)-(6.112) and Eqs.(6.113)-(6.114).

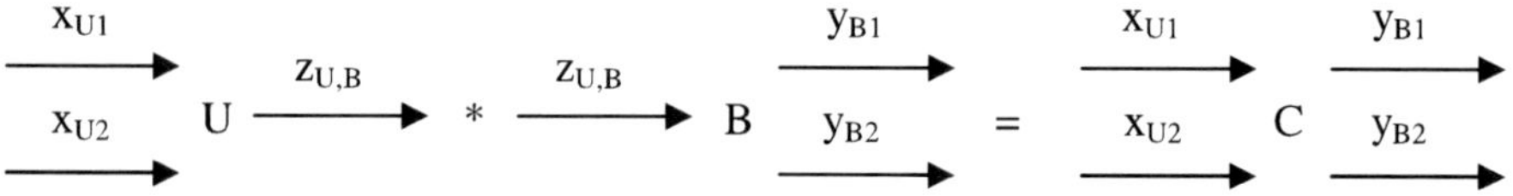

Fig. 6.25 FN with nodes U and B horizontally merged into node C

$$[U]\ (x_{U1}, x_{U2} \mid z_{U,B}) * [B]\ (z_{U,B} \mid y_{B1}, y_{B2}) = [C]\ (x_{U1}, x_{U2} \mid y_{B1}, y_{B2}) \qquad (6.110)$$

B: (6.111)

y_{B1}, y_{B2} / $z_{U,B}$	11	12	21	22
1	1	0	0	0
2	0	0	0	0
3	0	0	0	1

$$B: \{(1, 11), (3, 22)\} \tag{6.112}$$

C: (6.113)

y_{B1}, y_{B2} / x_{U1}, x_{U2}	11	12	21	22
11	0	0	0	1
12	0	0	0	1
21	1	0	0	0
22	1	0	0	0

$$C: \{(11, 22), (12, 22), (21, 11), (22, 11)\} \tag{6.114}$$

The rows of the Boolean matrix for the unknown node U can be found from Eq.(6.108) which is presented in a detailed form for this example by Eqs.(6.115)-(6.118).

$$\begin{aligned}
&\max\ [\min(u_{11}, 1), \min(u_{12}, 0), \min(u_{13}, 0)] = 0\\
&\max\ [\min(u_{11}, 0), \min(u_{12}, 0), \min(u_{13}, 0)] = 0\\
&\max\ [\min(u_{11}, 0), \min(u_{12}, 0), \min(u_{13}, 0)] = 0\\
&\max\ [\min(u_{11}, 0), \min(u_{12}, 0), \min(u_{13}, 1)] = 1
\end{aligned} \tag{6.115}$$

$$\begin{aligned}
&\max\ [\min(u_{21}, 1), \min(u_{22}, 0), \min(u_{23}, 0)] = 0\\
&\max\ [\min(u_{21}, 0), \min(u_{22}, 0), \min(u_{23}, 0)] = 0\\
&\max\ [\min(u_{21}, 0), \min(u_{22}, 0), \min(u_{23}, 0)] = 0\\
&\max\ [\min(u_{21}, 0), \min(u_{22}, 0), \min(u_{23}, 1)] = 1
\end{aligned} \tag{6.116}$$

$$\begin{aligned}
&\max\ [\min(u_{31}, 1), \min(u_{32}, 0), \min(u_{33}, 0)] = 1\\
&\max\ [\min(u_{31}, 0), \min(u_{32}, 0), \min(u_{33}, 0)] = 0\\
&\max\ [\min(u_{31}, 0), \min(u_{32}, 0), \min(u_{33}, 0)] = 0\\
&\max\ [\min(u_{31}, 0), \min(u_{32}, 0), \min(u_{33}, 1)] = 0
\end{aligned} \tag{6.117}$$

$$\begin{aligned}
&\max\ [\min(u_{41}, 1), \min(u_{42}, 0), \min(u_{43}, 0)] = 1\\
&\max\ [\min(u_{41}, 0), \min(u_{42}, 0), \min(u_{43}, 0)] = 0\\
&\max\ [\min(u_{41}, 0), \min(u_{42}, 0), \min(u_{43}, 0)] = 0\\
&\max\ [\min(u_{41}, 0), \min(u_{42}, 0), \min(u_{43}, 1)] = 0
\end{aligned} \tag{6.118}$$

The potential variation in the number of solutions for the set of 4 systems of Boolean equations in Eqs.(6.115)-(6.118) with 4 equations and 3 unknowns each can be found from Eq.(6.109) which is presented in a specific form for this example by Eq.(6.119).

$$(2^4)^1 = 16 \tag{6.119}$$

The solution for node U is given by the Boolean matrices and the binary relations in Eqs.(6.120)-(6.151). In this case, each subscript i = 1,16 for U represents an individual solution from the solution set. If an individual solution represents a inconsistent rule base, i.e. a rule base whose Boolean matrix has more than one non-zero element in at least one row, then this solution is discarded. Therefore, the only admissible solution for node U is U_1.

U_1: (6.120)

x_{U1}, x_{U2} \ $z_{U,B}$	1	2	3
11	0	0	1
12	0	0	1
21	1	0	0
22	1	0	0

$$U_1: \{(11, 3), (12, 3), (21, 1), (22, 1)\} \tag{6.121}$$

U_2: (6.122)

x_{U1}, x_{U2} \ $z_{U,B}$	1	2	3
11	0	1	1
12	0	0	1
21	1	0	0
22	1	0	0

$$U_2: \{(11, 2), (11, 3), (12, 3), (21, 1), (22, 1)\} \tag{6.123}$$

U_3: (6.124)

x_{U1}, x_{U2} \ $z_{U,B}$	1	2	3
11	0	0	1
12	0	1	1
21	1	0	0
22	1	0	0

$$U_3: \{(11, 3), (12, 2), (12, 3), (21, 1), (22, 1)\} \tag{6.125}$$

U_4: (6.126)

x_{U1}, x_{U2} \ $z_{U,B}$	1	2	3
11	0	0	1
12	0	0	1
21	1	1	0
22	1	0	0

$$U_4: \{(11, 3), (12, 3), (21, 1), (21, 2), (22, 1)\} \tag{6.127}$$

U_5: (6.128)

x_{U1}, x_{U2} \ $z_{U,B}$	1	2	3
11	0	0	1
12	0	0	1
21	1	0	0
22	1	1	0

$$U_5: \{(11, 3), (12, 3), (21, 1), (22, 1), (22, 2)\} \tag{6.129}$$

U_6: (6.130)

x_{U1}, x_{U2} \ $z_{U,B}$	1	2	3
11	0	1	1
12	0	1	1
21	1	0	0
22	1	0	0

$$U_6: \{(11, 2), (11, 3), (12, 2), (12, 3), (21, 1), (22, 1)\} \tag{6.131}$$

U_7: (6.132)

x_{U1}, x_{U2} \ $z_{U,B}$	1	2	3
11	0	1	1
12	0	0	1
21	1	1	0
22	1	0	0

$$U_7: \{(11, 2), (11, 3), (12, 3), (21, 1), (21, 2), (22, 1)\} \tag{6.133}$$

U_8: (6.134)

x_{U1}, x_{U2} \ $z_{U,B}$	1	2	3
11	0	1	1
12	0	0	1
21	1	0	0
22	1	1	0

$$U_8: \{(11, 2), (11, 3), (12, 3), (21, 1), (22, 1), (22, 2)\} \tag{6.135}$$

U_9: (6.136)

x_{U1}, x_{U2} \ $z_{U,B}$	1	2	3
11	0	0	1
12	0	1	1
21	1	1	0
22	1	0	0

$$U_9: \{(11, 3), (12, 2), (12, 3), (21, 1), (21, 2), (22, 1)\} \tag{6.137}$$

U_{10}: (6.138)

x_{U1}, x_{U2} \ $z_{U,B}$	1	2	3
11	0	0	1
12	0	1	1
21	1	0	0
22	1	1	0

$$U_{10}: \{(11, 3), (12, 2), (12, 3), (21, 1), (22, 1), (22, 2)\} \quad (6.139)$$

U_{11}: (6.140)

x_{U1}, x_{U2} \ $z_{U,B}$	1	2	3
11	0	0	1
12	0	0	1
21	1	1	0
22	1	1	0

$$U_{11}: \{(11, 3), (12, 3), (21, 1), (21, 2), (22, 1), (22, 2)\} \quad (6.141)$$

U_{12}: (6.142)

x_{U1}, x_{U2} \ $z_{U,B}$	1	2	3
11	0	1	1
12	0	1	1
21	1	1	0
22	1	0	0

$$U_{12}: \{(11, 2), (11, 3), (12, 2), (12, 3), (21, 1), (21, 2), (22, 1)\} \quad (6.143)$$

U_{13}: (6.144)

x_{U1}, x_{U2} \ $z_{U,B}$	1	2	3
11	0	1	1
12	0	1	1
21	1	0	0
22	1	1	0

$$U_{13}: \{(11, 2), (11, 3), (12, 2), (12, 3), (21, 1), (22, 1), (22, 2)\} \quad (6.145)$$

U_{14}: (6.146)

x_{U1}, x_{U2} \ $z_{U,B}$	1	2	3
11	0	1	1
12	0	0	1
21	1	1	0
22	1	1	0

$$U_{14}: \{(11, 2), (11, 3), (12, 3), (21, 1), (21, 2), (22, 1), (22, 2)\} \quad (6.147)$$

U_{15}: (6.148)

x_{U1}, x_{U2} \ $z_{U,B}$	1	2	3
11	0	0	1
12	0	1	1
21	1	1	0
22	1	1	0

U_{15}: {(11, 3), (12, 2), (12, 3), (21, 1), (21, 2), (22, 1), (22, 2)} (6.149)

U_{16}: (6.150)

x_{U1}, x_{U2} \ $z_{U,B}$	1	2	3
11	0	1	1
12	0	1	1
21	1	1	0
22	1	1	0

U_{16}: {(11,2), (11, 3), (12, 2), (12, 3), (21, 1), (21, 2), (22, 1), (22, 2)} (6.151)

Problem 6.3

In this problem, node identification is considered in the context of horizontal merging of three nodes A, U and B into a node C whereby only A, B and C are given. Therefore, it is necessary to identify the unknown node U on the basis of the known nodes A, B and C. This problem can be described in a general form by the Boolean matrix equation in Eq.(6.152) where A, U, B and C represent the Boolean matrices for the above four nodes. The dimensions and the detailed descriptions of these Boolean matrices are given in Eqs.(6.153)-(6.156).

$$A * U * B = C \qquad (6.152)$$

$$A_p \times_q = \begin{matrix} a_{11} \ldots\ldots a_{1q} \\ \ldots\ldots\ldots\ldots\ldots \\ a_{p1} \ldots\ldots a_{pq} \end{matrix} \qquad (6.153)$$

$$U_q \times_r = \begin{matrix} u_{11} \ldots\ldots u_{1r} \\ \ldots\ldots\ldots\ldots\ldots \\ u_{q1} \ldots\ldots u_{qr} \end{matrix} \qquad (6.154)$$

$$B_r \times_s = \begin{matrix} b_{11} \ldots\ldots b_{1s} \\ \ldots\ldots\ldots\ldots\ldots \\ b_{r1} \ldots\ldots b_{rs} \end{matrix} \qquad (6.155)$$

$$C_p \times_s = \begin{matrix} c_{11} \ldots\ldots c_{1s} \\ \ldots\ldots\ldots\ldots\ldots \\ c_{p1} \ldots\ldots c_{ps} \end{matrix} \qquad (6.156)$$

The Boolean matrix equation in Eq.(6.152) can be solved in two different ways. In either case, the solution can be found on the basis of the solutions for Problems 6.1-6.2.

In the first case, Eq.(6.152) is presented as a system of two Boolean matrix equations. This case is shown by Eqs.(6.157)-(6.158) whereby D is an additional unknown Boolean matrix given by Eq.(6.159). The solution for this system of Boolean matrix equations can be found by first solving Eq.(6.158) with respect to D and then solving Eq.(6.157) with respect to U.

$$A * U = D \tag{6.157}$$

$$D * B = C \tag{6.158}$$

$$D_p \times_r = \begin{matrix} d_{11} \;\ldots\ldots\; d_{1r} \\ \ldots\ldots\ldots\ldots\ldots \\ d_{p1} \;\ldots\ldots\; d_{pr} \end{matrix} \tag{6.159}$$

In the second case, Eq.(6.152) is also presented as a system of two Boolean matrix equations. This case is shown by Eqs.(6.160)-(6.161) whereby E is an additional unknown Boolean matrix given by Eq.(6.162). The solution for this system of Boolean matrix equations can be found by first solving Eq.(6.161) with respect to E and then solving Eq.(6.160) with respect to U.

$$U * B = E \tag{6.160}$$

$$A * E = C \tag{6.161}$$

$$E_q \times_s = \begin{matrix} e_{11} \;\ldots\ldots\; e_{1s} \\ \ldots\ldots\ldots\ldots\ldots \\ e_{q1} \;\ldots\ldots\; e_{qs} \end{matrix} \tag{6.162}$$

Example 6.18

This example considers a FN with three sequential nodes A, U and B whereby $\{x_{A1}, x_{A2}\}$ is the input set for A, $z_{A,U}$ and $z_{U,B}$ are the intermediate variables for U and $\{y_{B1}, y_{B2}\}$ is the output set for B. These nodes are horizontally merged into node C that represents the equivalent fuzzy system for this FN, as shown by the block-scheme in Figure 6.26 and the topological expression in Eq.(6.163). The known nodes A, B and C are described by the Boolean matrices and the binary relations in Eqs.(6.85)-(6.86) from Example 6.16, Eqs.(6.111)-(6.112) from Example 6.17 and Eqs.(6.164)-(6.165).

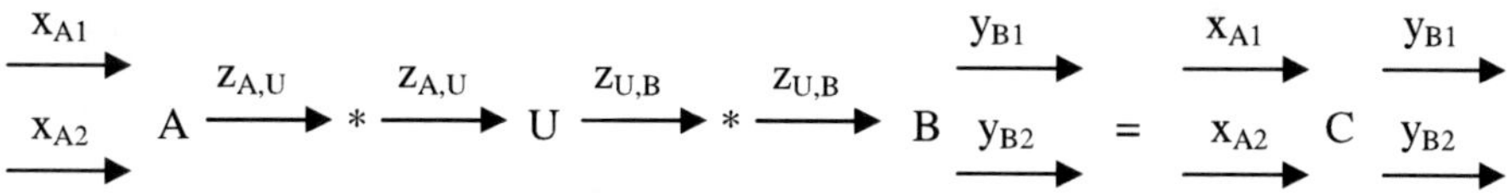

Fig. 6.26 FN with nodes A, U and B horizontally merged into node C

$$[A](x_{A1},x_{A2}\,|\,z_{A,U})*[U](z_{A,U}\,|\,z_{U,B})*[B](z_{U,B}\,|\,y_{B1},y_{B2}) = [C](x_{A1},x_{A2}\,|\,y_{B1},y_{B2}) \quad (6.163)$$

C: (6.164)

x_{A1}, x_{A2} \ y_{B1}, y_{B2}	11	12	21	22
11	0	0	0	1
12	0	0	0	1
21	1	0	0	0
22	1	0	0	0

C: {(11, 22), (12, 22), (21, 11), (22, 11)} (6.165)

In the first case, the only admissible solution for node D is D_1, as shown by the Boolean matrix and the binary relation in Eqs.(6.166)-(6.167). The remaining 15 potential solutions are discarded as none of them represents a consistent rule base.

D_1: (6.166)

x_{A1}, x_{A2} \ $z_{U,B}$	1	2	3
11	0	0	1
12	0	0	1
21	1	0	0
22	1	0	0

D_1: {(11, 3), (12, 3), (21, 1), (22, 1)} (6.167)

The solution set for node U is $\{U_1, U_2, U_3, U_4\}$, as shown by the Boolean matrices and the binary relations in Eqs.(6.168)-(6.175). Each of these solutions represents a consistent rule base.

U_1: (6.168)

$z_{A,U}$ \ $z_{U,B}$	1	2	3
1	0	0	1
2	0	0	0
3	1	0	0

U_1: {(1, 3), (3, 1)} (6.169)

$U_2:$ (6.170)

$z_{A,U}$ \ $z_{U,B}$	1	2	3
1	0	0	1
2	1	0	0
3	1	0	0

$$U_1: \{(1,3), (2,1), (3,1)\} \quad (6.171)$$

$U_3:$ (6.172)

$z_{A,U}$ \ $z_{U,B}$	1	2	3
1	0	0	1
2	0	1	0
3	1	0	0

$$U_3: \{(1,3), (2,2), (3,1)\} \quad (6.173)$$

$U_4:$ (6.174)

$z_{A,U}$ \ $z_{U,B}$	1	2	3
1	0	0	1
2	0	0	1
3	1	0	0

$$U_4: \{(1,3), (2,3), (3,1)\} \quad (6.175)$$

In the second case, the solution set for node E is $\{E_1, E_2, E_3, E_4, E_5\}$, as shown by the Boolean matrices and the binary relations in Eqs.(6.176)-(6.185). Each of these solutions represents a consistent rule base.

$E_1:$ (6.176)

$z_{A,U}$ \ y_{B1}, x_{B2}	11	12	21	22
1	0	0	0	1
2	0	0	0	0
3	1	0	0	0

$$E_1: \{(1,22), (3,11)\} \quad (6.177)$$

$E_2:$ (6.178)

$z_{A,U}$ \ y_{B1}, x_{B2}	11	12	21	22
1	0	0	0	1
2	1	0	0	0
3	1	0	0	0

$$E_2: \{(1,22), (2,11), (3,11)\} \quad (6.179)$$

E_3: (6.180)

$z_{A,U}$ \ y_{B1}, x_{B2}	11	12	21	22
1	0	0	0	1
2	0	1	0	0
3	1	0	0	0

$$E_3 : \{(1, 22), (2, 12), (3, 11)\} \quad (6.181)$$

E_4: (6.182)

$z_{A,U}$ \ y_{B1}, x_{B2}	11	12	21	22
1	0	0	0	1
2	0	0	1	0
3	1	0	0	0

$$E_4 : \{(1, 22), (2, 21), (3, 11)\} \quad (6.183)$$

E_5: (6.184)

$z_{A,U}$ \ y_{B1}, x_{B2}	11	12	21	22
1	0	0	0	1
2	0	0	0	1
3	1	0	0	0

$$E_5 : \{(1, 22), (2, 22), (3, 11)\} \quad (6.185)$$

The solution set for node U is also $\{U_1, U_2, U_3, U_4\}$, as shown by the Boolean matrices and the binary relations in Eqs.(6.168)-(6.175). Each of these individual solutions for U follows from an individual solution for E and represents a consistent rule base. In particular, the admissible solutions U_1 and U_3 follow from the basic solution E_1 whereas the additional solutions U_2 and U_4 follow from the non-basic solutions E_2 and E_5, respectively. There are no solutions for U that follow from the non-basic solutions E_3 or E_4. In this context, the term 'basic solution' refers to a Boolean matrix with at least one zero row whereas the term 'non-basic solution' refers to a Boolean matrix without any zero rows.

6.6 Node Identification in Vertical Merging

Node identification is usually applied in vertical merging when one or more nodes in the same layer of a FN are unknown but the node for the equivalent fuzzy system for this layer is given. In this case, it is necessary to find the unknown nodes from the other nodes in this layer and the node for the equivalent fuzzy system. The purpose of this type of node identification is to ensure that once the unknown nodes have been identified and vertically merged with the known nodes, the resultant node is identical with the one given in advance. In this context, node identification in vertical merging can not always be applied as there is no guarantee for a solution to exist in accordance with the above requirement. However, when a solution can be found, it is usually unique.

When Boolean matrices are used as formal models during node identification in vertical merging, this process is based on examining the structure of the known matrices and the given matrix. In this case, a location based correspondence is sought between the non-zero elements of the known Boolean matrices and any identical non-zero blocks of the given Boolean matrix. If such a correspondence is to be found, then the unknown Boolean matrix is equal to these non-zero blocks or to a compressed image of the given Boolean matrix whereby all non-zero and zero blocks are represented by 1's and 0's, respectively.

Node identification can also be applied in vertical merging in the context of binary relations when such relations are used as formal models for the known, given and unknown nodes. In this case, a location based correspondence is sought between the individual pairs of the known relations and any blocks of pairs with similar pattern from the given relation, such that the first and the second elements in the corresponding pairs in each block are identical. If such a correspondence is to be found, then the unknown binary relation can be derived from the similarity between the blocks of pairs in the given relation and the individual pairs in the known relation.

Problem 6.4

In this problem, node identification is considered in the context of vertical merging of two nodes A and U into a node C whereby only A and C are given. Therefore, it is necessary to identify the unknown node U on the basis of the known nodes A and C. This problem can be described in a general form by the Boolean matrix equation in Eq.(6.186) where A, U and C represent the Boolean matrices for the above three nodes. The dimensions and the detailed descriptions of these Boolean matrices are given in Eqs.(6.187)-(6.189).

$$A + U = C \tag{6.186}$$

$$A_p{\times}_q = \begin{matrix} a_{11} \;\ldots\ldots\; a_{1q} \\ \ldots\ldots\ldots\ldots\ldots \\ a_{p1} \;\ldots\ldots\; a_{pq} \end{matrix} \tag{6.187}$$

$$U_r{\times}_s = \begin{matrix} u_{11} \;\ldots\ldots\; u_{1s} \\ \ldots\ldots\ldots\ldots\ldots \\ u_{r1} \;\ldots\ldots\; u_{rs} \end{matrix} \tag{6.188}$$

$$C_{p.r}{\times}_{q.s} = \begin{matrix} c_{11} \;\ldots\ldots\ldots\; c_{1,r.s} \\ \ldots\ldots\ldots\ldots\ldots\ldots \\ c_{p.r,1} \;\ldots\ldots c_{p.r,q.s} \end{matrix} \tag{6.189}$$

If the Boolean matrix C in Eq.(6.189) contains only one set of not more than 'p' identical non-zero blocks C^k, $1 \leq k \leq p$ whereby there is not more than one such block in any block row of this matrix, the elements with a location in A corresponding to the location of non-zero blocks C^k in C are all equal to '1' and all other elements with a location in A corresponding to the location of zero blocks C^0 in C are equal to '0', then the Boolean matrix U is equal to C^k. In this case, the location of the elements of the non-zero identical blocks C^k in C can be described by Eq.(6.190) which also shows the admissible initial values for the subscripts i and j of the elements of C^k. As far as the zero blocks C^0 are concerned, they are with the same number of rows and columns as the non-zero blocks C^k.

$$C^k = \begin{matrix} c_{ij}^{\ k} \ldots\ldots\ldots c_{i,j+s-1}^{\ k} \\ \ldots\ldots\ldots\ldots\ldots\ldots \\ c_{i+r-1,j}^{\ k} \ldots\ldots c_{i+r-1,j+s-1}^{\ k} \end{matrix}, \; i=1,1+r,\ldots,1+(p-1).r, \; j=1,1+s,\ldots,1+(q-1).s \qquad (6.190)$$

Example 6.19

This example considers a FN with two parallel nodes A and U whereby x_A is the input to A, y_A is the output from A, x_U is the input to U and y_U is the output from U. These nodes are vertically merged into node C that represents the equivalent fuzzy system for this FN, as shown by the block-scheme in Figure 6.27 and the topological expression in Eq.(6.191). The known nodes A and C are described by the Boolean matrices and the binary relations in Eqs.(6.192)-(6.193) and Eqs.(6.194)-(6.195).

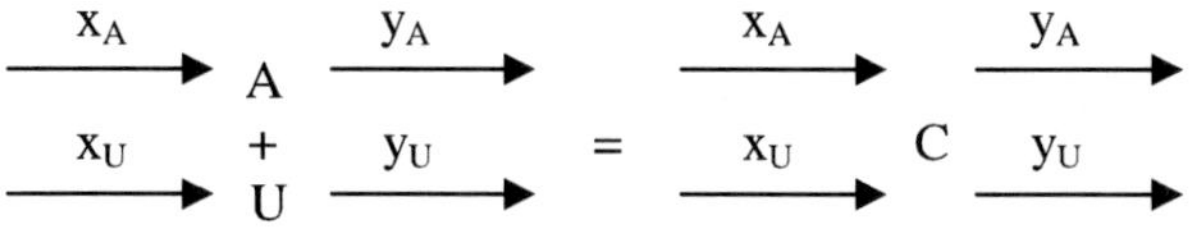

Fig. 6.27 FN with nodes A and U vertically merged into node C

$$[A]\,(x_A \mid y_A) + [U]\,(x_U \mid y_U) = [C]\,(x_A, x_U \mid y_A, y_U) \qquad (6.191)$$

A: (6.192)

x_A \ y_A	1	2	3
1	0	1	0
2	1	0	0
3	0	0	1

A: {(1, 2), (2, 1), (3, 3)} (6.193)

C: (6.194)

x_A, x_U \ y_A, y_U	11	12	13	21	22	23	31	32	33
11	0	0	0	1	0	0	0	0	0
12	0	0	0	0	0	1	0	0	0
13	0	0	0	0	1	0	0	0	0
21	1	0	0	0	0	0	0	0	0
22	0	0	1	0	0	0	0	0	0
23	0	1	0	0	0	0	0	0	0
31	0	0	0	0	0	0	1	0	0
32	0	0	0	0	0	0	0	0	1
33	0	0	0	0	0	0	0	1	0

C: {(11, 21), (12, 23), (13, 22),
(21, 11), (22, 13), (23, 12),
(31, 31), (32, 33), (33, 32)} (6.195)

The Boolean matrix C in Eq.(6.194) contains only one set of 3 identical non-zero blocks C^k whereby there is not more than one such block in any block row of this matrix. Also, the elements with a location in A corresponding to the location of non-zero blocks C^k in C are all equal to '1' and all other elements with a location in A corresponding to the location of zero blocks C^0 in C are equal to '0'.

Therefore, the Boolean matrix U is equal to C^k, as shown by the Boolean matrix and the binary relation in Eqs.(6.196)-(6.197). In this case, the location of the elements of C^k, k=1,2,3 in C is given by Eqs.(6.198)-(6.200).

U: (6.196)

x_U \ y_U	1	2	3
1	1	0	0
2	0	0	1
3	0	1	0

U: {(1, 1), (2, 3), (3, 2)} (6.197)

$$C^1 = \begin{matrix} c_{14}^{\;1} & c_{15}^{\;1} & c_{16}^{\;1} \\ c_{24}^{\;1} & c_{25}^{\;1} & c_{26}^{\;1} \\ c_{34}^{\;1} & c_{35}^{\;1} & c_{36}^{\;1} \end{matrix} \qquad (6.198)$$

$$C^2 = \begin{matrix} c_{41}^{\;2} & c_{42}^{\;2} & c_{43}^{\;2} \\ c_{51}^{\;2} & c_{52}^{\;2} & c_{53}^{\;2} \\ c_{61}^{\;2} & c_{62}^{\;2} & c_{63}^{\;2} \end{matrix} \qquad (6.199)$$

$$C^3 = \begin{matrix} c_{77}^{\;3} & c_{78}^{\;3} & c_{79}^{\;3} \\ c_{87}^{\;3} & c_{88}^{\;3} & c_{89}^{\;3} \\ c_{97}^{\;3} & c_{98}^{\;3} & c_{99}^{\;3} \end{matrix} \qquad (6.200)$$

Problem 6.5

In this problem, node identification is considered in the context of vertical merging of two nodes U and B into a node C whereby only B and C are given. Therefore, it is necessary to identify the unknown node U on the basis of the known nodes B and C. This problem can be described in a general form by the Boolean matrix equation in Eq.(6.201) where U, B and C represent the Boolean matrices for the above three nodes. The dimensions and the detailed descriptions of these Boolean matrices are given in Eqs.(6.202)-(6.203) and Eq.(6.189).

$$U + B = C \tag{6.201}$$

$$U_p \times_q = \begin{matrix} u_{11} \ldots\ldots u_{1q} \\ \ldots\ldots\ldots\ldots\ldots \\ u_{p1} \ldots\ldots u_{pq} \end{matrix} \tag{6.202}$$

$$B_r \times_s = \begin{matrix} b_{11} \ldots\ldots b_{1s} \\ \ldots\ldots\ldots\ldots\ldots \\ b_{r1} \ldots\ldots b_{rs} \end{matrix} \tag{6.203}$$

If the Boolean matrix C in Eq.(6.189) contains only one set of not more than ‘p’ identical non-zero blocks C^k, $1 \leq k \leq p$ whereby there is not more than one such block in any block row of this matrix and the Boolean matrix B is equal to C^k, then the elements with a location in U corresponding to the location of non-zero blocks C^k in C are all equal to ‘1’ and all other elements with a location in U corresponding to the location of zero blocks C^0 in C are equal to ‘0’. In this case, a non-zero identical block C^k can be described by Eq.(6.190) which also shows the admissible initial values for the subscripts i and j of the elements of C^k. As far as the zero blocks C^0 are concerned, they are with the same number of rows and columns as the non-zero blocks C^k.

Example 6.20

This example considers a FN with two parallel nodes U and B whereby x_U is the input to U, y_U is the output from U, x_B is the input to B and y_B is the output from B. These nodes are vertically merged into node C that represents the equivalent fuzzy system for this FN, as shown by the block-scheme in Figure 6.28 and the topological expression in Eq.(6.204). The known nodes B and C are described by the Boolean matrices and the binary relations in Eqs.(6.205)-(6.206) and Eqs.(6.207)-(6.208).

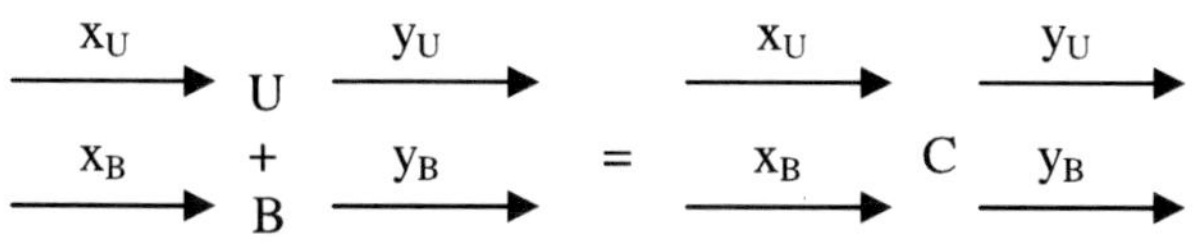

Fig. 6.28 FN with nodes U and B vertically merged into node C

$$[U]\,(x_U \mid y_U) + [B]\,(x_B \mid y_B) = [C]\,(x_U, x_B \mid y_U, y_B) \tag{6.204}$$

B: (6.205)

x_B \ y_B	1	2	3
1	0	1	0
2	0	0	1
3	1	0	0

B: {(1, 2), (2, 3), (3, 1)} (6.206)

C: (6.207)

x_U, x_B \ y_U, y_B	11	12	13	21	22	23	31	32	33
11	0	0	0	0	0	0	0	1	0
12	0	0	0	0	0	0	0	0	1
13	0	0	0	0	0	0	1	0	0
21	0	1	0	0	0	0	0	0	0
22	0	0	1	0	0	0	0	0	0
23	1	0	0	0	0	0	0	0	0
31	0	0	0	0	1	0	0	0	0
32	0	0	0	0	0	1	0	0	0
33	0	0	0	1	0	0	0	0	0

C: {(11, 32), (12, 33), (13, 31),
(21, 12), (22, 13), (23, 11),
(31, 22), (32, 23), (33, 21)} (6.208)

The Boolean matrix C in Eq.(6.207) contains only one set of 3 identical non-zero blocks C^k whereby there is not more than one such block in any block row of this matrix. Also, the Boolean matrix B is equal to C^k.

Therefore, the elements with a location in U corresponding to the location of non-zero blocks C^k in C are all equal to '1' and all other elements with a location in U corresponding to the location of zero blocks C^0 in C are equal to '0', as shown by the Boolean matrix and the binary relation in Eqs.(6.209)-(6.210). In this case, the location of the elements of C^k, k=1,2,3 in C is given by Eq.(6.211), Eq.(6.199) and Eq.(2.212).

U: (6.209)

x_U \ y_U	1	2	3
1	0	0	1
2	1	0	0
3	0	1	0

U: {(1, 3), (2, 1), (3, 2)} (6.210)

$$C^1 = \begin{matrix} c_{17}^{\,1} & c_{18}^{\,1} & c_{19}^{\,1} \\ c_{27}^{\,1} & c_{28}^{\,1} & c_{29}^{\,1} \\ c_{37}^{\,1} & c_{38}^{\,1} & c_{39}^{\,1} \end{matrix} \tag{6.211}$$

$$C^3 = \begin{matrix} c_{74}^{3} & c_{75}^{3} & c_{76}^{3} \\ c_{84}^{3} & c_{85}^{3} & c_{86}^{3} \\ c_{94}^{3} & c_{95}^{3} & c_{96}^{3} \end{matrix} \quad (6.212)$$

Problem 6.6

In this problem, node identification is considered in the context of vertical merging of three nodes A, U and B into a node C whereby only A, B and C are given. Therefore, it is necessary to identify the unknown node U on the basis of the known nodes A, B and C. This problem can be described in a general form by the Boolean matrix equation in Eq.(6.213) where A, U, B and C represent the Boolean matrices for the above four nodes. The dimensions and the detailed descriptions of these Boolean matrices are given in Eq.(6.187), Eq.(6.214), Eq.(6.203) and Eq.(6.215).

$$A + U + B = C \quad (6.213)$$

$$U_f \times_g = \begin{matrix} u_{11} \ \ldots\ldots \ u_{1g} \\ \ldots\ldots\ldots\ldots\ldots \\ u_{f1} \ \ldots\ldots\ldots \ u_{fg} \end{matrix} \quad (6.214)$$

$$C_{p.f.r} \times_{q.g.s} = \begin{matrix} c_{11} \ \ldots\ldots\ldots \ c_{1,q.g.s} \\ \ldots\ldots\ldots\ldots\ldots\ldots\ldots \\ c_{p.f.r,1} \ \ldots\ldots \ c_{p.f.r,q,g,s} \end{matrix} \quad (6.215)$$

The Boolean matrix equation in Eq.(6.213) can be solved in two different ways. In either case, the solution can be found on the basis of the solutions for Problems 6.4-6.5.

In the first case, Eq.(6.213) is presented as a system of two Boolean matrix equations. This case is shown by Eqs.(6.216)-(6.217) whereby D is an additional unknown Boolean matrix given by Eq.(6.218). The solution for this system of Boolean matrix equations can be found by first solving Eq.(6.217) with respect to D and then solving Eq.(6.216) with respect to U.

$$A * U = D \quad (6.216)$$

$$D * B = C \quad (6.217)$$

$$D_p \times_g = \begin{matrix} d_{11} \ \ldots\ldots \ d_{1g} \\ \ldots\ldots\ldots\ldots\ldots \\ d_{p1} \ \ldots\ldots \ d_{pg} \end{matrix} \quad (6.218)$$

In the second case, Eq.(6.213) is also presented as a system of two Boolean matrix equations. This case is shown by Eqs.(6.219)-(6.220) whereby E is an additional unknown Boolean matrix given by Eq.(6.221). The solution for this system of Boolean matrix equations can be found by first solving Eq.(6.220) with respect to E and then solving Eq.(6.219) with respect to U.

$$U + B = E \tag{6.219}$$

$$A + E = C \tag{6.220}$$

$$E_f{\times}_s = \begin{matrix} e_{11} \;\ldots\ldots\; e_{1s} \\ \ldots\ldots\ldots\ldots\ldots \\ e_{f1} \;\ldots\ldots\; e_{fs} \end{matrix} \tag{6.221}$$

Example 6.21

This example considers a FN with three parallel nodes A, U and B whereby x_A is the input to A, y_A is the output from A, x_U is the input to U, y_U is the output from U, x_B is the input to B and y_B is the output from B. These nodes are vertically merged into node C that represents the equivalent fuzzy system for this FN, as shown by the block-scheme in Figure 6.29 and the topological expression in Eq.(6.222). The known nodes A, B and C are described by the Boolean matrices and the binary relations in Eqs.(6.192)-(6.193), Eqs.(6.205)-(6.206) and Eqs.(6.223)-(6.224). In this case, the labels and the elements of the Boolean matrix in Eq.(6.223) are represented by a compact notation. In particular, each of the capital letters A, B, C, D, E, F, G, H, I stands for three sequential rows and columns as indicated in brackets, 1_3 denotes the Boolean matrix for node B from Eq.(6.205) and 0_3 denotes a zero Boolean matrix of dimension equal to that for 1_3.

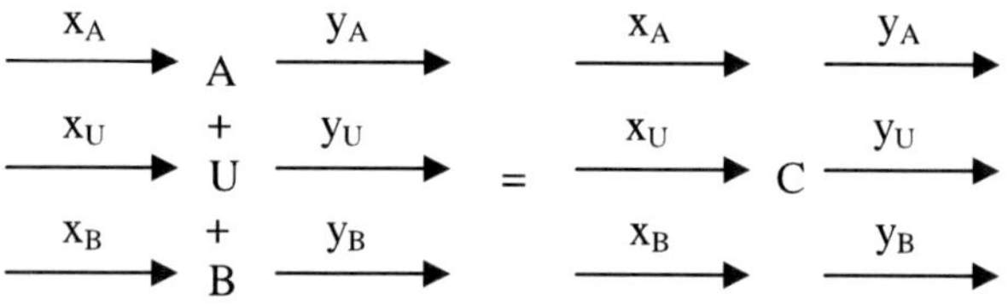

Fig. 6.29 FN with nodes A, U and B vertically merged into node C

$$[A]\,(x_A \mid y_A) + [U]\,(x_U \mid y_U) + [B]\,(x_B \mid y_B) = [C]\,(x_A, x_U, x_B \mid y_A, y_U, y_B) \tag{6.222}$$

C: (6.223)

x_A, x_U, x_B \ y_A, y_U, y_B	A	B	C	D	E	F	G	H	I
A (111-113)	0_3	0_3	0_3	0_3	0_3	1_3	0_3	0_3	0_3
B (121-123)	0_3	0_3	0_3	0_3	1_3	0_3	0_3	0_3	0_3
C (131-133)	0_3	0_3	0_3	1_3	0_3	0_3	0_3	0_3	0_3
D (211-213)	0_3	0_3	1_3	0_3	0_3	0_3	0_3	0_3	0_3
E (221-223)	0_3	1_3	0_3	0_3	0_3	0_3	0_3	0_3	0_3
F (231-233)	1_3	0_3	0_3	0_3	0_3	0_3	0_3	0_3	0_3
G (311-313)	0_3	0_3	0_3	0_3	0_3	0_3	0_3	0_3	1_3
H (321-323)	0_3	0_3	0_3	0_3	0_3	0_3	0_3	1_3	0_3
I (331-333)	0_3	0_3	0_3	0_3	0_3	0_3	1_3	0_3	0_3

C: {(111, 232), (112, 233), (113, 231), (6.224)
(121, 222), (122, 223), (123, 221),
(131, 212), (132, 213), (133, 211),
(211, 132), (212, 133), (213, 131),
(221, 122), (222, 123), (223, 121),
(231, 112), (232, 113), (233, 111),
(311, 332), (312, 333), (313, 331),
(321, 322), (322, 323), (323, 321),
(331, 312), (332, 313), (333, 311)}

In the first case, the solution for node D is given by the Boolean matrix and the binary relation in Eqs.(6.225)-(6.226) whereas the solution for node U is given by the Boolean matrix and the binary relation in Eqs.(6.227)-(6.228).

D: (6.225)

x_A, x_U \ y_A, y_U	11	12	13	21	22	23	31	32	33
11	0	0	0	0	0	1	0	0	0
12	0	0	0	0	1	0	0	0	0
13	0	0	0	1	0	0	0	0	0
21	0	0	1	0	0	0	0	0	0
22	0	1	0	0	0	0	0	0	0
23	1	0	0	0	0	0	0	0	0
31	0	0	0	0	0	0	0	0	1
32	0	0	0	0	0	0	0	1	0
33	0	0	0	0	0	0	1	0	0

D: {(11, 23), (12, 22), (13, 21), (6.226)
(21, 13), (22, 12), (23, 11),
(31, 33), (32, 32), (33, 31)}

U: (6.227)

x_U \ y_U	1	2	3
1	0	0	1
2	0	1	0
3	1	0	0

U: {(1, 3), (2, 2), (3, 1)} (6.228)

In the second case, the solution for node E is given by the Boolean matrix and the binary relation in Eqs.(6.229)-(6.230) whereas the solution for node U is also given by the Boolean matrix and the binary relation in Eqs.(6.227)-(6.228).

E: (6.229)

x_U, x_B \ y_U, y_B	11	12	13	21	22	23	31	32	33
11	0	0	0	0	0	0	0	1	0
12	0	0	0	0	0	0	0	0	1
13	0	0	0	0	0	0	1	0	0
21	0	0	0	0	1	0	0	0	0
22	0	0	0	0	0	1	0	0	0
23	0	0	0	1	0	0	0	0	0
31	0	1	0	0	0	0	0	0	0
32	0	0	1	0	0	0	0	0	0
33	1	0	0	0	0	0	0	0	0

E: {(11, 32), (12, 33), (13, 31),
(21, 22), (22, 23), (23, 21),
(31, 12), (32, 13), (33, 11)} (6.230)

6.7 Node Identification in Output Merging

Node identification is usually applied in output merging when one or more nodes with common inputs in the same layer of a FN are unknown but the node for the equivalent fuzzy system for this layer is given. In this case, it is necessary to find the unknown nodes from the other nodes in this layer and the node for the equivalent fuzzy system. The purpose of this type of node identification is to ensure that once the unknown nodes have been identified and output merged with the known nodes, the resultant node is identical with the one given in advance. In this context, node identification in output merging can not always be applied as there is no guarantee for a solution to exist in accordance with the above requirement. However, when a solution can be found, it is usually unique.

When Boolean matrices are used as formal models during node identification in output merging, this process is based on examining the structure of the known matrices and the given matrix. In this case, a location based correspondence is sought between the non-zero elements of the known Boolean matrices and any non-zero row blocks of the given Boolean matrix. If such a correspondence is to be found, then the rows of the unknown Boolean matrix are equal to the non-zero row blocks or to a compressed image of the given Boolean matrix whereby all non-zero and zero row blocks are represented by 1's and 0's, respectively.

Node identification can also be applied in output merging in the context of binary relations when such relations are used as formal models for the known, given and unknown nodes. In this case, a location based correspondence is sought between the pairs of the known relations and any pairs with similar pattern from the given relation. If such a correspondence is to be found, then the unknown binary relation can be derived from the similarity between the pairs in the given relation and the pairs in the known relations.

Problem 6.7

In this problem, node identification is considered in the context of output merging of two nodes A and U with common inputs into a node C whereby only A and C are given. Therefore, it is necessary to identify the unknown node U on the basis of the known nodes A and C. This problem can be described in a general form by the Boolean matrix equation in Eq.(6.231) where A, U and C represent the Boolean matrices for the above three nodes. The dimensions and the detailed descriptions of these Boolean matrices are given in Eqs.(6.232)-(6.234).

$$A \; ; \; U = C \tag{6.231}$$

$$A_p \times_q = \begin{matrix} a_{11} \;\ldots\ldots\; a_{1q} \\ \ldots\ldots\ldots\ldots\ldots \\ a_{p1} \;\ldots\ldots\; a_{pq} \end{matrix} \tag{6.232}$$

$$U_p \times_r = \begin{matrix} u_{11} \;\ldots\ldots\; u_{1r} \\ \ldots\ldots\ldots\ldots\ldots \\ u_{p1} \;\ldots\ldots\; u_{pr} \end{matrix} \tag{6.233}$$

$$C_p \times_{q.r} = \begin{matrix} c_{11} \;\ldots\ldots\ldots\; c_{1,q.r} \\ \ldots\ldots\ldots\ldots\ldots\ldots \\ c_{p,1} \;\ldots\ldots\ldots\; c_{p,q.r} \end{matrix} \tag{6.234}$$

If the Boolean matrix C in Eq.(6.234) contains only one set of not more than ‘p’ non-zero row blocks C^k, $1 \leq k \leq p$ whereby there is not more than one such block in any row of this matrix, the elements with a location in A corresponding to the location of non-zero row blocks C^k in C are all equal to ‘1’ and all other elements with a location in A corresponding to the location of zero row blocks C^0 in C are equal to ‘0’, then the rows of the Boolean matrix U are equal to C^k. In this case, the location of the elements of the non-zero row blocks C^k in C can be described by Eq.(6.235) which also shows the admissible initial values for the subscripts i and j of the elements of C^k. As far as the zero row blocks C^0 are concerned, they are with the same number of elements as the non-zero row blocks C^k.

$$C^k = c_{ij}^{\;k} \;\ldots\ldots\; c_{i+r-1,j}^{\;k} \,,\; i=1,p,\; j=1,1+r,\ldots,1+(q-1).r \tag{6.235}$$

Example 6.22

This example considers a FN with two nodes A and U with common input $x_{A,U}$ whereby y_A is the output from A and y_U is the output from U. These nodes are output merged into node C that represents the equivalent fuzzy system for this FN, as shown by the block-scheme in Figure 6.30 and the topological expression in Eq.(6.236). The known nodes A and C are described by the Boolean matrices and the binary relations in Eqs.(6.237)-(6.238) and Eqs.(6.239)-(6.240).

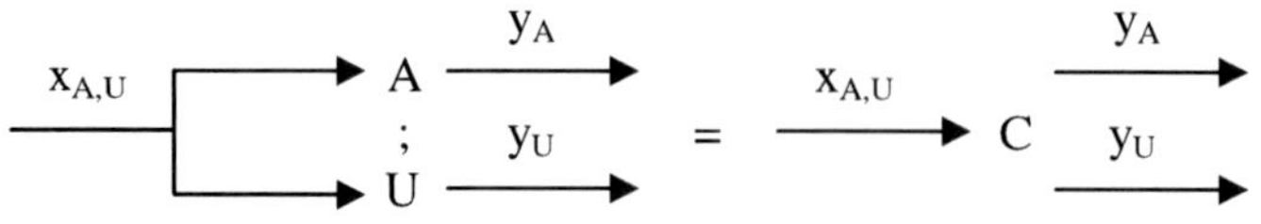

Fig. 6.30 FN with nodes A and U output merged into node C

$$[A]\ (x_{A,U} \mid y_A)\ ;\ [U]\ (x_{A,U} \mid y_U) = [C]\ (x_{A,U} \mid y_A, y_U) \tag{6.236}$$

A: (6.237)

$x_{A,U}$ \ y_A	1	2	3
1	0	1	0
2	1	0	0
3	0	0	1

A: {(1, 2), (2, 1), (3, 3)} (6.238)

C: (6.239)

$x_{A,U}$ \ y_A, y_U	11	12	13	21	22	23	31	32	33
1	0	0	0	1	0	0	0	0	0
2	0	0	1	0	0	0	0	0	0
3	0	0	0	0	0	0	0	1	0

C: {(1, 21), (2, 13), (3, 32), (6.240)

The Boolean matrix C in Eq.(6.239) contains only 3 non-zero row blocks C^k whereby there is not more than one such block in any row of this matrix. Also, the elements with a location in A corresponding to the location of non-zero row blocks C^k in C are all equal to '1' and all other elements with a location in A corresponding to the location of zero row blocks C^0 in C are equal to '0'.

Therefore, the rows of the Boolean matrix U are equal to C^k, as shown by the Boolean matrix and the binary relation in Eqs.(6.241)-(6.242). In this case, the location of the elements of C^k, k=1,2,3 in C is given by Eqs.(6.243)-(6.245).

U: (6.241)

$x_{A,U}$ \ y_U	1	2	3
1	1	0	0
2	0	0	1
3	0	1	0

U: {(1, 1), (2, 3), (3, 2)} (6.242)

$$C^1 = c_{14}{}^1 \quad c_{15}{}^1 \quad c_{16}{}^1 \tag{6.243}$$

$$C^2 = c_{21}{}^2 \quad c_{22}{}^2 \quad c_{23}{}^2 \tag{6.244}$$

$$C^3 = c_{37}{}^3 \quad c_{38}{}^3 \quad c_{39}{}^3 \tag{6.245}$$

Problem 6.8

In this problem, node identification is considered in the context of output merging of two nodes U and B with common inputs into a node C whereby only B and C are given. Therefore, it is necessary to identify the unknown node U on the basis of the known nodes B and C. This problem can be described in a general form by the Boolean matrix equation in Eq.(6.246) where U, B and C represent the Boolean matrices for the above three nodes. The dimensions and the detailed descriptions of these Boolean matrices are given in Eqs.(6.247)-(6.248) and Eq.(6.234).

$$U \,;\, B = C \tag{6.246}$$

$$U_p \times_q = \begin{matrix} u_{11} \;\ldots\ldots\; u_{1q} \\ \ldots\ldots\ldots\ldots\ldots \\ u_{p1} \;\ldots\ldots\; u_{pq} \end{matrix} \tag{6.247}$$

$$B_p \times_r = \begin{matrix} b_{11} \;\ldots\ldots\; b_{1r} \\ \ldots\ldots\ldots\ldots\ldots \\ b_{p1} \;\ldots\ldots\; b_{pr} \end{matrix} \tag{6.248}$$

If the Boolean matrix C in Eq.(6.246) contains only one set of not more than ‘p’ non-zero row blocks C^k, $1 \leq k \leq p$ whereby there is not more than one such block in any row of this matrix, and the rows of the Boolean matrix B are equal to C^k, then the elements with a location in U corresponding to the location of non- zero row blocks C^k in C are all equal to ‘1’ and all other elements with a location in U corresponding to the location of zero row blocks C^0 in C are equal to ‘0’. In this case, the location of the elements of a non-zero row block C^k can be described by Eq.(6.235) which also shows the admissible initial values for the subscripts i and j of the elements of C^k. As far as the zero row blocks C^0 are concerned, they are with the same number of elements as the non-zero row blocks C^k.

Example 6.23

This example considers a FN with two nodes U and B with common input $x_{U,B}$ whereby y_U is the output from U and y_B is the output from B. These nodes are output merged into node C that represents the equivalent fuzzy system for this FN, as shown by the block-scheme in Figure 6.31 and the topological expression in Eq.(6.249). The known nodes B and C are described by the Boolean matrices and the binary relations in Eqs.(6.250)-(6.251) and Eqs.(6.252)-(6.253).

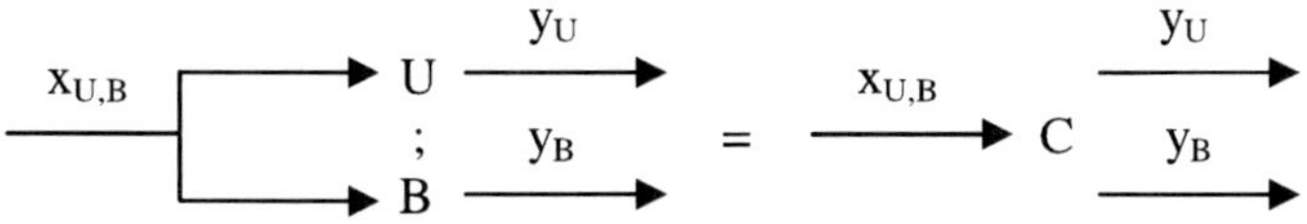

Fig. 6.31 FN with nodes U and B output merged into node C

$$[U]\ (x_{U,B} \mid y_U)\ ;\ [B]\ (x_{U,B} \mid y_B) = [C]\ (x_{U,B} \mid y_U, y_B) \quad (6.249)$$

B : (6.250)

$x_{U,B}$ \ y_B	1	2	3
1	0	1	0
2	0	0	1
3	1	0	0

B: {(1, 2), (2, 3), (3, 1)} (6.251)

C: (6.252)

$x_{U,B}$ \ y_U, y_B	11	12	13	21	22	23	31	32	33
1	0	0	0	0	0	0	0	1	0
2	0	0	1	0	0	0	0	0	0
3	0	0	0	1	0	0	0	0	0

C: {(1, 32), (2, 13), (3, 21), (6.253)

The Boolean matrix C in Eq.(6.252) contains only 3 non-zero blocks C^k whereby there is not more than one such block in any row of this matrix. Also, the rows of the Boolean matrix B are equal to C^k.

Therefore, the elements with a location in U corresponding to the location of non-zero row blocks C^k in C are all equal to '1' and all other elements with a location in U corresponding to the location of zero row blocks C^0 in C are equal to '0', as shown by the Boolean matrix and the binary relation in Eqs.(6.254)-(6.255). In this case, the location of the elements of C^k, k=1,2,3 in C is given by Eq.(6.256), Eq.(6.244) and Eq.(2.257).

U : (6.254)

$x_{U,B}$ \ y_U	1	2	3
1	0	0	1
2	1	0	0
3	0	1	0

U: {(1, 3), (2, 1), (3, 2)} (6.255)

$$C^1 = c_{17}^{\ 1} \quad c_{18}^{\ 1} \quad c_{19}^{\ 1} \quad (6.256)$$

$$C^3 = c_{34}^{\ 3} \quad c_{35}^{\ 3} \quad c_{36}^{\ 3} \quad (6.257)$$

Problem 6.9

In this problem, node identification is considered in the context of output merging of three nodes A, U and B into a node C whereby only A, B and C are given. Therefore, it is necessary to identify the unknown node U on the basis of the known nodes A, B and C. This problem can be described in a general form by the Boolean matrix equation in Eq.(6.258) where A, U, B and C represent the Boolean matrices for the above four nodes. The dimensions and the detailed descriptions of these Boolean matrices are given in Eqs.(6.232)-(6.233) and Eqs.(6.259)-(6.260).

$$A \,;\, U \,;\, B = C \tag{6.258}$$

$$B_p X_s = \begin{matrix} b_{11} \;\ldots\ldots\; b_{1s} \\ \ldots\ldots\ldots\ldots\ldots \\ b_{p1} \;\ldots\ldots\; b_{ps} \end{matrix} \tag{6.259}$$

$$C_p X_{q.r.s} = \begin{matrix} c_{11} \;\ldots\ldots\ldots\; c_{1,q.r.s} \\ \ldots\ldots\ldots\ldots\ldots\ldots \\ c_{p,1} \;\ldots\ldots\ldots\; c_{p,q.r.s} \end{matrix} \tag{6.260}$$

The Boolean matrix equation in Eq.(6.258) can be solved in two different ways. In either case, the solution can be found on the basis of the solutions for Problems 6.7-6.8.

In the first case, Eq.(6.258) is presented as a system of two Boolean matrix equations. This case is shown by Eqs.(6.261)-(6.262) whereby D is an additional unknown Boolean matrix given by Eq.(6.263). The solution for this system of Boolean matrix equations can be found by first solving Eq.(6.262) with respect to D and then solving Eq.(6.261) with respect to U.

$$A \,;\, U = D \tag{6.261}$$

$$D \,;\, B = C \tag{6.262}$$

$$D_p X_{q.r} = \begin{matrix} d_{11} \;\ldots\ldots\; d_{1,q.r} \\ \ldots\ldots\ldots\ldots\ldots \\ d_{p1} \;\ldots\ldots\; d_{p,q.r} \end{matrix} \tag{6.263}$$

In the second case, Eq.(6.258) is also presented as a system of two Boolean matrix equations. This case is shown by Eqs.(6.264)-(6.265) whereby E is an additional unknown Boolean matrix given by Eq.(6.266). The solution for this system of Boolean matrix equations can be found by first solving Eq.(6.265) with respect to E and then solving Eq.(6.264) with respect to U.

$$U \,;\, B = E \tag{6.264}$$

$$A \,;\, E = C \tag{6.265}$$

$$E_p X_{r.s} = \begin{matrix} e_{11} \;\ldots\ldots\; e_{1,r.s} \\ \ldots\ldots\ldots\ldots\ldots \\ e_{p1} \;\ldots\ldots\; e_{p,r.s} \end{matrix} \tag{6.266}$$

Example 6.24

This example considers a FN with three nodes A, U and B with common input $x_{A,U,B}$ whereby y_A is the output from A, y_U is the output from U and y_B is the output from B. These nodes are output merged into node C that represents the equivalent fuzzy system for this FN, as shown by the block-scheme in Figure 6.32 and the topological expression in Eq.(6.267). The known nodes A, B and C are described by the Boolean matrices and the binary relations in Eqs.(6.237)-(6.238), Eqs.(6.250)-(6.251) and Eqs.(6.268)-(6.269). In this case, the labels and the elements of the Boolean matrix in Eq.(6.268) are represented by a compact notation. In particular, each of the capital letters A, B, C, D, E, F, G, H, I stands for three sequential columns in accordance with Eq.(6.223), 1_j , j=1,3 denotes the j-th Boolean row in the Boolean matrix for node B from Eq.(6.250) and 0_3 denotes a zero Boolean row of dimension equal to that for 1_j.

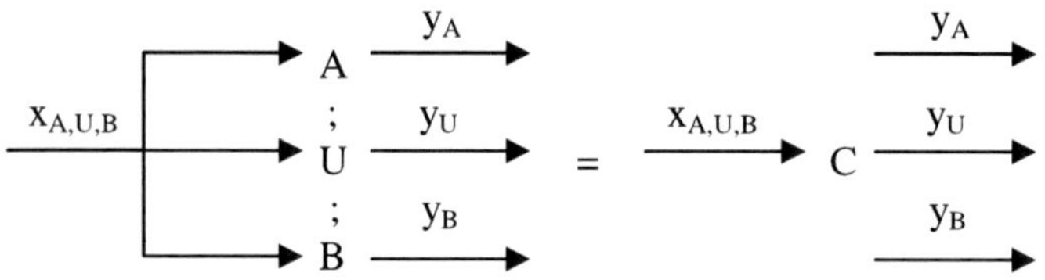

Fig. 6.32 FN with nodes A, U and B output merged into node C

$$[A]\,(x_{A,U,B} \mid y_A)\ ;\ [U]\,(x_{A,U,B} \mid y_U)\ ;\ [B]\,(x_{A,U,B} \mid y_B) = [C]\,(x_{A,U,B} \mid y_A,\, y_U,\, y_B) \qquad (6.267)$$

C: (6.268)

$x_{A,U,B}$ \ y_A, y_U, y_B	A	B	C	D	E	F	G	H	I
1	0_3	0_3	0_3	0_3	0_3	1_1	0_3	0_3	0_3
2	0_3	1_2	0_3	0_3	0_3	0_3	0_3	0_3	0_3
3	0_3	0_3	0_3	0_3	0_3	0_3	1_3	0_3	0_3

C: {(1, 232), (2, 123), (3, 311)} (6.269)

In the first case, the solution for node D is given by the Boolean matrix and the binary relation in Eqs.(6.270)-(6.271) whereas the solution for node U is given by the Boolean matrix and the binary relation in Eqs.(6.272)-(6.273).

D: (6.270)

$x_{A,U,B}$ \ y_A, y_U	11	12	13	21	22	23	31	32	33
1	0	0	0	0	0	1	0	0	0
2	0	1	0	0	0	0	0	0	0
3	0	0	0	0	0	0	1	0	0

D: {(1, 23), (2, 12), (3, 31)} (6.271)

U: (6.272)

$x_{A,U,B}$ \ y_U	1	2	3
1	0	0	1
2	0	1	0
3	1	0	0

$$U: \{(1, 3), (2, 2), (3, 1)\} \tag{6.273}$$

In the second case, the solution for node E is given by the Boolean matrix and the binary relation in Eqs.(6.274)-(6.275) whereas the solution for node U is also given by the Boolean matrix and the binary relation in Eqs.(6.272)-(6.273).

E: (6.274)

$x_{A,U,B}$ \ y_U, y_B	11	12	13	21	22	23	31	32	33
1	0	0	0	0	0	0	0	1	0
2	0	0	0	0	0	1	0	0	0
3	1	0	0	0	0	0	0	0	0

$$E: \{(1, 32), (2, 23), (3, 11)\} \tag{6.275}$$

6.8 Comparison of Advanced Operations

The advanced operations introduced in this chapter are central to the linguistic composition approach used in the book. This applies particularly to node transformation for input augmentation, output permutation and feedback equivalence which complement the basic operations of horizontal, vertical and output merging for composing the networked rule bases within a FN into a linguistically equivalent single rule base for a fuzzy system. The other advanced operations such as node identification in horizontal, vertical and output merging are aimed at decomposing a single rule base for a fuzzy system into linguistically equivalent networked rule bases within a FN. However, in some cases advanced operations for decomposition may be used together with advanced operations for composition in the context of the linguistic composition approach and this is shown by some examples further in the book.

The solution to some advanced operations always exists. The exceptions are node identification operations that may not have a solution. In this context, when a solution exists, node identification in horizontal merging is likely to have multiple solutions whereas node identification in vertical and output merging usually has a unique solution. Similar observations apply to node transformation for input augmentation, output permutation and feedback equivalence which often has a unique solution.

The characteristics of solutions for different types of advanced operations in FNs are summarised in Table 6.1.

Table 6.1 Solution characteristics for advanced operations in FNs

Advanced operation	Composition	Existence	Uniqueness
Node transformation for input augmentation	Yes	Yes	Yes
Node transformation for output permutation	Yes	Yes	Yes
Node transformation for feedback equivalence	Yes	Yes	Yes
Node identification in horizontal merging	No	No	No
Node identification in vertical merging	No	No	Yes
Node identification in output merging	No	No	Yes

The next chapter shows some applications of the theoretical results from Chapters 4-6 to FNs. In particular, several basic types of feedforward FNs are considered.

Chapter 7
Feedforward Fuzzy Networks

7.1 Preliminaries on Feedforward Fuzzy Networks

The basic operations, their structural properties and the advanced operations introduced in Chapters 4-6 are illustrated there mainly on fairly simple FNs with interconnected nodes. Although these networks are assumed to be part of the structure of more complex FNs, the latter are taken into account only implicitly in the considerations so far. Therefore, it is necessary to show explicitly the application of the above operations and their properties to the overall structure of fairly complex FNs.

The current chapter describes the application of basic operations, their properties and advanced operations in feedforward FNs. The latter are FNs all of whose connections are only in a forward direction, i.e. from nodes residing in specific layers to nodes residing in subsequent layers. This feedforward characteristic is reflected by right-sided arrows in the corresponding block scheme for the FN under consideration. In this context, a right-sided arrow represents an output from a node that is fed forward as an input to another node.

Four types of feedforward FNs are considered in the context of both analysis and design. The analysis part is presented first and is then followed by the design part. In the case of analysis, all network nodes are known and the aim is to derive a formula for the single node representing the linguistically equivalent fuzzy system. In the case of design, each network node is unknown at a time with all other network nodes known and the aim is to derive an algorithm for the unknown node from the known network nodes and the single node representing the linguistically equivalent fuzzy system. The design task can be easily extended to cases with more than one unknown node.

The four types of feedforward FNs represent different network topologies with respect to single or multiple levels and layers in the underlying grid structure for the FN under consideration. Each of these types is illustrated with several examples that are presented at a fairly high level of abstraction using mainly block schemes and topological expressions. These formal models for FNs are at network level and they both lend themselves easily to advanced manipulations in the context of the linguistic composition approach. Boolean matrices are used as formal models only implicitly in block schemes and topological expressions as well as in design tasks.

A. Gegov: Fuzzy Networks for Complex Systems, STUDFUZZ 259, pp. 161–200.
springerlink.com

All presented examples are for feedforward FNs with a fairly small number of nodes but the extension of these examples to networks with a larger number of nodes is straightforward. The only difference in this extension is the higher complexity of the formulas for the derivation of the single node representing the linguistically equivalent fuzzy system in the case of analysis and the algorithms for the derivation of the unknown node in the case of design.

7.2 Networks with Single Level and Single Layer

The simplest type of FN is the one with single level and single layer. This network has only one node residing in the single level and the single layer of the underlying grid structure. Due to the absence of other nodes, there are not any feedforward connections from and to this node. Moreover, there are not any connections between the outputs from and the inputs to this node as such connections are only of feedback type and are therefore outside the scope of the current chapter.

Therefore, a FN with single level and single layer is a single node network that is identical to a fuzzy system with single rule base. This implies that a fuzzy system is a simple FN, i.e. a special case of a FN with single node. Similarly, a FN can be viewed as a complex fuzzy system, i.e. a general case of a fuzzy system with networked rule bases.

As the focus of this book is on FNs, FNs with single level and single layer which are identical to fuzzy systems are considered in the current section very briefly only for completeness and consistency. In this context, the other three types of feedforward FNs which are an extension of fuzzy systems are considered in much more detail in the following sections.

7.3 Networks with Single Level and Multiple Layers

A more complex type of FN is the one with single level and multiple layers. This network has at least two nodes residing in the single level and the multiple layers of the underlying grid structure, i.e. it is identical to a queue of fuzzy systems with single rule bases. Due to the presence of multiple nodes, there are feedforward connections from and to at least some of these nodes. However, there are not any connections between the outputs from and the inputs to the same or other nodes as such connections are of feedback type and outside the scope of the current chapter.

Example 7.1

This example considers a FN with nodes N_{11} and N_{12} where x_{11} is an input for N_{11}, y_{12} is an output for N_{12}, $z_{11,12}^{1,2}$ is the connection from the first output for N_{11} to the second input for N_{12} and $z_{11,12}^{2,1}$ is the connection from the second output for N_{11} to the first input for N_{12}. This initial FN can be described by the block-scheme in Fig.7.1 and the topological expression in Eq.(7.1) from where it can be seen that the connections have crossing paths.

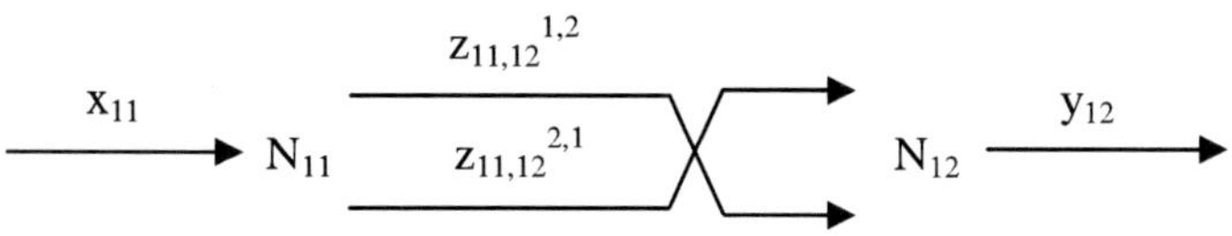

Fig. 7.1 Initial FN for Example 7.1

$$[N_{11}]\ (x_{11} \mid z_{11,12}^{1,2}, z_{11,12}^{2,1}) * [N_{12}]\ (z_{11,12}^{2,1}, z_{11,12}^{1,2} \mid y_{12}) \tag{7.1}$$

In order to merge horizontally the nodes N_{11} and N_{12} of the initial FN, it is necessary to remove the crossing of the connection paths. This can be done by permuting the connections $z_{11,12}^{1,2}$ and $z_{11,12}^{2,1}$ at their output points in node N_{11}. This permutation operation transforms the initial FN into an interim FN with nodes N_{11}^{PO} and N_{12} whereby the permutation of outputs is reflected by the replacement of N_{11} with N_{11}^{PO}. This interim FN can be described by the block-scheme in Fig.7.2 and the topological expression in Eq.(7.2) from where it can be seen that the connections have parallel paths.

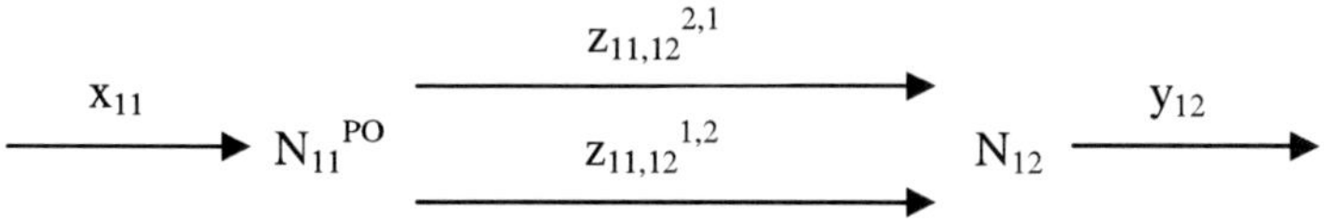

Fig. 7.2 Interim FN for Example 7.1

$$[N_{11}^{PO}]\ (x_{11} \mid z_{11,12}^{2,1}, z_{11,12}^{1,2}) * [N_{12}]\ (z_{11,12}^{2,1}, z_{11,12}^{1,2} \mid y_{12}) \tag{7.2}$$

The nodes N_{11}^{PO} and N_{12} of the interim FN can be merged horizontally due to the parallel connection paths. This merging operation transforms the interim FN into a final FN with a single equivalent node whereby the merging of nodes N_{11}^{PO} and N_{12} is reflected by their replacement with node $N_{11}^{PO} * N_{12}$. This final FN can be described by the block-scheme in Fig.7.3 and the topological expression in Eq.(7.3) from where it can be seen that the two original nodes are implicit in the single equivalent node and the second node is unchanged in relation to the initial FN.

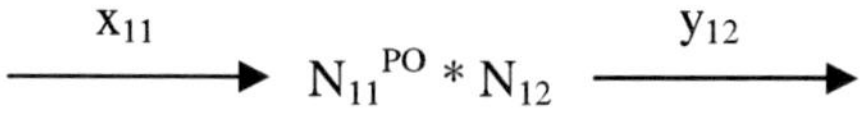

Fig. 7.3 Final FN for Example 7.1

$$[N_{11}^{PO} * N_{12}]\ (x_{11} \mid y_{12}) \tag{7.3}$$

The considerations in Example 7.1 are concerned with network analysis when all network nodes are known. In the case of network design, at least one node is unknown. In this context, Algorithms 7.1-7.2 describe the process of deriving an unknown node in the initial FN from Fig.7.1 when the other node and the single equivalent node N_E are known. In this case, node N_E is given by the Boolean matrix equation in Eq.(7.4).

$$N_E = N_{11}^{PO} * N_{12} \tag{7.4}$$

Algorithm 7.1

1. Define N_E and N_{12}.
2. Derive N_{11}^{PO} from Eq.(7.4), if possible.
3. Find N_{11} by inverse output permutation of N_{11}^{PO}.

Algorithm 7.2

1. Define N_E and N_{11}.
2. Find N_{11}^{PO} by output permutation of N_{11}.
3. Derive N_{12} from Eq.(7.4), if possible.

Example 7.2

This example considers a FN with nodes N_{11}, N_{12} and N_{13} where x_{11} is an input for N_{11}, y_{13} is an output for N_{13}, $z_{11,12}^{1,2}$ is the connection from the first output for N_{11} to the second input for N_{12}, $z_{11,12}^{2,1}$ is the connection from the second output for N_{11} to the first input for N_{12}, $z_{12,13}^{1,2}$ is the connection from the first output for N_{12} to the second input for N_{13} and $z_{12,13}^{2,1}$ is the connection from the second output for N_{12} to the first input for N_{13}. This initial FN can be described by the block-scheme in Fig.7.4 and the topological expression in Eq.(7.5) from where it can be seen that the connections have crossing paths.

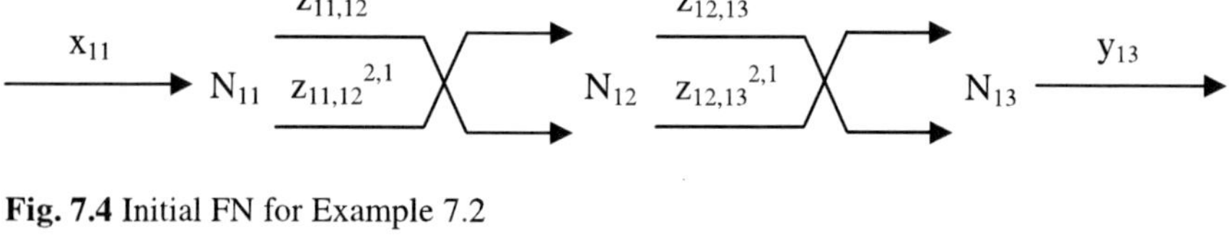

Fig. 7.4 Initial FN for Example 7.2

$$[N_{11}]\,(x_{11} \mid z_{11,12}^{1,2}, z_{11,12}^{2,1}) * [N_{12}]\,(z_{11,12}^{2,1}, z_{11,12}^{1,2} \mid z_{12,13}^{1,2}, z_{12,13}^{2,1}) *$$

$$[N_{13}]\,(z_{12,13}^{2,1}, z_{12,13}^{1,2} \mid y_{13}) \tag{7.5}$$

In order to merge horizontally the nodes N_{11}, N_{12} and N_{13} of the initial FN, it is necessary to remove the crossing of the connection paths. This can be done by first permuting the connections $z_{11,12}^{1,2}$, $z_{11,12}^{2,1}$ at their output points in node N_{11}

and then permuting the connections $z_{12,13}^{1,2}$, $z_{12,13}^{2,1}$ at their output points in node N_{12}. This permutation operation transforms the initial FN into an interim FN with nodes N_{11}^{PO}, N_{12}^{PO} and N_{13} whereby the permutation of outputs is reflected by the replacement of N_{11} and N_{12} with N_{11}^{PO} and N_{12}^{PO}, respectively. This interim FN can be described by the block-scheme in Fig.7.5 and the topological expression in Eq.(7.6) from where it can be seen that the connections have parallel paths.

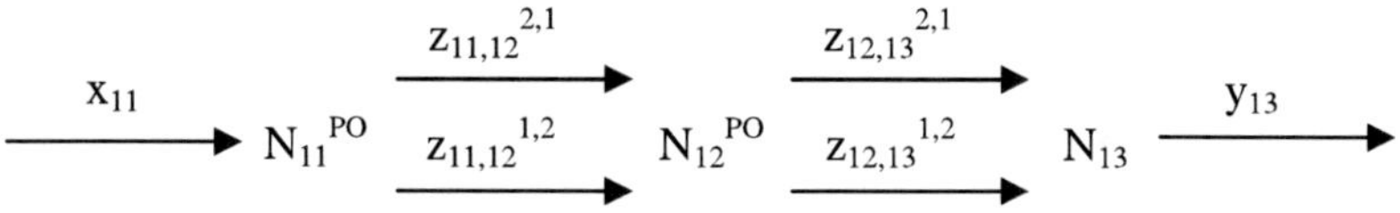

Fig. 7.5 Interim FN for Example 7.2

$$[N_{11}^{PO}]\,(x_{11} \,|\, z_{11,12}^{2,1}, z_{11,12}^{1,2}) * [N_{12}^{PO}]\,(z_{11,12}^{2,1}, z_{11,12}^{1,2} \,|\, z_{12,13}^{2,1}, z_{12,13}^{1,2}) * \qquad (7.6)$$

$$[N_{13}]\,(z_{12,13}^{2,1}, z_{12,13}^{1,2} \,|\, y_{13})$$

The nodes N_{11}^{PO}, N_{12}^{PO} and N_{13} of the interim FN can be merged horizontally due to the parallel connection paths. This merging operation transforms the interim FN into a final FN with a single equivalent node whereby the merging of nodes N_{11}^{PO}, N_{12}^{PO} and N_{13} is reflected by their replacement with node $N_{11}^{PO} * N_{12}^{PO} * N_{13}$. This final FN can be described by the block-scheme in Fig.7.6 and the topological expression in Eq.(7.7) from where it can be seen that the three original nodes are implicit in the single equivalent node and the third node is unchanged in relation to the initial FN.

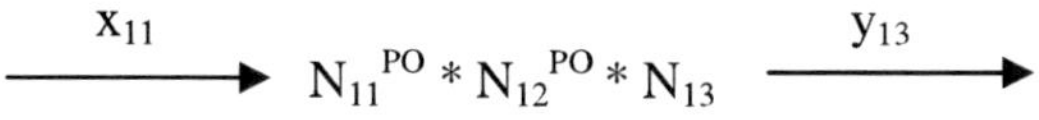

Fig. 7.6 Final FN for Example 7.2

$$[N_{11}^{PO} * N_{12}^{PO} * N_{13}]\,(x_{11} \,|\, y_{13}) \qquad (7.7)$$

The considerations in Example 7.2 are concerned with network analysis when all network nodes are known. In the case of network design, at least one node is unknown. In this context, Algorithms 7.3-7.5 describe the process of deriving an unknown node in the initial FN from Fig.7.4 when the other nodes and the single equivalent node N_E are known. In this case, node N_E is given by the Boolean matrix equation in Eq.(7.8).

$$N_E = N_{11}^{PO} * N_{12}^{PO} * N_{13} \qquad (7.8)$$

Algorithm 7.3

1. Define N_E, N_{12} and N_{13}.
2. Find N_{12}^{PO} by output permutation of N_{12}.
3. Derive N_{11}^{PO} from Eq.(7.8), if possible.
4. Find N_{11} by inverse output permutation of N_{11}^{PO}.

Algorithm 7.4

1. Define N_E, N_{11} and N_{13}.
2. Find N_{11}^{PO} by output permutation of N_{11}.
3. Derive N_{12}^{PO} from Eq.(7.8), if possible.
4. Find N_{12} by inverse output permutation of N_{12}^{PO}.

Algorithm 7.5

1. Define N_E, N_{11} and N_{12}.
2. Find N_{11}^{PO} by output permutation of N_{11}.
3. Find N_{12}^{PO} by output permutation of N_{12}.
4. Derive N_{13} from Eq.(7.4), if possible.

Example 7.3

This example considers a FN with nodes N_{11}, N_{12} and N_{13} where x_{11} is an input for N_{11}, y_{13} is an output for N_{13}, $z_{11,13}^{1,2}$ is the connection from the first output for N_{11} to the second input for N_{13}, $z_{11,12}^{2,1}$ is the connection from the second output for N_{11} to the first and only input for N_{12} and $z_{12,13}^{1,1}$ is the connection from the first and only output for N_{12} to the first input for N_{13}. This initial FN can be described by the block-scheme in Fig.7.7 and the topological expression in Eq.(7.9) from where it can be seen that there is a crossing connection path at the bottom propagating through the second layer of the FN within a virtual second level.

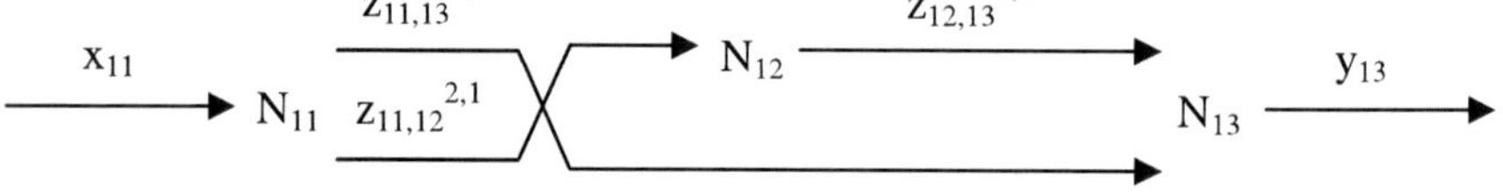

Fig. 7.7 Initial FN for Example 7.3

$$[N_{11}]\ (x_{11} \mid z_{11,13}^{1,2},\ z_{11,12}^{2,1})\ *\ [N_{12}]\ (z_{11,12}^{2,1} \mid z_{12,13}^{1,1})\ * \qquad (7.9)$$

$$[N_{13}]\ (z_{12,13}^{1,1},\ z_{11,13}^{1,2} \mid y_{13})$$

In order to merge horizontally the nodes N_{11}, N_{12} and N_{13} of the initial FN, it is necessary to remove the crossing of the connection path at the bottom. This can be done by permuting the connections $z_{11,13}^{1,2}$ and $z_{11,12}^{2,1}$ at their output points in node N_{11}. This permutation operation transforms the initial FN into a first interim FN with nodes N_{11}^{PO}, N_{12} and N_{13} whereby the permutation of outputs is reflected by the replacement of N_{11} with N_{11}^{PO}. This first interim FN can be described by the block-scheme in Fig.7.8 and the topological expression in Eq.(7.10) from where it can be seen that there is a parallel connection path at the bottom propagating through the second layer of the FN within a virtual second level.

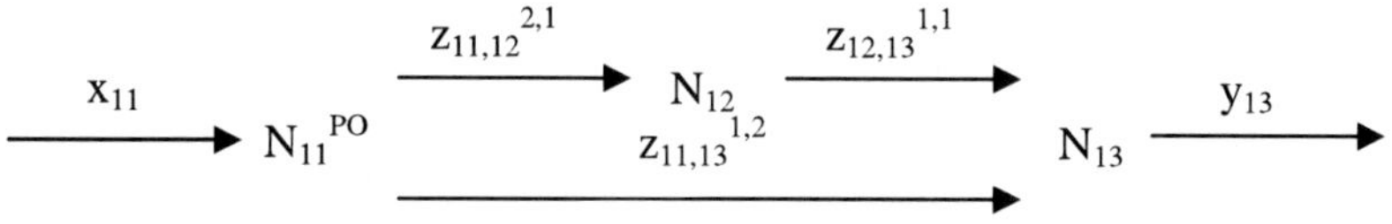

Fig. 7.8 First interim FN for Example 7.3

$$[N_{11}^{PO}]\,(x_{11} \mid z_{11,12}^{2,1}, z_{11,13}^{1,2}) * [N_{12}]\,(z_{11,12}^{2,1} \mid z_{12,13}^{1,1}) * [N_{13}]\,(z_{12,13}^{1,1}, z_{11,13}^{1,2} \mid y_{13}) \quad (7.10)$$

Further on, it is also necessary to represent the parallel connection path at the bottom by inserting an implicit identity node I_{22}. This insertion transforms the first interim FN into a second interim FN with nodes N_{11}^{PO}, N_{12}, N_{13} and I_{22}. This second interim FN can be described by the block-scheme in Fig.7.9 and the topological expression in Eq.(7.11) from where it can be seen that the parallel connection path at the bottom propagating through the second layer of the FN within a virtual second level is preserved.

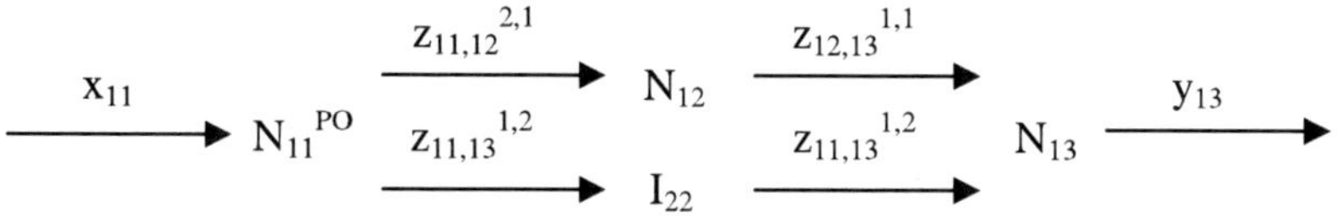

Fig. 7.9 Second interim FN for example 7.3

$$[N_{11}^{PO}]\,(x_{11} \mid z_{11,12}^{2,1}, z_{11,13}^{1,2}) * \{[N_{12}]\,(z_{11,12}^{2,1} \mid z_{12,13}^{1,1}) + [I_{22}]\,(z_{11,13}^{1,2} \mid z_{11,13}^{1,2})\} * [N_{13}]\,(z_{12,13}^{1,1}, z_{11,13}^{1,2} \mid y_{13}) \quad (7.11)$$

Nodes N_{12} and I_{22} of the interim FN can be merged vertically into a temporary node $N_{12} + I_{22}$. This node can be further merged horizontally with node N_{11}^{PO} on the left and node N_{13} on the right. These merging operations transform the interim FN into a final FN with a single equivalent node whereby the horizontal merging of nodes N_{11}^{PO}, $N_{12} + I_{22}$ and N_{13} is reflected by their replacement with node $N_{11}^{PO} * (N_{12} + I_{22}) * N_{13}$. This final FN can be described by the block-scheme in Fig.7.10 and the topological expression in Eq.(7.12) from where it can be seen that the identity node from the parallel connection path at the bottom and the three original nodes are implicit in the single equivalent node whereby the last two of these nodes are unchanged in relation to the initial FN.

$$\xrightarrow{x_{11}} N_{11}{}^{PO} * (N_{12} + I_{22}) * N_{13} \xrightarrow{y_{13}}$$

Fig. 7.10 Final FN for Example 7.3

$$[N_{11}{}^{PO} * (N_{12} + I_{22}) * N_{13}]\ (x_{11} \mid y_{13}) \quad (7.12)$$

The considerations in Example 7.3 are concerned with network analysis when all network nodes are known. In the case of network design, at least one node is unknown. In this context, Algorithms 7.6-7.8 describe the process of deriving an unknown node in the initial FN from Fig.7.7 when the other nodes, the implicit identity node I_{22} and the single equivalent node N_E are known. In this case, node N_E is given by the Boolean matrix equation in Eq.(7.13).

$$N_E = N_{11}{}^{PO} * (N_{12} + I_{22}) * N_{13} \quad (7.13)$$

Algorithm 7.6

1. Define N_E, N_{12}, N_{13} and I_{22}.
2. Find $N_{12} + I_{22}$ by vertical merging of N_{12} and I_{22}.
3. Derive $N_{11}{}^{PO}$ from Eq.(7.13), if possible.
4. Find N_{11} by inverse output permutation of $N_{11}{}^{PO}$.

Algorithm 7.7

1. Define N_E, N_{11}, N_{13} and I_{22}.
2. Find $N_{11}{}^{PO}$ by output permutation of N_{11}.
3. Derive $N_{12} + I_{22}$ from Eq.(7.13), if possible.
4. Derive N_{12} from $N_{12} + I_{22}$, if possible.

Algorithm 7.8

1. Define N_E, N_{11}, N_{12} and I_{22}.
2. Find $N_{11}{}^{PO}$ by output permutation of N_{11}.
3. Find $N_{12} + I_{22}$ by vertical merging of N_{12} and I_{22}.
4. Derive N_{13} from Eq.(7.13), if possible.

Example 7.4

This example considers a FN with nodes N_{11}, N_{12} and N_{13} where x_{11} is an input for N_{11}, y_{13} is an output for N_{13}, $z_{11,12}{}^{1,1}$ is the connection from the first output for N_{11} to the first and only input for N_{12}, $z_{11,13}{}^{2,1}$ is the connection from the second output for N_{11} to the first input for N_{13} and $z_{12,13}^{1,2}$ is the connection from the first and only output for N_{12} to the second input for N_{13}. This initial FN can be described by the block-scheme in Fig.7.11 and the topological expression in Eq.(7.14) from where it can be seen that there is a crossing connection path at the top propagating through the second layer of the FN within a virtual zero level.

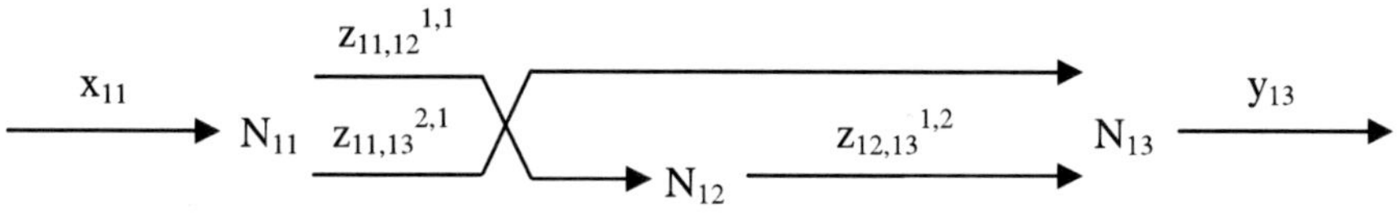

Fig. 7.11 Initial FN for Example 7.4

$$[N_{11}]\ (x_{11} \mid z_{11,12}^{1,1},\ z_{11,13}^{2,1}) * [N_{12}]\ (z_{11,12}^{1,1} \mid z_{12,13}^{1,2}) * \qquad (7.14)$$

$$[N_{13}]\ (z_{11,13}^{2,1},\ z_{12,13}^{1,2} \mid y_{13})$$

In order to merge horizontally the nodes N_{11}, N_{12} and N_{13} of the initial FN, it is necessary to remove the crossing of the connection path at the top. This can be done by permuting the connections $z_{11,12}^{1,1}$ and $z_{11,13}^{2,1}$ at their output points in node N_{11}. This permutation operation transforms the initial FN into a first interim FN with nodes N_{11}^{PO}, N_{12} and N_{13} whereby the permutation of outputs is reflected by the replacement of N_{11} with N_{11}^{PO}. This first interim FN can be described by the block-scheme in Fig.7.12 and the topological expression in Eq.(7.15) from where it can be seen that there is a parallel connection path at the top propagating through the second layer of the FN within a virtual zero level.

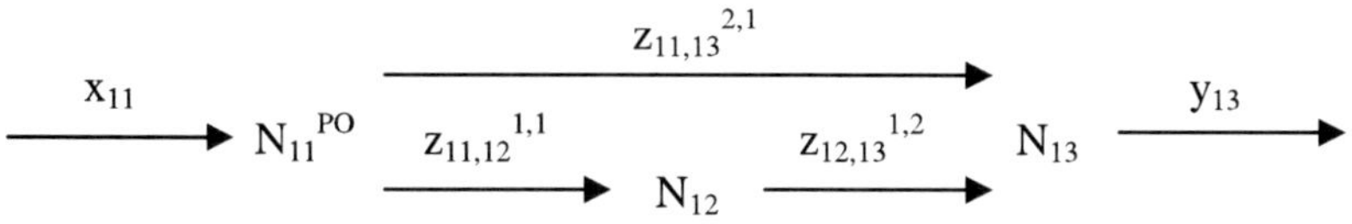

Fig. 7.12 First interim FN for Example 7.4

$$[N_{11}^{PO}]\ (x_{11} \mid z_{11,13}^{2,1},\ z_{11,12}^{1,1}) * [N_{12}]\ (z_{11,12}^{1,1} \mid z_{12,13}^{1,2})\} * \qquad (7.15)$$

$$[N_{13}]\ (z_{11,13}^{2,1},\ z_{12,13}^{1,2} \mid y_{13})$$

Further on, it is also necessary to represent the parallel connection path at the top by inserting an implicit identity node I_{02}. This insertion transforms the first interim FN into a second interim FN with nodes N_{11}^{PO}, N_{12}, N_{13} and I_{02}. This second interim FN can be described by the block-scheme in Fig.7.13 and the topological expression in Eq.(7.16) from where it can be seen that the parallel connection path at the top propagating through the second layer of the FN within a virtual zero level is preserved.

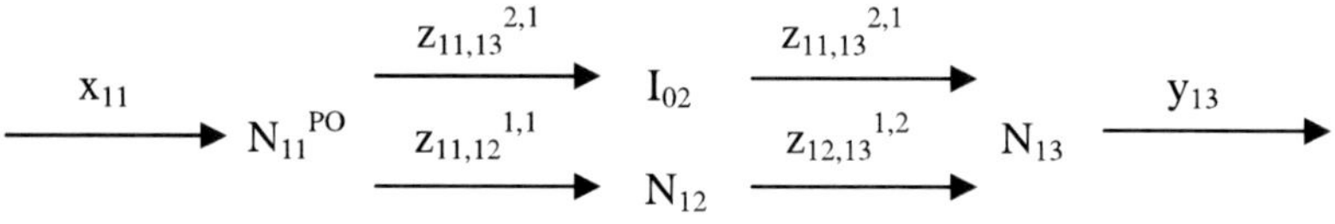

Fig. 7.13 Second interim FN for Example 7.4

$$[N_{11}^{PO}]\ (x_{11} \mid z_{11,13}^{2,1}, z_{11,12}^{1,1}) \ * \tag{7.16}$$

$$\{[I_{02}]\ (z_{11,13}^{2,1} \mid z_{11,13}^{2,1}) + [N_{12}]\ (z_{11,12}^{1,1} \mid z_{12,13}^{1,2})\} * [N_{13}]\ (z_{11,13}^{2,1}, z_{12,13}^{1,2} \mid y_{13})$$

Nodes I_{02} and N_{12} of the interim FN can be merged vertically into a temporary node $I_{02} + N_{12}$. This node can be further merged horizontally with node N_{11}^{PO} on the left and node N_{13} on the right. These merging operations transform the interim FN into a final FN with a single equivalent node whereby the horizontal merging of nodes N_{11}^{PO}, $I_{02} + N_{12}$ and N_{13} is reflected by their replacement with node $N_{11}^{PO} * (I_{02} + N_{12}) * N_{13}$. This final FN can be described by the block-scheme in Fig.7.14 and the topological expression in Eq.(7.17). It can be seen from there that the identity node from the parallel connection path at the top and the three original nodes are implicit in the single equivalent node whereby the last two of these nodes are unchanged in relation to the initial FN.

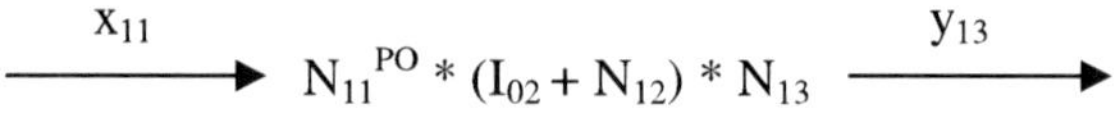

Fig. 7.14 Final FN for Example 7.4

$$[N_{11}^{PO} * (I_{02} + N_{12}) * N_{13}]\ (x_{11} \mid y_{13}) \tag{7.17}$$

The considerations in Example 7.4 are concerned with network analysis when all network nodes are known. In the case of network design, at least one node is unknown. In this context, Algorithms 7.9-7.11 describe the process of deriving an unknown node in the initial FN from Fig.7.11 when the other nodes, the implicit identity node I_{02} and the single equivalent node N_E are known. In this case, node N_E is given by the Boolean matrix equation in Eq.(7.18).

$$N_E = N_{11}^{PO} * (I_{02} + N_{12}) * N_{13} \tag{7.18}$$

Algorithm 7.9

1. Define N_E, N_{12}, N_{13} and I_{02}.
2. Find $I_{02} + N_{12}$ by vertical merging of I_{02} and N_{12}.
3. Derive N_{11}^{PO} from Eq.(7.18), if possible.
4. Find N_{11} by inverse output permutation of N_{11}^{PO}.

Algorithm 7.10

1. Define N_E, N_{11}, N_{13} and I_{02}.
2. Find N_{11}^{PO} by output permutation of N_{11}.
3. Derive $I_{02} + N_{12}$ from Eq.(7.18), if possible.
4. Derive N_{12} from $I_{02} + N_{12}$, if possible.

Algorithm 7.11

1. Define N_E, N_{11}, N_{12} and I_{02}.
2. Find N_{11}^{PO} by output permutation of N_{11}.
3. Find $I_{02} + N_{12}$ by vertical merging of I_{02} and N_{12}.
4. Derive N_{13} from Eq.(7.18), if possible.

7.4 Networks with Multiple Levels and Single Layer

Another more complex type of FN is the one with multiple levels and single layer. This network has at least two nodes residing in the multiple levels and the single layer of the underlying grid structure, i.e. it is identical to a stack of fuzzy systems with single rule bases. Due to the presence of multiple nodes, there may be common inputs to at least some of these nodes. However, there are not any connections between the outputs from any nodes and the inputs to the same or other nodes as such connections are of feedback type and outside the scope of the current chapter.

Example 7.5

This example considers a FN with nodes N_{11} and N_{21} where $x_{11,21}^{1,1}$ is a common input that is the first input for N_{11} and the first and only input for N_{21}, x_{11}^{2} is the second input for N_{11}, y_{11} is an output for N_{11} and y_{21} is an output for N_{21}. This initial FN can be described by the block-scheme in Fig.7.15 and the topological expression in Eq.(7.19). It can be seen from there that one of the inputs for the top node is not an input for the bottom node.

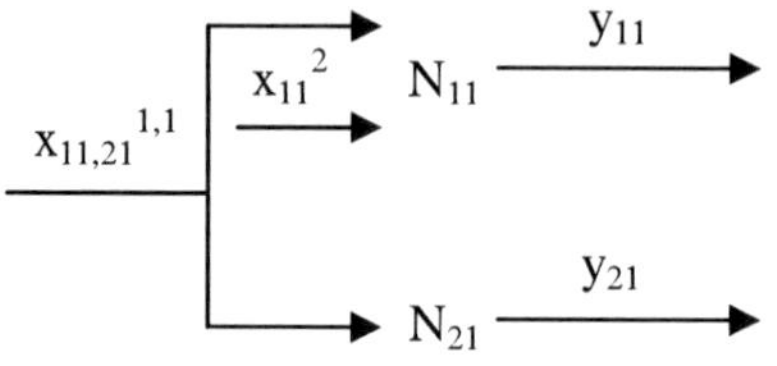

Fig. 7.15 Initial FN for Example 7.5

$$[N_{11}]\ (x_{11,21}^{1,1}, x_{11}^{2} \mid y_{11})\ ;\ [N_{21}]\ (x_{11,21}^{1,1} \mid y_{21}) \tag{7.19}$$

In order to merge the outputs of the nodes N_{11} and N_{21} of the initial FN, it is necessary to make the uncommon input x_{11}^{2} for the top node common. This can be done by augmenting the bottom node with the same input. This augmentation operation transforms the initial FN into an interim FN with nodes N_{11} and N_{21}^{AI} whereby the augmentation of inputs is reflected by the replacement of N_{21} with N_{21}^{AI}. This interim FN can be described by the block-scheme in Fig.7.16 and the topological expression in Eq.(7.20). It can be seen from there that all the inputs for the top node are common as they are also inputs for the bottom node.

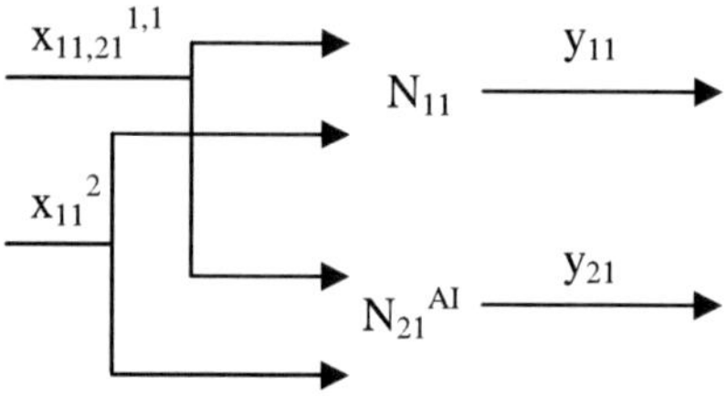

Fig. 7.16 Interim FN for Example 7.5

$$[N_{11}]\ (x_{11,21}^{1,1}, x_{11}^{2} \mid y_{11})\ ;\ [N_{21}^{AI}]\ (x_{11,21}^{1,1}, x_{11}^{2} \mid y_{21}) \tag{7.20}$$

The outputs of the nodes N_{11} and N_{21}^{AI} of the interim FN can be merged due to the common inputs. This merging operation transforms the interim FN into a final FN with a single equivalent node whereby the merging of nodes N_{11} and N_{21}^{AI} is reflected by their replacement with node N_{11} ; N_{21}^{AI}. This final FN can be described by the block-scheme in Fig.7.17 and the topological expression in Eq.(7.21). It can be seen from there that the two original nodes are implicit in the single equivalent node and the top node is unchanged in relation to the initial FN.

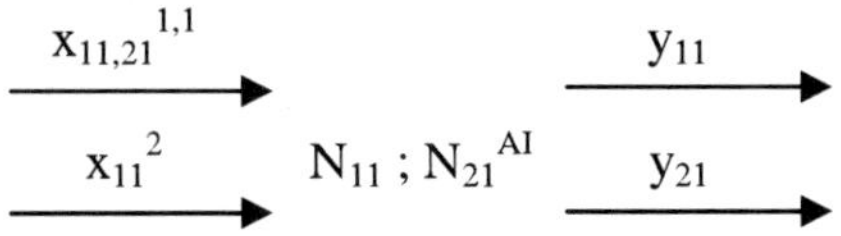

Fig. 7.17 Final FN for Example 7.5

$$[N_{11} ; N_{21}^{AI}] (x_{11,21}^{1,1}, x_{11}^{2} \mid y_{11}, y_{21}) \tag{7.21}$$

The considerations in Example 7.5 are concerned with network analysis when all network nodes are known. In the case of network design, at least one node is unknown. In this context, Algorithms 7.12-7.13 describe the process of deriving an unknown node in the initial FN from Fig.7.15 when the other node and the single equivalent node N_E are known. In this case, node N_E is given by the Boolean matrix equation in Eq.(7.22).

$$N_E = N_{11} ; N_{21}^{AI} \tag{7.22}$$

Algorithm 7.12

1. Define N_E and N_{21}.
2. Find N_{21}^{AI} by input augmentation of N_{21}.
3. Derive N_{11} from Eq.(7.22), if possible.

Algorithm 7.13

1. Define N_E and N_{11}.
2. Derive N_{21}^{AI} from Eq.(7.22), if possible.
3. Find N_{21} by inverse input augmentation of N_{21}^{AI}.

Example 7.6

This example considers a FN with nodes N_{11} and N_{21} where $x_{11,21}^{1,1}$ is a common input that is the first and only input for N_{11} and the first input for N_{21}, x_{21}^{2} is the second input for N_{21}, y_{11} is an output for N_{11} and y_{21} is an output for N_{21}. This initial FN can be described by the block-scheme in Fig.7.18 and the topological expression in Eq.(7.23). It can be seen from there that one of the inputs for the bottom node is uncommon as it is not an input for the top node.

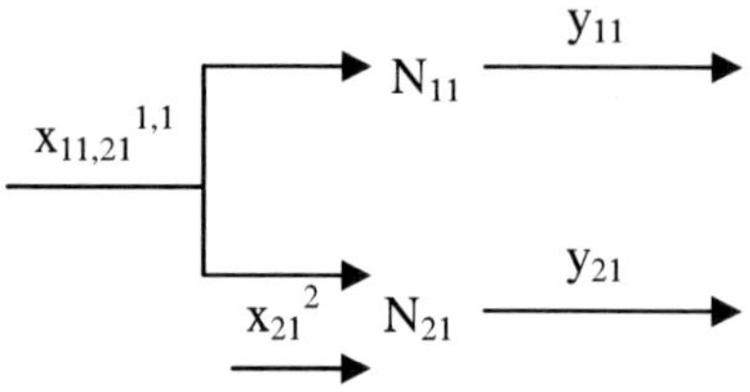

Fig. 7.18 Initial FN for Example 7.6

$$[N_{11}]\ (x_{11,21}^{1,1} \mid y_{11})\ ;\ [N_{21}]\ (x_{11,21}^{1,1}, x_{21}^{2} \mid y_{21}) \tag{7.23}$$

In order to merge the outputs of the nodes N_{11} and N_{21} of the initial FN, it is necessary to make the uncommon input x_{21}^{2} for the bottom node common. This can be done by augmenting the top node with the same input. This augmentation operation transforms the initial FN into an interim FN with nodes N_{11}^{AI} and N_{21} whereby the augmentation of inputs is reflected by the replacement of N_{11} with N_{11}^{AI}. This interim FN can be described by the block-scheme in Fig.7.19 and the topological expression in Eq.(7.24). It can be seen from there that all the inputs for the bottom node are common as they are also inputs for the top node.

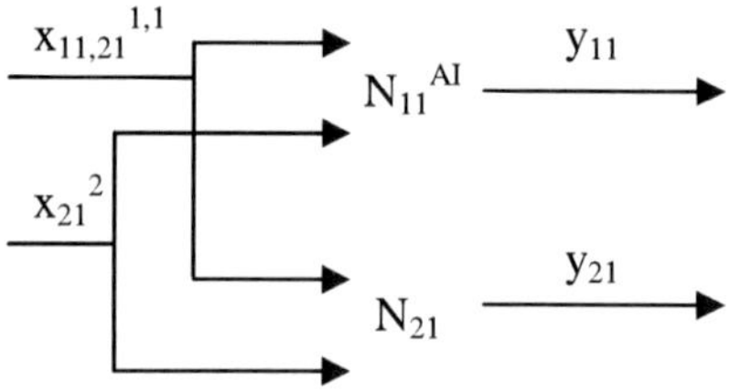

Fig. 7.19 Interim FN for Example 7.6

$$[N_{11}^{AI}]\ (x_{11,21}^{1,1}, x_{21}^{2} \mid y_{11})\ ;\ [N_{21}]\ (x_{11,21}^{1,1}, x_{21}^{2} \mid y_{21}) \tag{7.24}$$

The outputs of the nodes N_{11}^{AI} and N_{21} of the interim FN can be merged due to the common inputs. This merging operation transforms the interim FN into a final FN with a single equivalent node whereby the merging of nodes N_{11}^{AI} and N_{21} is reflected by their replacement with node N_{11}^{AI} ; N_{21}. This final FN can be described by the block-scheme in Fig.7.20 and the topological expression in Eq.(7.25) from where it can be seen that the two original nodes are implicit in the single equivalent node and the bottom node is unchanged in relation to the initial FN.

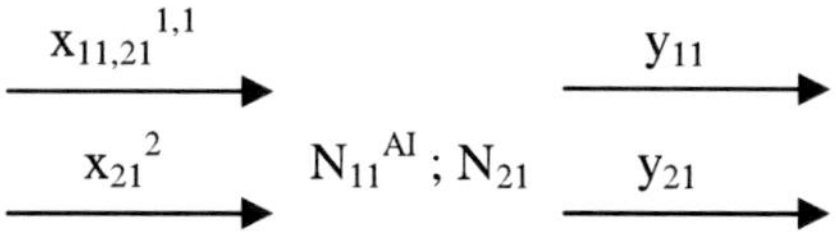

Fig. 7.20 Final FN for Example 7.6

$$[N_{11}^{AI} ; N_{21}] \, (x_{11,21}^{1,1}, x_{21}^{2} \mid y_{11}, y_{21}) \tag{7.25}$$

The considerations in Example 7.6 are concerned with network analysis when all network nodes are known. In the case of network design, at least one node is unknown. In this context, Algorithms 7.14-7.15 describe the process of deriving an unknown node in the initial FN from Fig.7.18 when the other node and the single equivalent node N_E are known. In this case, node N_E is given by the Boolean matrix equation in Eq.(7.26).

$$N_E = N_{11}^{AI} ; N_{21} \tag{7.26}$$

Algorithm 7.14

1. Define N_E and N_{21}.
2. Derive N_{11}^{AI} from Eq.(7.26), if possible.
3. Find N_{11} by inverse input augmentation of N_{11}^{AI}.

Algorithm 7.15

1. Define N_E and N_{11}.
2. Find N_{11}^{AI} by input augmentation of N_{11}.
3. Derive N_{21} from Eq.(7.26), if possible.

Example 7.7

This example considers a FN with nodes N_{11}, N_{21} and N_{31} where $x_{11,21,31}^{1,1,1}$ is a common input that is the first input for N_{11} and the first and only input for both N_{21} and N_{31}, x_{11}^{2} is the second input for N_{11}, y_{11} is an output for N_{11}, y_{21} is an output for N_{21} and y_{31} is an output for N_{31}. This initial FN can be described by the block-scheme in Fig.7.21 and the topological expression in Eq.(7.27). It can be seen from there that one of the inputs for the top node is uncommon as it is not an input for the middle and the bottom node.

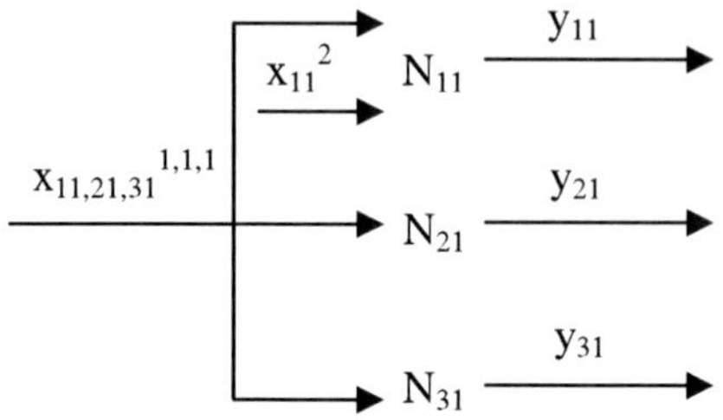

Fig. 7.21 Initial FN for Example 7.7

$$[N_{11}]\ (x_{11,21,31}^{1,1,1}, x_{11}^{2} \mid y_{11})\ ;\ [N_{21}]\ (x_{11,21,31}^{1,1,1} \mid y_{21})\ ;\ [N_{31}]\ (x_{11,21,31}^{1,1,1} \mid y_{31}) \quad (7.27)$$

In order to merge the outputs of the nodes N_{11}, N_{21} and N_{31} of the initial FN, it is necessary to make the uncommon input x_{11}^{2} for the top node common. This can be done by augmenting the middle and the bottom node with the same input. This augmentation operation transforms the initial FN into an interim FN with nodes N_{11}, N_{21}^{AI} and N_{31}^{AI} whereby the augmentation of inputs is reflected by the replacement of N_{21} and N_{31} with N_{21}^{AI} and N_{31}^{AI}, respectively. This interim FN can be described by the block-scheme in Fig.7.22 and the topological expression in Eq.(7.28). It can be seen from there that all the inputs for the top node are common as they are also inputs for the middle and the bottom node.

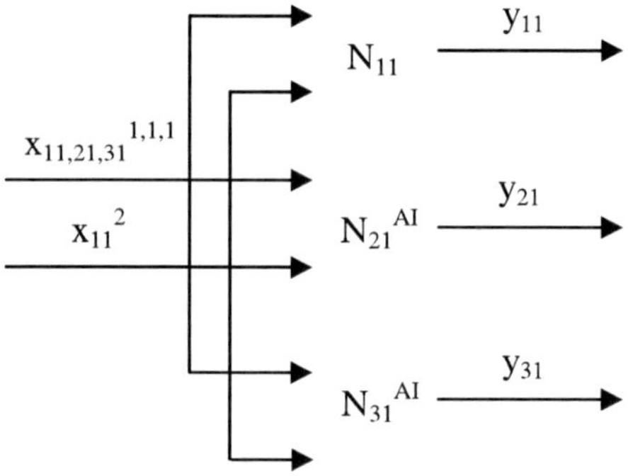

Fig. 7.22 Interim FN for Example 7.7

$$[N_{11}]\ (x_{11,21,31}^{1,1,1}, x_{11}^{2} \mid y_{11})\ ;\ [N_{21}^{AI}]\ (x_{11,21,31}^{1,1,1}, x_{11}^{2} \mid y_{21})\ ; \quad (7.28)$$

$$[N_{31}^{AI}]\ (x_{11,21,31}^{1,1,1}, x_{11}^{2} \mid y_{31})$$

The outputs of the nodes N_{11}, N_{21}^{AI} and N_{31}^{AI} of the interim FN can be merged due to the common inputs. This merging operation transforms the interim FN into a final FN with a single equivalent node whereby the merging of nodes N_{11}, N_{21}^{AI} and

N_{31}^{AI} is reflected by their replacement with node N_{11} ; N_{21}^{AI} ; N_{31}^{AI}. This final FN can be described by the block-scheme in Fig.7.23 and the topological expression in Eq.(7.29) from where it can be seen that the three original nodes are implicit in the single equivalent node and the top node is unchanged in relation to the initial FN.

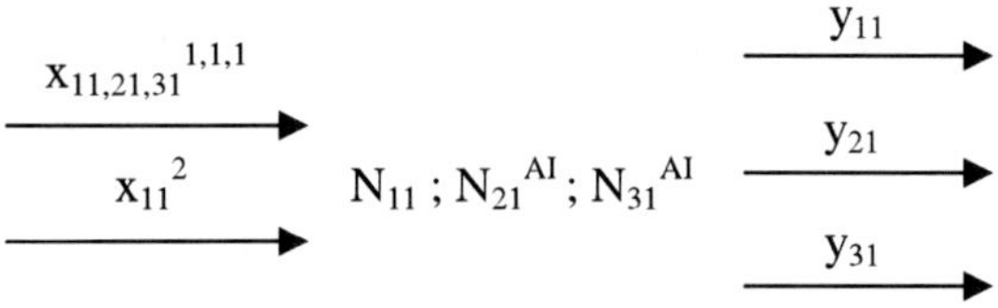

Fig. 7.23 Final FN for Example 7.7

$$[N_{11} ; N_{21}^{AI} ; N_{31}^{AI}] (x_{11,21,31}^{1,1,1}, x_{11}^{2} \mid y_{11}, y_{21}, y_{31}) \quad (7.29)$$

The considerations in Example 7.7 are concerned with network analysis when all network nodes are known. In the case of network design, at least one node is unknown. In this context, Algorithms 7.16-7.18 describe the process of deriving an unknown node in the initial FN from Fig.7.21 when the other nodes and the single equivalent node N_E are known. In this case, node N_E is given by the Boolean matrix equation in Eq.(7.30).

$$N_E = N_{11} ; N_{21}^{AI} ; N_{31}^{AI} \quad (7.30)$$

Algorithm 7.16

1. Define N_E, N_{21} and N_{31}.
2. Find N_{21}^{AI} by input augmentation of N_{21}.
3. Find N_{31}^{AI} by input augmentation of N_{31}.
4. Derive N_{11} from Eq.(7.30), if possible.

Algorithm 7.17

1. Define N_E, N_{11} and N_{31}.
2. Find N_{31}^{AI} by input augmentation of N_{31}.
3. Derive N_{21}^{AI} from Eq.(7.30), if possible.
4. Find N_{21} by inverse input augmentation of N_{21}^{AI}.

Algorithm 7.18

1. Define N_E, N_{11} and N_{21}.
2. Find N_{21}^{AI} by input augmentation of N_{21}.
3. Derive N_{31}^{AI} from Eq.(7.30), if possible.
4. Find N_{31} by inverse input augmentation of N_{31}^{AI}.

Example 7.8

This example considers a FN with nodes N_{11}, N_{21} and N_{31} where $x_{11,21,31}^{1,1,1}$ is a common input that is the first input for N_{21} and the first and only input for both N_{11} and N_{31}, x_{21}^{2} is the second input for N_{21}, y_{11} is an output for N_{11}, y_{21} is an output for N_{21} and y_{31} is an output for N_{31}. This initial FN can be described by the block-scheme in Fig.7.24 and the topological expression in Eq.(7.31). It can be seen from there that one of the inputs for the middle node is uncommon as it is not an input for the top and the bottom node.

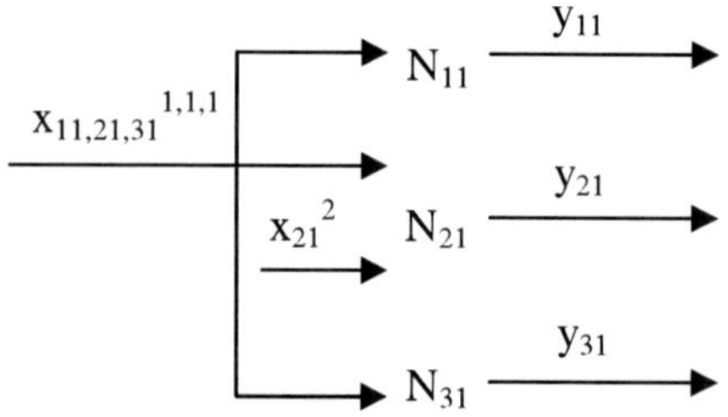

Fig. 7.24 Initial FN for Example 7.8

$$[N_{11}]\,(x_{11,21,31}^{1,1,1} \mid y_{11})\ ;\ [N_{21}]\,(x_{11,21,31}^{1,1,1}, x_{21}^{2} \mid y_{21})\ ;\ [N_{31}]\,(x_{11,21,31}^{1,1,1} \mid y_{31}) \quad (7.31)$$

In order to merge the outputs of the nodes N_{11}, N_{21} and N_{31} of the initial FN, it is necessary to make the uncommon input x_{21}^{2} for the middle node common. This can be done by augmenting the top and the bottom node with the same input. This augmentation operation transforms the initial FN into an interim FN with nodes N_{11}^{AI}, N_{21} and N_{31}^{AI} whereby the augmentation of inputs is reflected by the replacement of N_{11} and N_{31} with N_{11}^{AI} and N_{31}^{AI}, respectively. This interim FN can be described by the block-scheme in Fig.7.25 and the topological expression in Eq.(7.32). It can be seen from there that all the inputs for the middle node are common as they are also inputs for the top and the bottom node.

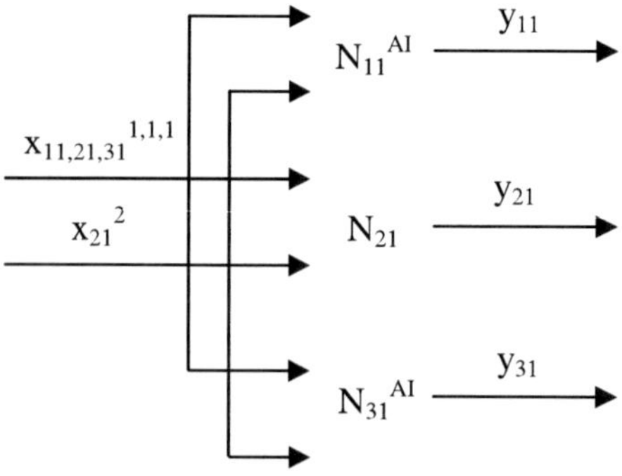

Fig. 7.25 Interim FN for Example 7.8

$$[N_{11}^{AI}]\ (x_{11,21,31}^{1,1,1}, x_{21}^{2} \mid y_{11})\ ;\ [N_{21}]\ (x_{11,21,31}^{1,1,1}, x_{21}^{2} \mid y_{21})\ ; \tag{7.32}$$

$$[N_{31}^{AI}]\ (x_{11,21,31}^{1,1,1}, x_{21}^{2} \mid y_{31})$$

The outputs of the nodes N_{11}^{AI}, N_{21} and N_{31}^{AI} of the interim FN can be merged due to the common inputs. This merging operation transforms the interim FN into a final FN with a single equivalent node whereby the merging of nodes N_{11}^{AI}, N_{21} and N_{31}^{AI} is reflected by their replacement with node N_{11}^{AI} ; N_{21} ; N_{31}^{AI}. This final FN can be described by the block-scheme in Fig.7.26 and the topological expression in Eq.(7.33) from where it can be seen that the three original nodes are implicit in the single equivalent node and the middle node is unchanged in relation to the initial FN.

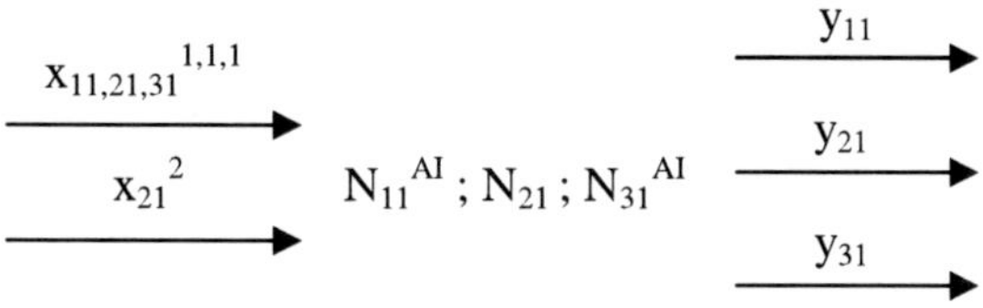

Fig. 7.26 Final FN for Example 7.8

$$[N_{11}^{AI}\ ;\ N_{21}\ ;\ N_{31}^{AI}]\ (x_{11,21,31}^{1,1,1}, x_{21}^{2} \mid y_{11}, y_{21}, y_{31}) \tag{7.33}$$

The considerations in Example 7.8 are concerned with network analysis when all network nodes are known. In the case of network design, at least one node is unknown. In this context, Algorithms 7.19-7.21 describe the process of deriving an unknown node in the initial FN from Fig.7.24 when the other nodes and the single equivalent node N_E are known. In this case, node N_E is given by the Boolean matrix equation in Eq.(7.34).

$$N_E = N_{11}^{AI}\ ;\ N_{21}\ ;\ N_{31}^{AI} \tag{7.34}$$

Algorithm 7.19

1. Define N_E, N_{21} and N_{31}.
2. Find N_{31}^{AI} by input augmentation of N_{31}.
3. Derive N_{11}^{AI} from Eq.(7.34), if possible.
4. Find N_{11} by inverse input augmentation of N_{11}^{AI}.

Algorithm 7.20

1. Define N_E, N_{11} and N_{31}.
2. Find N_{11}^{AI} by input augmentation of N_{11}.
3. Find N_{31}^{AI} by input augmentation of N_{31}.
4. Derive N_{21} from Eq.(7.34), if possible.

Algorithm 7.21

1. Define N_E, N_{11} and N_{21}.
2. Find N_{11}^{AI} by input augmentation of N_{11}.
3. Derive N_{31}^{AI} from Eq.(7.34), if possible.
4. Find N_{31} by inverse input augmentation of N_{31}^{AI}.

Example 7.9

This example considers a FN with nodes N_{11}, N_{21} and N_{31} where $x_{11,21,31}^{1,1,1}$ is a common input that is the first input for N_{31} and the first and only input for both N_{11} and N_{21}, x_{31}^{2} is the second input for N_{31}, y_{11} is an output for N_{11}, y_{21} is an output for N_{21} and y_{31} is an output for N_{31}. This initial FN can be described by the block-scheme in Fig.7.27 and the topological expression in Eq.(7.35). It can be seen from there that one of the inputs for the bottom node is uncommon as it is not an input for the top and the middle node.

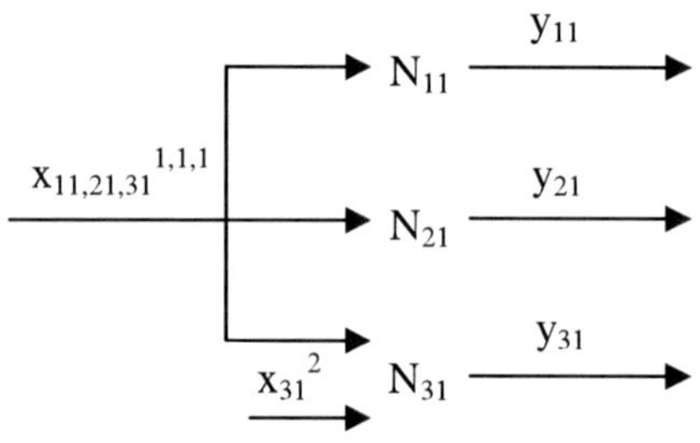

Fig. 7.27 Initial FN for Example 7.9

$$[N_{11}]\,(x_{11,21,31}^{1,1,1} \mid y_{11})\ ;\ [N_{21}]\,(x_{11,21,31}^{1,1,1} \mid y_{21})\ ;\ [N_{31}]\,(x_{11,21,31}^{1,1,1}, x_{31}^{2} \mid y_{31}) \quad (7.35)$$

In order to merge the outputs of the nodes N_{11}, N_{21} and N_{31} of the initial FN, it is necessary to make the uncommon input x_{31}^{2} for the bottom node common. This can be done by augmenting the top and the middle node with the same input. This augmentation operation transforms the initial FN into an interim FN with nodes N_{11}^{AI}, N_{21}^{AI} and N_{31} whereby the augmentation of inputs is reflected by the replacement of N_{11} and N_{21} with N_{11}^{AI} and N_{21}^{AI}, respectively This interim FN can be described by the block-scheme in Fig.7.28 and the topological expression in Eq.(7.36). It can be seen from there that all the inputs for the bottom node are common as they are also inputs for the top and the middle node.

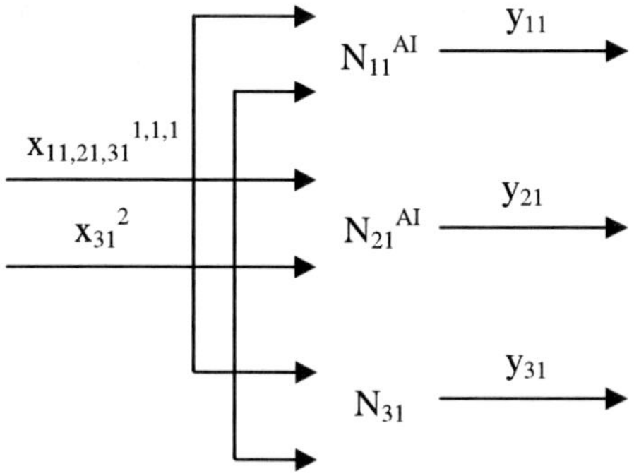

Fig. 7.28 Interim FN for Example 7.9

$$[N_{11}^{AI}]\,(x_{11,21,31}^{1,1,1}, x_{31}^{2} \mid y_{11})\,;\,[N_{21}^{AI}]\,(x_{11,21,31}^{1,1,1}, x_{31}^{2} \mid y_{21})\,; \quad (7.36)$$

$$[N_{31}]\,(x_{11,21,31}^{1,1,1}, x_{31}^{2} \mid y_{31})$$

The outputs of the nodes N_{11}^{AI}, N_{21}^{AI} and N_{31} of the interim FN can be merged due to the common inputs. This merging operation transforms the interim FN into a final FN with a single equivalent node whereby the merging of nodes N_{11}^{AI}, N_{21}^{AI} and N_{31} is reflected by their replacement with node $N_{11}^{AI}\,;\,N_{21}^{AI}\,;\,N_{31}$. This final FN can be described by the block-scheme in Fig.7.29 and the topological expression in Eq.(7.37) from where it can be seen that the three original nodes are implicit in the single equivalent node and the bottom node is unchanged in relation to the initial FN.

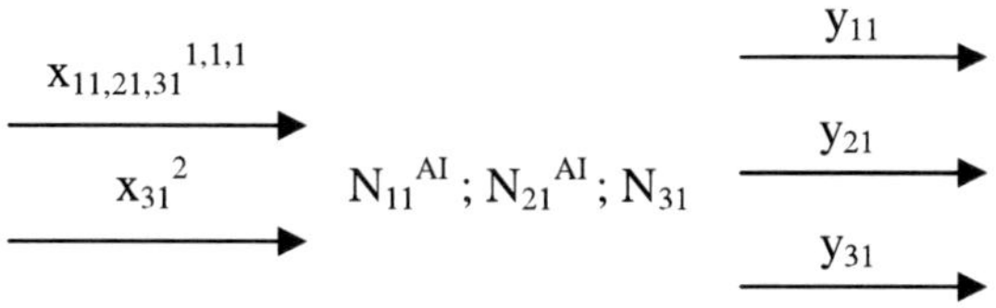

Fig. 7.29 Final FN for Example 7.9

$$[N_{11}^{AI}\,;\,N_{21}^{AI}\,;\,N_{31}]\,(x_{11,21,31}^{1,1,1}, x_{31}^{2} \mid y_{11}, y_{21}, y_{31}) \quad (7.37)$$

The considerations in Example 7.9 are concerned with network analysis when all network nodes are known. In the case of network design, at least one node is unknown. In this context, Algorithms 7.22-7.24 describe the process of deriving an unknown node in the initial FN from Fig.7.27 when the other nodes and the single equivalent node N_E are known. In this case, node N_E is given by the Boolean matrix equation in Eq.(7.38).

$$N_E = N_{11} \,;\, N_{21}^{AI} ;\, N_{31}^{AI} \tag{7.38}$$

Algorithm 7.22

1. Define N_E, N_{21} and N_{31}.
2. Find N_{21}^{AI} by input augmentation of N_{21}.
3. Find N_{31}^{AI} by input augmentation of N_{31}.
4. Derive N_{11} from Eq.(7.38), if possible.

Algorithm 7.23

1. Define N_E, N_{11} and N_{31}.
2. Find N_{31}^{AI} by input augmentation of N_{31}.
3. Derive N_{21}^{AI} from Eq.(7.38), if possible.
4. Find N_{21} by inverse input augmentation of N_{21}^{AI}.

Algorithm 7.24

1. Define N_E, N_{11} and N_{21}.
2. Find N_{21}^{AI} by input augmentation of N_{21}.
3. Derive N_{31}^{AI} from Eq.(7.38), if possible.
4. Find N_{31} by inverse input augmentation of N_{31}^{AI}.

7.5 Networks with Multiple Levels and Multiple Layers

The most complex type of FN is the one with multiple levels and multiple layers. This network has at least two nodes residing in the multiple levels and layers of the underlying grid structure, i.e. it is identical to a grid of fuzzy systems with single rule bases. Due to the presence of multiple nodes in the layers, there are feedforward connections from and to at least some of these nodes. Also, due to the presence of multiple nodes in the levels, there may be common inputs to at least some of these nodes. However, there are not any connections between the outputs from and the inputs to the same or other nodes as such connections are of feedback type and outside the scope of the current chapter.

Example 7.10

This example considers a FN with nodes N_{11}, N_{21}, N_{12} and N_{22} where x_{11} is an input for N_{11}, x_{21} is an input for N_{21}, y_{12} is an output for N_{12}, y_{22} is an output for N_{22}, $z_{11,12}^{1,1}$ is the connection from the first output for N_{11} to the first input for N_{12}, $z_{11,22}^{2,1}$ is the connection from the second output for N_{11} to the first and only input for N_{22} and $z_{21,12}^{1,2}$ is the connection from the first and only output for N_{21} to the second input for N_{12}. This initial FN can be described by the block-scheme in Fig.7.30 and the topological expression in Eq.(7.39) from where it can be seen that the connections in the middle and at the bottom have crossing paths.

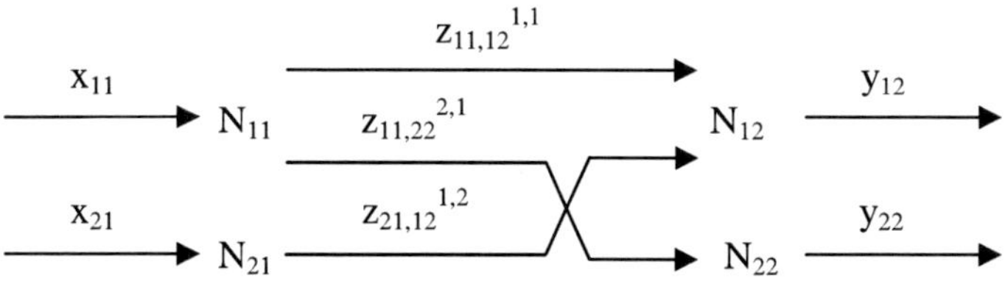

Fig. 7.30 Initial FN for Example 7.10

$$\{[N_{11}]\ (x_{11} \mid z_{11,12}^{1,1}, z_{11,22}^{2,1}) + [N_{21}]\ (x_{21} \mid z_{21,12}^{1,2})\} \ * \qquad (7.39)$$

$$\{[N_{12}]\ (z_{11,12}^{1,1}, z_{21,12}^{1,2} \mid y_{12}) + [N_{22}]\ (z_{11,22}^{2,1} \mid y_{22})\}$$

The initial FN has four nodes which can be merged vertically in pairs, i.e. N_{11} with N_{21} and N_{12} with N_{22}. This merging operation transforms the initial FN into a first interim FN with nodes $N_{11} + N_{21}$ and $N_{12} + N_{22}$ whereby the merging of nodes is reflected by the replacement of N_{11} and N_{21} with $N_{11} + N_{21}$ and the replacement of N_{12} and N_{22} with $N_{12} + N_{22}$. This first interim FN can be described by the block-scheme in Fig.7.31 and the topological expression in Eq.(7.40) from where it can be seen that the connections in the middle and at the bottom still have crossing paths.

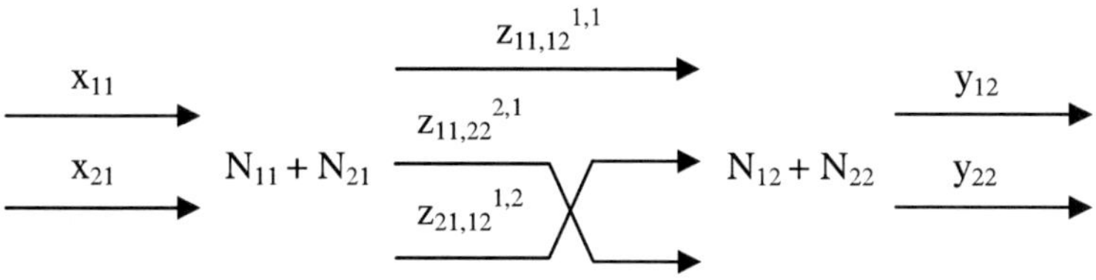

Fig. 7.31 First interim FN for Example 7.10

$$[N_{11} + N_{21}]\ (x_{11}, x_{21} \mid z_{11,12}^{1,1}, z_{11,22}^{2,1}, z_{21,12}^{1,2}) \ * \qquad (7.40)$$

$$[N_{12} + N_{22}]\ (z_{11,12}^{1,1}, z_{21,12}^{1,2}, z_{11,22}^{2,1} \mid y_{12}, y_{22})$$

In order to merge horizontally the nodes $N_{11} + N_{21}$ and $N_{12} + N_{22}$ of the first interim FN, it is necessary to remove the crossing of the connection paths. This can be done by permuting the connections $z_{11,22}^{2,1}$ and $z_{11,12}^{2,1}$ at their output points in node $N_{11} + N_{21}$. This permutation operation transforms the first interim FN into a second interim FN with nodes $(N_{11} + N_{21})^{PO}$ and $N_{12} + N_{22}$ whereby the permutation of outputs is reflected by the replacement of $N_{11} + N_{21}$ with $(N_{11} + N_{21})^{PO}$. This second interim FN can be described by the block-scheme in Fig.7.32 and the topological expression in Eq.(7.41 from where it can be seen that all connections already have parallel paths.

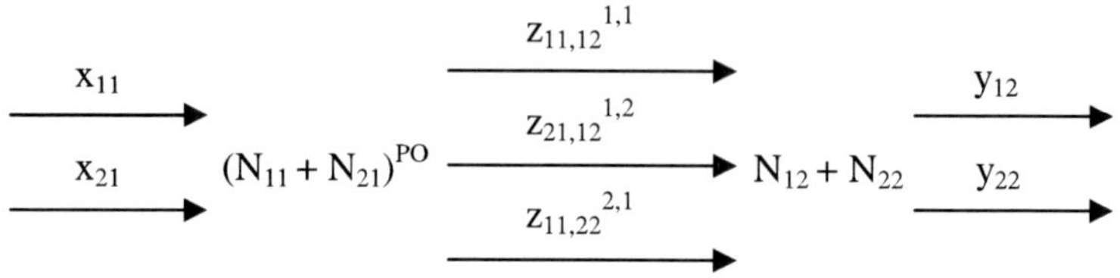

Fig. 7.32 Second interim FN for Example 7.10

$$[(N_{11} + N_{21})^{PO}] (x_{11}, x_{21} \mid z_{11,12}^{1,1}, z_{21,12}^{1,2}, z_{11,22}^{2,1}) * \tag{7.41}$$

$$[N_{12} + N_{22}] (z_{11,12}^{1,1}, z_{21,12}^{1,2}, z_{11,22}^{2,1} \mid y_{12}, y_{22})$$

The nodes $(N_{11} + N_{21})^{PO}$ and $N_{12} + N_{22}$ of the second interim FN can be merged horizontally due to the parallel connection paths. This merging operation transforms the second interim FN into a final FN with a single equivalent node whereby the merging of nodes $(N_{11} + N_{21})^{PO}$ and $N_{12} + N_{22}$ is reflected by their replacement with node $(N_{11} + N_{21})^{PO} * (N_{12} + N_{22})$. This final FN can be described by the block-scheme in Fig.7.33 and the topological expression in Eq.(7.42) from where it can be seen that the two composite nodes are implicit in the single equivalent node and the first composite node has permuted outputs.

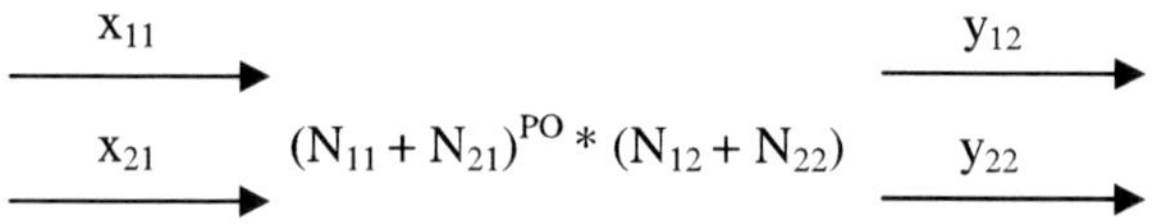

Fig. 7.33 Final FN for Example 7.10

$$[(N_{11} + N_{21})^{PO} * (N_{12} + N_{22})] (x_{11}, x_{21} \mid y_{12}, y_{22}) \tag{7.42}$$

The considerations in Example 7.10 are concerned with network analysis when all network nodes are known. In the case of network design, at least one node is unknown. In this context, Algorithms 7.25-7.28 describe the process of deriving an unknown node in the initial FN from Fig.7.30 when the other nodes and the single equivalent node N_E are known. In this case, node N_E is given by the Boolean matrix equation in Eq.(7.43).

$$N_E = (N_{11} + N_{21})^{PO} * (N_{12} + N_{22}) \tag{7.43}$$

Algorithm 7.25

1. Define N_E, N_{21}, N_{12} and N_{22}.
2. Find $N_{12} + N_{22}$ by vertical merging of $N_{12} + N_{22}$.
3. Derive $(N_{11} + N_{21})^{PO}$ from Eq.(7.43), if possible.
4. Find $N_{11} + N_{21}$ by inverse output permutation of $(N_{11} + N_{21})^{PO}$.
5. Derive N_{11} from $N_{11} + N_{21}$, if possible.

Algorithm 7.26

1. Define N_E, N_{11}, N_{12} and N_{22}.
2. Find $N_{12} + N_{22}$ by vertical merging of $N_{12} + N_{22}$.
3. Derive $(N_{11} + N_{21})^{PO}$ from Eq.(7.43), if possible.
4. Find $N_{11} + N_{21}$ by inverse output permutation of $(N_{11} + N_{21})^{PO}$.
5. Derive N_{21} from $N_{11} + N_{21}$, if possible.

Algorithm 7.27

1. Define N_E, N_{11}, N_{21} and N_{22}.
2. Find $N_{11} + N_{21}$ by vertical merging of $N_{11} + N_{21}$.
3. Find $(N_{11} + N_{21})^{PO}$ by output permutation of $N_{11} + N_{21}$.
4. Derive $N_{12} + N_{22}$ from Eq.(7.43), if possible.
5. Derive N_{12} from $N_{12} + N_{22}$, if possible.

Algorithm 7.28

1. Define N_E, N_{11}, N_{21} and N_{12}.
2. Find $N_{11} + N_{21}$ by vertical merging of $N_{11} + N_{21}$.
3. Find $(N_{11} + N_{21})^{PO}$ by output permutation of $N_{11} + N_{21}$.
4. Derive $N_{12} + N_{22}$ from Eq.(7.43), if possible.
5. Derive N_{22} from $N_{12} + N_{22}$, if possible.

Example 7.11

This example considers a FN with nodes N_{11}, N_{21}, N_{12} and N_{22} where x_{11} is an input for N_{11}, x_{21} is an input for N_{21}, y_{12} is an output for N_{12}, y_{22} is an output for N_{22}, $z_{11,22}^{1,1}$ is the connection from the first and only output for N_{11} to the first input for N_{22}, $z_{21,12}^{1,1}$ is the connection from the first output for N_{21} to the first and only input for N_{12} and $z_{21,22}^{2,2}$ is the connection from the second output for N_{21} to the second input for N_{22}. This initial FN can be described by the block-scheme in Fig.7.34 and the topological expression in Eq.(7.44) from where it can be seen that the connections at the top and in the middle have crossing paths.

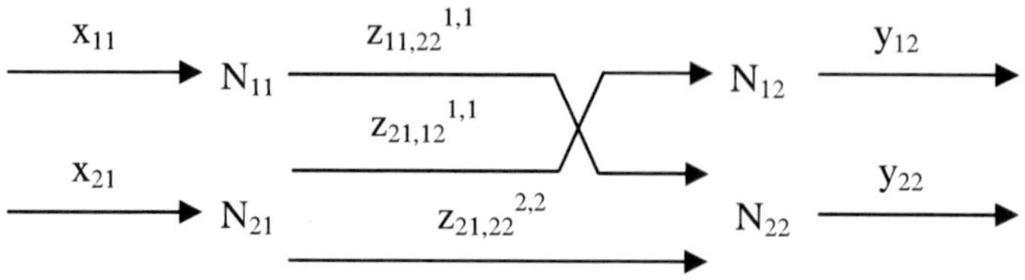

Fig. 7.34 Initial FN for Example 7.11

$$\{[N_{11}]\ (x_{11} \mid z_{11,22}^{1,1}) + [N_{21}]\ (x_{21} \mid z_{21,12}^{1,1},\ z_{21,22}^{2,2})\} \ * \tag{7.44}$$
$$\{[N_{12}]\ (z_{21,12}^{1,1} \mid y_{12}) + [N_{22}]\ (z_{11,22}^{1,1},\ z_{21,22}^{2,2} \mid y_{22})\}$$

The initial FN has four nodes which can be merged vertically in pairs, i.e. N_{11} with N_{21} and N_{12} with N_{22}. This merging operation transforms the initial FN into a first interim FN with nodes $N_{11} + N_{21}$ and $N_{12} + N_{22}$ whereby the merging of nodes is reflected by the replacement of N_{11} and N_{21} with $N_{11} + N_{21}$ and the replacement of N_{12} and N_{22} with $N_{12} + N_{22}$. This first interim FN can be described by the block-scheme in Fig.7.35 and the topological expression in Eq.(7.45) from where it can be seen that the connections at the top and in the middle still have crossing paths.

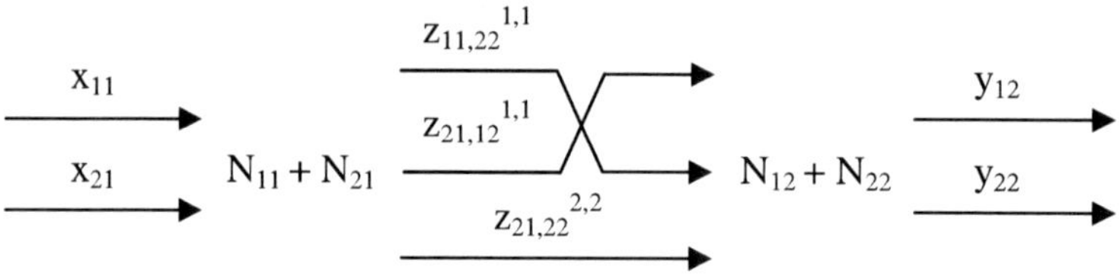

Fig. 7.35 First interim FN for Example 7.11

$$[N_{11} + N_{21}]\ (x_{11},\ x_{21} \mid z_{11,22}^{1,1},\ z_{21,12}^{1,1},\ z_{21,22}^{2,2}) \ * \tag{7.45}$$
$$[N_{12} + N_{22}]\ (z_{21,12}^{1,1},\ z_{11,22}^{1,1},\ z_{21,22}^{2,2} \mid y_{12},\ y_{22})$$

In order to merge horizontally the nodes $N_{11} + N_{21}$ and $N_{12} + N_{22}$ of the first interim FN, it is necessary to remove the crossing of the connection paths. This can be done by permuting the connections $z_{11,22}^{1,1}$ and $z_{21,12}^{1,1}$ at their output points in node $N_{11} + N_{21}$. This permutation operation transforms the first interim FN into a second interim FN with nodes $(N_{11} + N_{21})^{PO}$ and $N_{12} + N_{22}$ whereby the permutation of outputs is reflected by the replacement of $N_{11} + N_{21}$ with $(N_{11} + N_{21})^{PO}$. This second interim FN can be described by the block-scheme in Fig.7.36 and the topological expression in Eq.(7.46) from where it can be seen that all connections already have parallel paths.

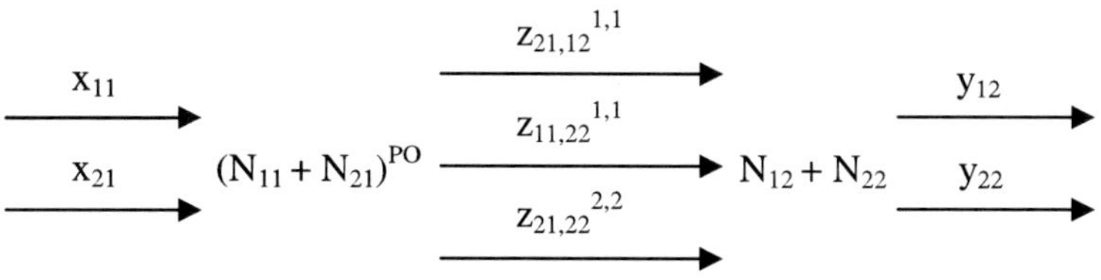

Fig. 7.36 Second interim FN for Example 7.11

$$[(N_{11} + N_{21})^{PO}] \, (x_{11}, x_{21} \mid z_{21,12}^{1,1}, z_{11,22}^{1,1}, z_{21,22}^{2,2}) \, * \tag{7.46}$$

$$[N_{12} + N_{22}] \, (z_{21,12}^{1,1}, z_{11,22}^{1,1}, z_{21,22}^{2,2} \mid y_{12}, y_{22})$$

The nodes $(N_{11} + N_{21})^{PO}$ and $N_{12} + N_{22}$ of the second interim FN can be merged horizontally due to the parallel connection paths. This merging operation transforms the second interim FN into a final FN with a single equivalent node whereby the merging of nodes $(N_{11} + N_{21})^{PO}$ and $N_{12} + N_{22}$ is reflected by their replacement with node $(N_{11} + N_{21})^{PO} * (N_{12} + N_{22})$. This final FN can be described by the block-scheme in Fig.7.33 and the topological expression in Eq.(7.42) from Example 7.10.

The considerations in Example 7.11 are concerned with network analysis when all network nodes are known. In the case of network design, at least one node is unknown. In this context, Algorithms 7.25-7.28 from Example 7.10 describe the process of deriving an unknown node in the initial FN from Fig.7.34 when the other nodes and the single equivalent node N_E are known. In this case, node N_E is given by the Boolean matrix equation in Eq.(7.43) from Example 7.10.

Example 7.12

This example considers a FN with nodes N_{11}, N_{21}, N_{12} and N_{22} where x_{11} is an input for N_{11}, x_{21} is an input for N_{21}, y_{12} is an output for N_{12}, y_{22} is an output for N_{22}, $z_{11,12}^{1,1}$ is the connection from the first and only output for N_{11} to the first input for N_{12} and $z_{21,12,22}^{1,2,1}$ is the connection from the first and only output for N_{21} to the second input for N_{12} and the second and only input for N_{22}. This initial FN can be described by the block-scheme in Fig.7.37 and the topological expression in Eq.(7.47) from where it can be seen that the connection at the top is an uncommon input for the nodes in the second layer.

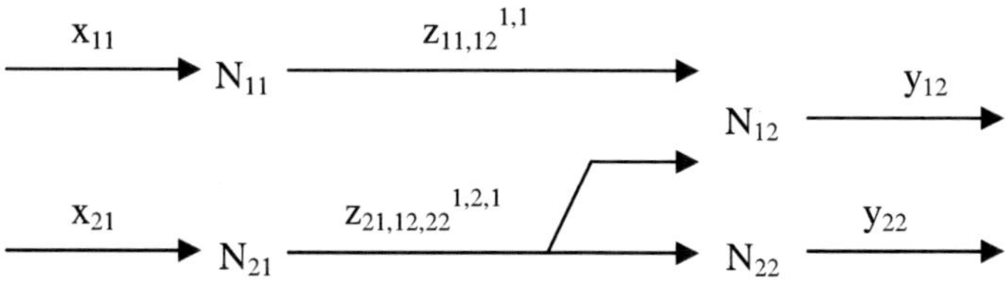

Fig. 7.37 Initial FN for Example 7.12

$$\{[N_{11}]\ (x_{11} \mid z_{11,12}^{1,1}) + [N_{21}]\ (x_{21} \mid z_{21,12,22}^{1,2,1})\} * \tag{7.47}$$

$$\{[N_{12}]\ (z_{11,12}^{1,1},\ z_{21,12,22}^{1,2,1} \mid y_{12}) + [N_{22}]\ (z_{21,12,22}^{1,2,1} \mid y_{22})\}$$

In order to merge the outputs of the nodes N_{12} and N_{22} of the initial FN, it is necessary to augment the top connection $z_{11,12}^{1,1}$, so that it becomes a common input for these two nodes. This augmentation operation transforms the initial FN into a first interim FN with nodes N_{11}, N_{21}, N_{12} and N_{22}^{AI} whereby the augmentation of inputs is reflected by the replacement of N_{22} with N_{22}^{AI}. This first interim FN can be described by the block-scheme in Fig.7.38 and the topological expression in Eq.(7.48) from where it can be seen that all inputs for the nodes in the second layer are already common.

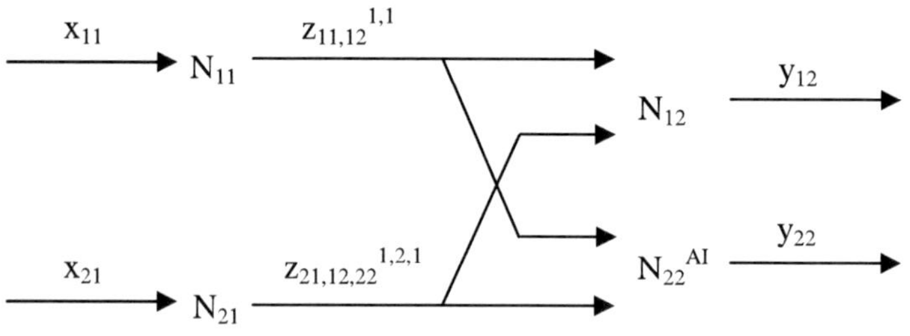

Fig. 7.38 First interim FN for Example 7.12

$$\{[N_{11}]\ (x_{11} \mid z_{11,12}^{1,1}) + [N_{21}]\ (x_{21} \mid z_{21,12,22}^{1,2,1})\} * \tag{7.48}$$

$$\{[N_{12}]\ (z_{11,12}^{1,1},\ z_{21,12,22}^{1,2,1} \mid y_{12}) + [N_{22}^{AI}]\ (z_{11,12}^{1,1},\ z_{21,12,22}^{1,2,1} \mid y_{22})\}$$

The first interim FN has four nodes whereby N_{11} can be vertically merged with N_{21} and N_{12} can be output merged with N_{22}^{AI}. These merging operations transform the first interim FN into a second interim FN with nodes $N_{11} + N_{21}$ and N_{12} ; N_{22}^{AI} whereby the merging of nodes is reflected by the replacement of N_{11} and N_{21} with $N_{11} + N_{21}$ and the replacement of N_{12} and N_{22} with N_{12} ; N_{22}^{AI}. This second interim FN can be described by the block-scheme in Fig.7.39 and the topological expression in Eq.(7.49) from where it can be seen that all connections already have parallel paths.

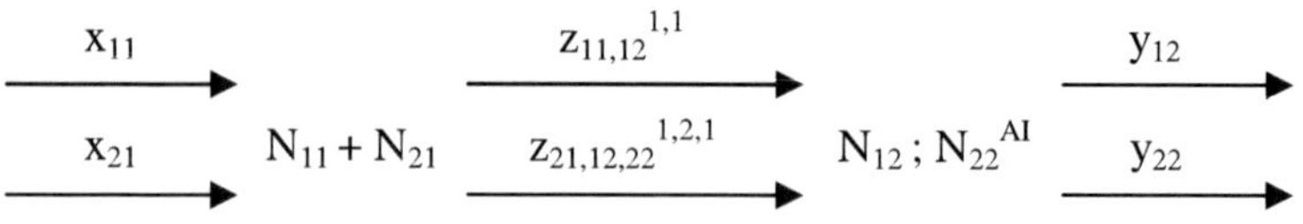

Fig. 7.39 Second interim FN for Example 7.12

$$[N_{11} + N_{21}]\ (x_{11}, x_{21} \mid z_{11,12}^{1,1}, z_{21,12,22}^{1,2,1}) \ * \tag{7.49}$$

$$[N_{12} ; N_{22}^{AI}]\ (z_{11,12}^{1,1}, z_{21,12,22}^{1,2,1} \mid y_{12}, y_{22})$$

The nodes $N_{11} + N_{21}$ and N_{12} ; N_{22}^{AI} of the second interim FN can be merged horizontally due to the parallel connection paths. This merging operation transforms the second interim FN into a final FN with a single equivalent node whereby the merging of nodes $N_{11} + N_{21}$ and N_{12} ; N_{22}^{AI} is reflected by their replacement with node $(N_{11} + N_{21}) * (N_{12} ; N_{22}^{AI})$. This final FN can be described by the block-scheme in Fig.7.40 and the topological expression in Eq.(7.50) from where it can be seen that the two composite nodes are implicit in the single equivalent node and the tail of the second composite node has augmented inputs.

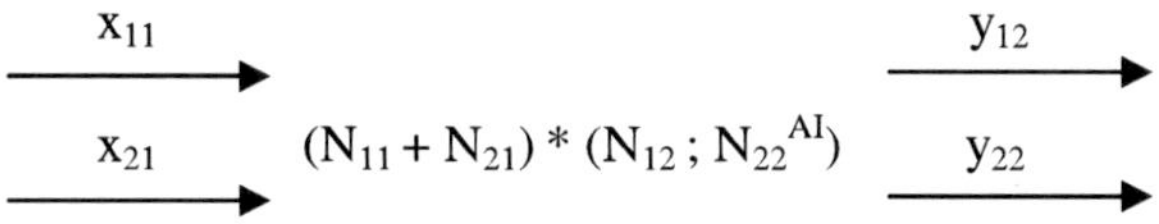

Fig. 7.40 Final FN for Example 7.12

$$[(N_{11} + N_{21}) * (N_{12} ; N_{22}^{AI})]\ (x_{11}, x_{21} \mid y_{12}, y_{22}) \tag{7.50}$$

The considerations in Example 7.12 are concerned with network analysis when all network nodes are known. In the case of network design, at least one node is unknown. In this context, Algorithms 7.29-7.32 describe the process of deriving an unknown node in the initial FN from Fig.7.37 when the other nodes and the single equivalent node N_E are known. In this case, node N_E is given by the Boolean matrix equation in Eq.(7.51).

$$N_E = (N_{11} + N_{21}) * (N_{12} ; N_{22}^{AI}) \tag{7.51}$$

Algorithm 7.29

1. Define N_E, N_{21}, N_{12} and N_{22}.
2. Find N_{22}^{AI} by input augmentation of N_{22}.
3. Find N_{12} ; N_{22}^{AI} by output merging of N_{12} and N_{22}^{AI}.
4. Derive $N_{11} + N_{21}$ from Eq.(7.51), if possible.
5. Derive N_{11} from $N_{11} + N_{21}$, if possible.

Algorithm 7.30

1. Define N_E, N_{11}, N_{12} and N_{22}.
2. Find N_{22}^{AI} by input augmentation of N_{22}.
3. Find N_{12} ; N_{22}^{AI} by output merging of N_{12} and N_{22}^{AI}.
4. Derive $N_{11} + N_{21}$ from Eq.(7.51), if possible.
5. Derive N_{21} from $N_{11} + N_{21}$, if possible.

Algorithm 7.31

1. Define N_E, N_{11}, N_{21} and N_{22}.
2. Find $N_{11} + N_{21}$ by vertical merging of $N_{11} + N_{21}$.
3. Derive $N_{12} ; N_{22}^{AI}$ from Eq.(7.51), if possible.
4. Find N_{22}^{AI} by input augmentation of N_{22}.
5. Derive N_{12} from $N_{12} ; N_{22}^{AI}$, if possible.

Algorithm 7.32

1. Define N_E, N_{11}, N_{21} and N_{12}.
2. Find $N_{11} + N_{21}$ by vertical merging of $N_{11} + N_{21}$.
3. Derive $N_{12} ; N_{22}^{AI}$ from Eq.(7.51), if possible.
4. Derive N_{22}^{AI} from $N_{12} ; N_{22}^{AI}$, if possible.
5. Find N_{22} by inverse input augmentation of N_{22}^{AI}.

Example 7.13

This example considers a FN with nodes N_{11}, N_{21}, N_{12} and N_{22} where x_{11} is an input for N_{11}, x_{21} is an input for N_{21}, y_{12} is an output for N_{12}, y_{22} is an output for N_{22}, $z_{11,12,22}^{1,1,1}$ is the connection from the first and only output for N_{11} to the first and only input for N_{12} and the first input for N_{22} and $z_{21,22}^{1,2}$ is the connection from the first and only output for N_{21} to the second input for N_{22}. This initial FN can be described by the block-scheme in Fig.7.41 and the topological expression in Eq.(7.52) from where it can be seen that the connection at the bottom is an uncommon input for the nodes in the second layer.

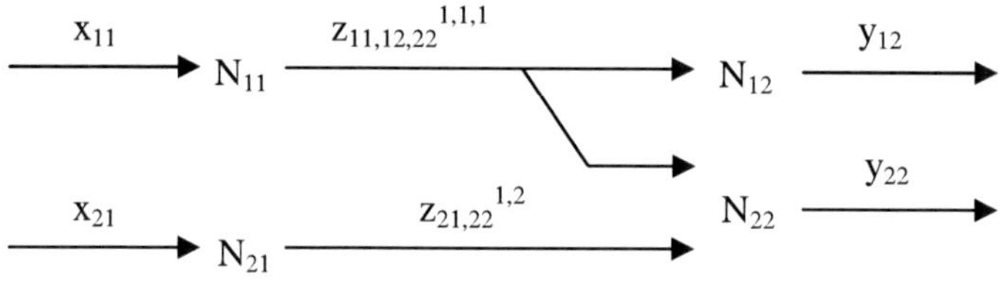

Fig. 7.41 Initial FN for Example 7.13

$$\{[N_{11}]\,(x_{11} \mid z_{11,12,22}^{1,1,1}) + [N_{21}]\,(x_{21} \mid z_{21,22}^{1,2})\} \; * \tag{7.52}$$

$$\{[N_{12}]\,(z_{11,12,22}^{1,1,1} \mid y_{12}) + [N_{22}]\,(z_{11,12,22}^{1,1,1},\, z_{21,22}^{1,2} \mid y_{22})\}$$

In order to merge the outputs of the nodes N_{12} and N_{22} of the initial FN, it is necessary to augment the bottom connection $z_{21,22}^{1,2}$, so that it becomes a common input for these two nodes. This augmentation operation transforms the initial FN into a first interim FN with nodes N_{11}, N_{21}, N_{12}^{AI} and N_{22} whereby the augmentation of inputs is reflected by the replacement of N_{12} with N_{12}^{AI}. This first interim FN can be described by the block-scheme in Fig.7.42 and the topological expression in

Eq.(7.53) from where it can be seen that all inputs for the nodes in the second layer are already common.

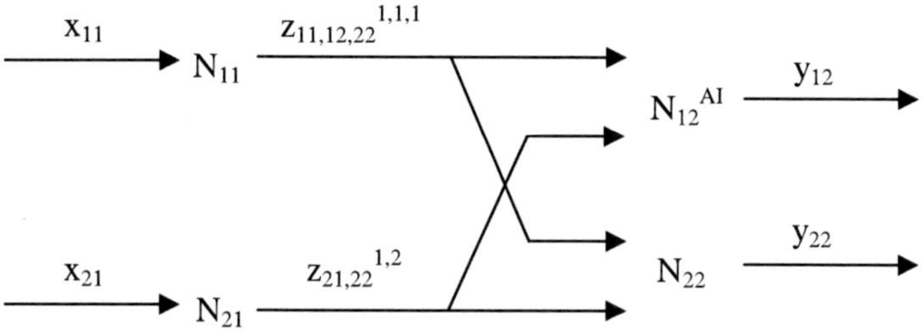

Fig. 7.42 First interim FN for Example 7.13

$$\{[N_{11}]\,(x_{11}\mid z_{11,12,22}{}^{1,1,1}) + [N_{21}]\,(x_{21}\mid z_{21,22}{}^{1,2})\} \; * \qquad (7.53)$$

$$\{[N_{12}{}^{AI}]\,(z_{11,12,22}{}^{1,1,1},\, z_{21,22}{}^{1,2}\mid y_{12}) + [N_{22}]\,(z_{11,12,22}{}^{1,1,1},\, z_{21,22}{}^{1,2}\mid y_{22})\}$$

The first interim FN has four nodes whereby N_{11} can be vertically merged with N_{21} and $N_{12}{}^{AI}$ can be output merged with N_{22}. These merging operations transform the first interim FN into a second interim FN with nodes $N_{11} + N_{21}$ and $N_{12}{}^{AI}$; N_{22} whereby the merging of nodes is reflected by the replacement of N_{11} and N_{21} with $N_{11} + N_{21}$ and the replacement of $N_{12}{}^{AI}$ and N_{22} with $N_{12}{}^{AI}$; N_{22}. This second interim FN can be described by the block-scheme in Fig.7.43 and the topological expression in Eq.(7.54) from where it can be seen that all connections already have parallel paths.

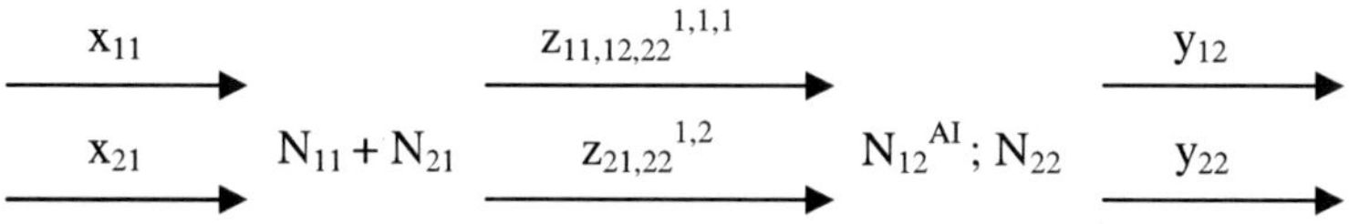

Fig. 7.43 Second interim FN for Example 7.13

$$[N_{11} + N_{21}]\,(x_{11}, x_{21}\mid z_{11,12,22}{}^{1,1,1},\, z_{21,22}{}^{1,2}) \; * \qquad (7.54)$$

$$[N_{12}{}^{AI} ; N_{22}]\,(z_{11,12,22}{}^{1,1,1},\, z_{21,22}{}^{1,2}\mid y_{12}, y_{22})$$

The nodes $N_{11} + N_{21}$ and $N_{12}{}^{AI}$; N_{22} of the second interim FN can be merged horizontally due to the parallel connection paths. This merging operation transforms the second interim FN into a final FN with a single equivalent node whereby the merging of nodes $N_{11} + N_{21}$ and $N_{12}{}^{AI}$; N_{22} is reflected by their replacement with node $(N_{11} + N_{21}) * (N_{12}{}^{AI} ; N_{22})$. This final FN can be described by the block-scheme in Fig.7.44 and the topological expression in Eq.(7.55) from where it can be seen that the two composite nodes are implicit in the single equivalent node and the head of the second composite node has augmented inputs.

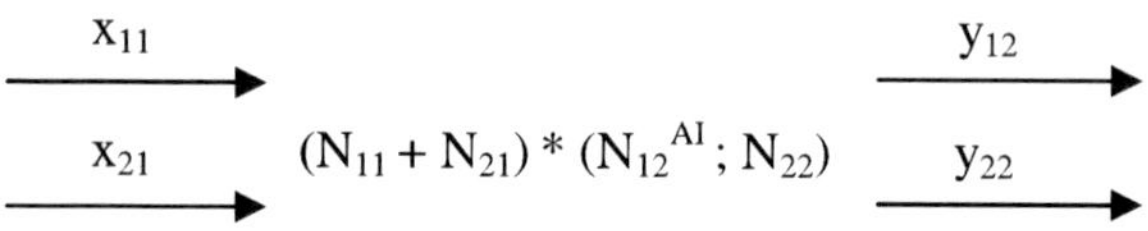

Fig. 7.44 Final FN for Example 7.13

$$[(N_{11} + N_{21}) * (N_{12}^{AI} ; N_{22})] (x_{11}, x_{21} \mid y_{12}, y_{22}) \quad (7.55)$$

The considerations in Example 7.13 are concerned with network analysis when all network nodes are known. In the case of network design, at least one node is unknown. In this context, Algorithms 7.33-7.36 describe the process of deriving an unknown node in the initial FN from Fig.7.41 when the other nodes and the single equivalent node N_E are known. In this case, node N_E is given by the Boolean matrix equation in Eq.(7.56).

$$N_E = (N_{11} + N_{21}) * (N_{12}^{AI} ; N_{22}) \quad (7.56)$$

Algorithm 7.33

1. Define N_E, N_{21}, N_{12} and N_{22}.
2. Find N_{12}^{AI} by input augmentation of N_{12}.
3. Find N_{12}^{AI} ; N_{22} by output merging of N_{12}^{AI} and N_{22}.
4. Derive $N_{11} + N_{21}$ from Eq.(7.56), if possible.
5. Derive N_{11} from $N_{11} + N_{21}$, if possible.

Algorithm 7.34

1. Define N_E, N_{11}, N_{12} and N_{22}.
2. Find N_{12}^{AI} by input augmentation of N_{12}.
3. Find N_{12}^{AI} ; N_{22}^{AI} by output merging of N_{12}^{AI} and N_{22}.
4. Derive $N_{11} + N_{21}$ from Eq.(7.56), if possible.
5. Derive N_{21} from $N_{11} + N_{21}$, if possible.

Algorithm 7.35

1. Define N_E, N_{11}, N_{21} and N_{22}.
2. Find $N_{11} + N_{21}$ by vertical merging of $N_{11} + N_{21}$.
3. Derive N_{12}^{AI} ; N_{22} from Eq.(7.56), if possible.
4. Derive N_{12}^{AI} from N_{12}^{AI} ; N_{22}, if possible.
5. Find N_{12} by inverse input augmentation of N_{12}^{AI}.

Algorithm 7.36

1. Define N_E, N_{11}, N_{21} and N_{12}.
2. Find $N_{11} + N_{21}$ by vertical merging of $N_{11} + N_{21}$.
3. Derive N_{12}^{AI} ; N_{22} from Eq.(7.56), if possible.
4. Find N_{12} by input augmentation of N_{12}^{AI}.
5. Derive N_{22} from N_{12}^{AI} ; N_{22}, if possible.

Example 7.14

This example considers a FN with nodes N_{11}, N_{21}, N_{12} and N_{22} where x_{11} is an input for N_{11}, $x_{21,12}^{1,2}$ is an input for N_{21} and the second input for N_{12}, y_{12} is an output for N_{12}, y_{22} is an output for N_{22}, $z_{11,12}^{1,1}$ is the connection from the first and only output for N_{11} to the first input for N_{12} and $z_{21,22}^{1,1}$ is the connection from the first and only output for N_{21} to the first and only input for N_{22}. This initial FN can be described by the block-scheme in Fig.7.45 and the topological expression in Eq.(7.57) from where it can be seen that the input at the bottom propagates through the first layer.

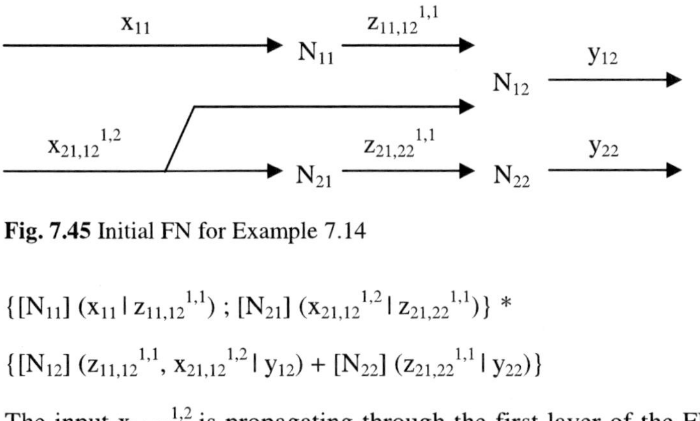

Fig. 7.45 Initial FN for Example 7.14

$$\{[N_{11}]\ (x_{11} \mid z_{11,12}^{1,1})\ ;\ [N_{21}]\ (x_{21,12}^{1,2} \mid z_{21,22}^{1,1})\}\ * \quad (7.57)$$

$$\{[N_{12}]\ (z_{11,12}^{1,1},\ x_{21,12}^{1,2} \mid y_{12}) + [N_{22}]\ (z_{21,22}^{1,1} \mid y_{22})\}$$

The input $x_{21,12}^{1,2}$ is propagating through the first layer of the FN within a virtual level between the first and the second level. This propagation can be represented by inserting an implicit identity node $I_{1.5,1}$. This representation transforms the initial FN into a first interim FN with nodes N_{11}, N_{21}, N_{12}, N_{22} and $I_{1.5,1}$. This first interim FN can be described by the block-scheme in Fig.7.46 and the topological expression in Eq.(7.58) from where it can be seen that the bottom input propagates through the first layer by means of the identity node.

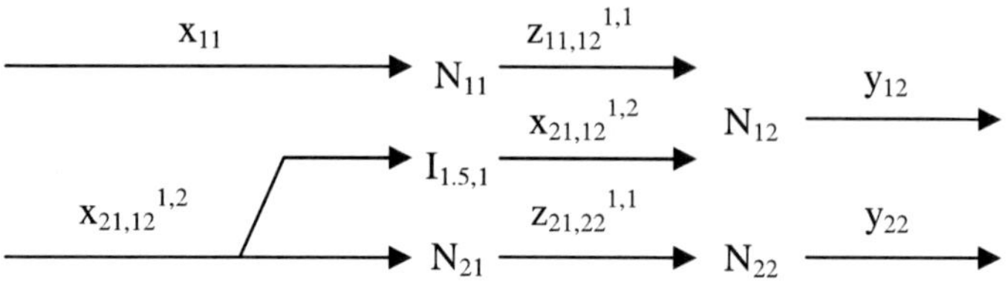

Fig. 7.46 First interim FN for Example 7.14

$$\{[N_{11}]\ (x_{11} \mid z_{11,12}^{1,1})\ ;\ [I_{1.5,1}]\ (x_{21,12}^{1,2} \mid x_{21,12}^{1,2})\ ;\ [N_{21}]\ (x_{21,12}^{1,2} \mid z_{21,22}^{1,1})\}\ * \tag{7.58}$$
$$\{[N_{12}]\ (z_{11,12}^{1,1},\ x_{21,12}^{1,2} \mid y_{12}) + [N_{22}]\ (z_{21,22}^{1,1} \mid y_{22})\}$$

In order to merge the outputs of the nodes N_{11}, $I_{1.5,1}$ and N_{21} of the first interim FN, it is necessary to augment the two inputs x_{11} and $x_{21,12}^{1,2}$ so that they become common inputs for these three nodes. This augmentation operation transforms the first interim FN into a second interim FN with nodes N_{11}^{AI}, $I_{1.5,1}^{AI}$, N_{21}^{AI}, N_{12} and N_{22} whereby the merging of nodes is reflected by the replacement of N_{11} with N_{11}^{AI}, $I_{1.5,1}$ with $I_{1.5,1}^{AI}$ and N_{21} with N_{21}^{AI}. This second interim FN can be described by the block-scheme in Fig.7.47 and the topological expression in Eq.(7.59).

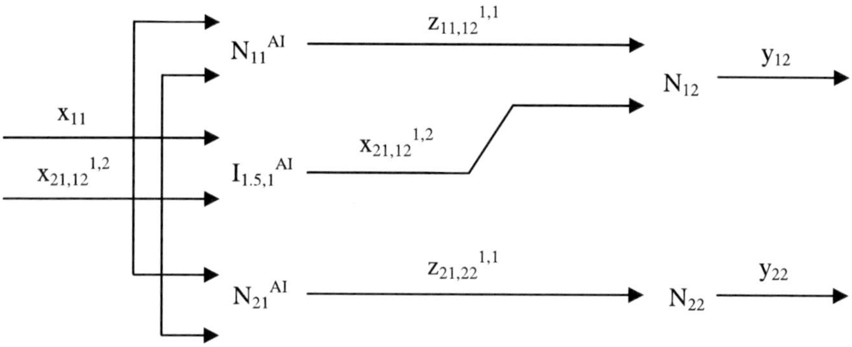

Fig. 7.47 Second interim FN for Example 7.14

$$\{[N_{11}^{AI}]\ (x_{11},\ x_{21,12}^{1,2} \mid z_{11,12}^{1,1})\ ;\ [I_{1.5,1}^{AI}]\ (x_{11},\ x_{21,12}^{1,2} \mid x_{21,12}^{1,2})\ ; \tag{7.59}$$
$$[N_{21}^{AI}]\ (x_{11},\ x_{21,12}^{1,2} \mid z_{21,22}^{1,1})\}\ *$$
$$\{[N_{12}]\ (z_{11,12}^{1,1},\ x_{21,12}^{1,2} \mid y_{12}) + [N_{22}]\ (z_{21,22}^{1,1} \mid y_{22})\}$$

The second interim FN has five nodes. In this case, N_{11}^{AI} can be output merged with $I_{1.5,1}^{AI}$ and N_{21}^{AI} whereas N_{12} can be vertically merged with N_{22}. These merging operations transform the second interim FN into a third interim FN with nodes N_{11}^{AI} ; $I_{1.5,1}^{AI}$; N_{21}^{AI} and $N_{12} + N_{22}$ whereby the merging of nodes is reflected by the replacement of N_{11}^{AI}, $I_{1.5,1}^{AI}$ and N_{21}^{AI} with N_{11}^{AI} ; $I_{1.5,1}^{AI}$; N_{21}^{AI} and the replacement of N_{12} and N_{22} with $N_{12} + N_{22}$. This third interim FN can be described by the block-scheme in Fig.7.48 and the topological expression in Eq.(7.60) from where it can be seen that all connections already have parallel paths.

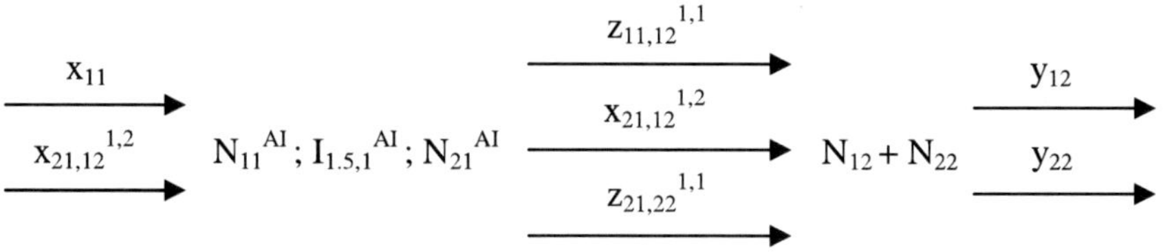

Fig. 7.48 Third interim FN for Example 7.14

$$[N_{11}^{AI}; I_{1.5,1}^{AI}; N_{21}^{AI}]\,(x_{11}, x_{21,12}^{1,2} \mid z_{11,12}^{1,1}, x_{21,12}^{1,2}, z_{21,22}^{1,1}) \ * \qquad (7.60)$$

$$[N_{12} + N_{22}]\,(z_{11,12}^{1,1}, x_{21,12}^{1,2}, z_{21,22}^{1,1} \mid y_{12}, y_{22})$$

The nodes N_{11}^{AI} ; $I_{1.5,1}^{AI}$; N_{21}^{AI} and $N_{12} + N_{22}$ of the third interim FN can be merged horizontally due to the parallel connection paths. This merging operation transforms the third interim FN into a final FN with a single equivalent node whereby the merging of nodes N_{11}^{AI} ; $I_{1.5,1}^{AI}$; N_{21}^{AI} and $N_{12} + N_{22}$ is reflected by their replacement with node $(N_{11}^{AI}; I_{1.5,1}^{AI}; N_{21}^{AI}) * (N_{12} + N_{22})$. This final FN can be described by the block-scheme in Fig.7.49 and the topological expression in Eq.(7.61) from where it can be seen that the two composite nodes are implicit in the single equivalent node and the nodes in the first composite node have augmented inputs.

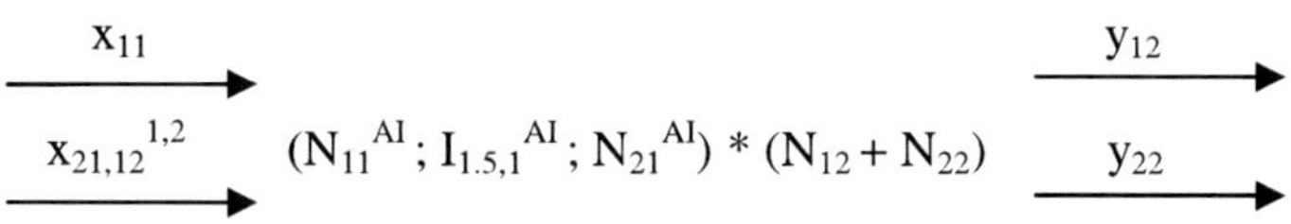

Fig. 7.49 Final FN for Example 7.14

$$[(N_{11}^{AI}; I_{1.5,1}^{AI}; N_{21}^{AI}) * (N_{12} + N_{22})]\,(x_{11}, x_{21,12}^{1,2} \mid y_{12}, y_{22}) \qquad (7.61)$$

The considerations in Example 7.14 are concerned with network analysis when all network nodes are known. In the case of network design, at least one node is unknown. In this context, Algorithms 7.37-7.40 describe the process of deriving an unknown node in the initial FN from Fig.7.45 when the other nodes, the implicit identity node $I_{1.5,1}^{AI}$ and the single equivalent node N_E are known. In this case, node N_E is given by the Boolean matrix equation in Eq.(7.62).

$$N_E = (N_{11}^{AI} ; I_{1.5,1}^{AI} ; N_{21}^{AI}) * (N_{12} + N_{22}) \quad (7.62)$$

Algorithm 7.37

1. Define N_E, N_{21}, N_{12}, N_{22} and $I_{1.5,1}^{AI}$.
2. Find $I_{1.5,1}^{AI}$ by input augmentation of $I_{1.5,1}$.
3. Find N_{21}^{AI} by input augmentation of N_{21}.
4. Find $N_{12} + N_{22}$ by vertical merging of N_{12} and N_{22}.
5. Derive $N_{11}^{AI} ; I_{1.5,1}^{AI} ; N_{21}^{AI}$ from Eq.(7.62), if possible.
6. Derive N_{11}^{AI} from $N_{11}^{AI} ; I_{1.5,1}^{AI} ; N_{21}^{AI}$, if possible.
7. Find N_{11} by inverse input augmentation of N_{11}^{AI}.

Algorithm 7.38

1. Define N_E, N_{11}, N_{12}, N_{22} and $I_{1.5,1}^{AI}$.
2. Find $I_{1.5,1}^{AI}$ by input augmentation of $I_{1.5,1}$.
3. Find N_{11}^{AI} by input augmentation of N_{11}.
4. Find $N_{12} + N_{22}$ by vertical merging of N_{12} and N_{22}.
5. Derive $N_{11}^{AI} ; I_{1.5,1}^{AI} ; N_{21}^{AI}$ from Eq.(7.62), if possible.
6. Derive N_{21}^{AI} from $N_{11}^{AI} ; I_{1.5,1}^{AI} ; N_{21}^{AI}$, if possible.
7. Find N_{21} by inverse input augmentation of N_{21}^{AI}.

Algorithm 7.39

1. Define N_E, N_{11}, N_{21}, N_{22} and $I_{1.5,1}^{AI}$.
2. Find N_{11}^{AI} by input augmentation of N_{11}.
3. Find $I_{1.5,1}^{AI}$ by input augmentation of $I_{1.5,1}$.
4. Find N_{21}^{AI} by input augmentation of N_{21}.
5. Find $N_{11}^{AI} ; I_{1.5,1}^{AI} ; N_{21}^{AI}$ by output merging of N_{11}^{AI}, $I_{1.5,1}^{AI}$ and N_{21}^{AI}.
6. Derive $N_{12} + N_{22}$ from Eq.(7.62), if possible.
7. Derive N_{12} from $N_{12} + N_{22}$, if possible.

Algorithm 7.40

1. Define N_E, N_{11}, N_{21}, N_{12} and $I_{1.5,1}^{AI}$.
2. Find N_{11}^{AI} by input augmentation of N_{11}.
3. Find $I_{1.5,1}^{AI}$ by input augmentation of $I_{1.5,1}$.
4. Find N_{21}^{AI} by input augmentation of N_{21}.
5. Find $N_{11}^{AI} ; I_{1.5,1}^{AI} ; N_{21}^{AI}$ by output merging of N_{11}^{AI}, $I_{1.5,1}^{AI}$ and N_{21}^{AI}.
6. Derive $N_{12} + N_{22}$ from Eq.(7.62), if possible.
7. Derive N_{22} from $N_{12} + N_{22}$, if possible.

Example 7.15

This example considers a FN with nodes N_{11}, N_{21}, N_{12} and N_{22} where $x_{11,22}^{1,1}$ is an input for N_{11} and the first input for N_{22}, x_{21} is an input for N_{21}, y_{12} is an output for N_{12}, y_{22} is an output for N_{22}, $z_{11,12}^{1,1}$ is the connection from the first and only output for N_{11} to the first and only input for N_{12} and $z_{21,22}^{1,2}$ is the connection from the second and only output for N_{21} to the second input for N_{22}. This initial FN can be described by the block-scheme in Fig.7.50 and the topological expression in Eq.(7.63) from where it can be seen that the input at the top propagates through the first layer.

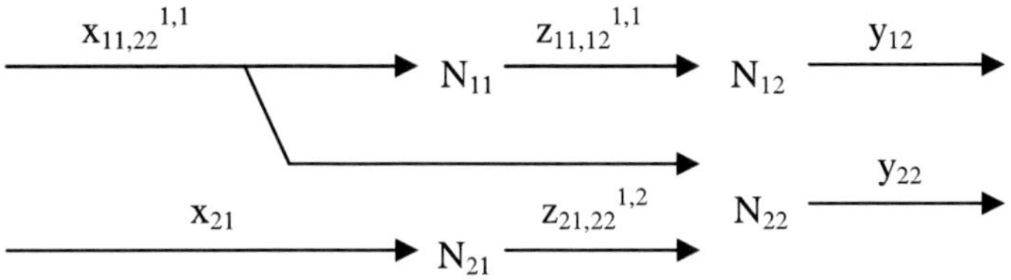

Fig. 7.50 Initial FN for Example 7.15

$$\{[N_{11}]\,(x_{11,22}^{1,1} \mid z_{11,12}^{1,1})\ ;\ [N_{21}]\,(x_{21} \mid z_{21,22}^{1,2})\}\ * \qquad (7.63)$$

$$\{[N_{12}]\,(z_{11,12}^{1,1},\, x_{11,22}^{1,1} \mid y_{12}) + [N_{22}]\,(z_{21,22}^{1,2} \mid y_{22})\}$$

The input $x_{11,22}^{1,1}$ is propagating through the first layer of the FN within a virtual level between the first and the second level. This propagation can be represented by inserting an implicit identity node $I_{1.5,1}$. This representation transforms the initial FN into a first interim FN with nodes N_{11}, N_{21}, N_{12}, N_{22} and $I_{1.5,1}$. This first interim FN can be described by the block-scheme in Fig.7.51 and the topological expression in Eq.(7.64) from where it can be seen that the top input propagates through the first layer by means of the identity input.

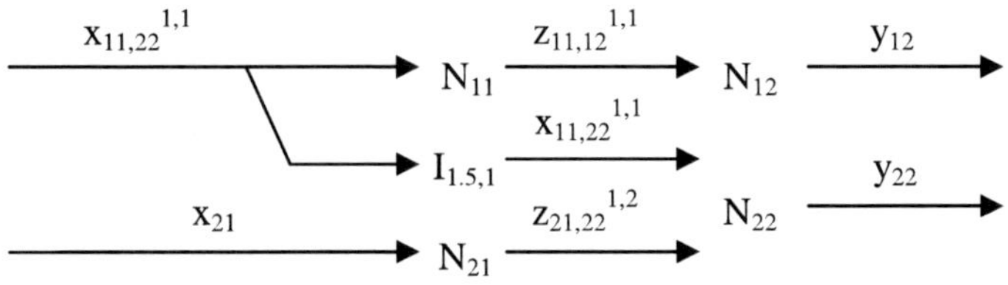

Fig. 7.51 First interim FN for Example 7.15

$$\{[N_{11}]\,(x_{11,22}^{1,1} \mid z_{11,12}^{1,1})\ ;\ [I_{1.5,1}]\,(x_{11,22}^{1,1} \mid x_{11,22}^{1,1})\ ;\ [N_{21}]\,(x_{21} \mid z_{21,22}^{1,2})\}\ * \qquad (7.64)$$

$$\{[N_{12}]\,(z_{11,12}^{1,1} \mid y_{12}) + [N_{22}]\,(x_{11,22}^{1,1},\, z_{21,22}^{1,2} \mid y_{22})\}$$

In order to merge the outputs of the nodes N_{11}, $I_{1.5,1}$ and N_{21} of the first interim FN, it is necessary to augment the two inputs $x_{11,22}^{1,1}$ and x_{21} so that they become common inputs for these three nodes. This augmentation operation transforms the first interim FN into a second interim FN with nodes N_{11}^{AI}, $I_{1.5,1}^{AI}$, N_{21}^{AI}, N_{12} and N_{22} whereby the merging of nodes is reflected by the replacement of N_{11} with N_{11}^{AI}, $I_{1.5,1}$ with $I_{1.5,1}^{AI}$ and N_{21} with N_{21}^{AI}. This second interim FN can be described by the block-scheme in Fig.7.52 and the topological expression in Eq.(7.65).

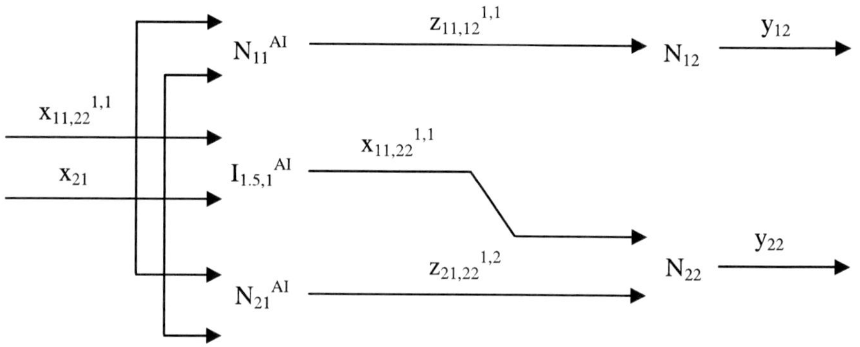

Fig. 7.52 Second interim FN for Example 7.15

$$\{[N_{11}^{AI}]\,(x_{11,22}^{1,1}, x_{21} \mid z_{11,12}^{1,1})\,;\,[I_{1.5,1}^{AI}]\,(x_{11,22}^{1,1}, x_{21} \mid x_{11,22}^{1,1})\,; \tag{7.65}$$

$$[N_{21}^{AI}]\,(x_{11,22}^{1,1}, x_{21} \mid z_{21,22}^{1,2})\} \;*$$

$$\{[N_{12}]\,(z_{11,12}^{1,1} \mid y_{12}) + [N_{22}]\,(x_{11,22}^{1,1}, z_{21,22}^{1,2} \mid y_{22})\}$$

The second interim FN has five nodes. In this case, N_{11}^{AI} can be output merged with $I_{1.5,1}^{AI}$ and N_{21}^{AI} whereas N_{12} can be vertically merged with N_{22}. These merging operations transform the second interim FN into a third interim FN with nodes N_{11}^{AI} ; $I_{1.5,1}^{AI}$; N_{21}^{AI} and $N_{12} + N_{22}$ whereby the merging of nodes is reflected by the replacement of N_{11}^{AI}, $I_{1.5,1}^{AI}$ and N_{21}^{AI} with N_{11}^{AI} ; $I_{1.5,1}^{AI}$; N_{21}^{AI} and the replacement of N_{12} and N_{22} with $N_{12} + N_{22}$. This third interim FN can be described by the block-scheme in Fig.7.53 and the topological expression in Eq.(7.66) from where it can be seen that all connections already have parallel paths.

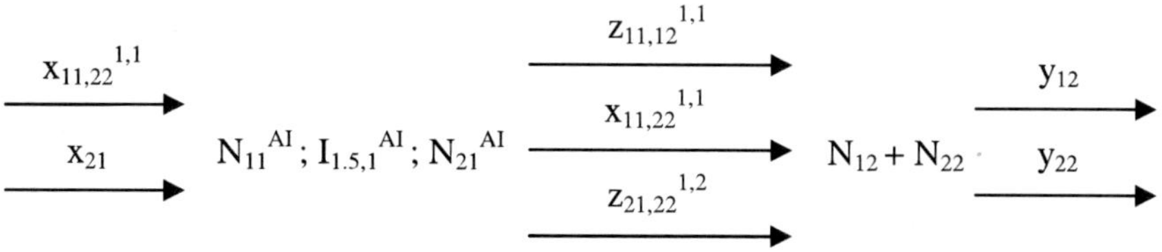

Fig. 7.53 Third interim FN for Example 7.15

$$[N_{11}^{AI}; I_{1.5,1}^{AI}; N_{21}^{AI}] (x_{11,22}^{1,1}, x_{21} \mid z_{11,12}^{1,1}, x_{11,22}^{1,1}, z_{21,22}^{1,2}) * \tag{7.66}$$

$$[N_{12} + N_{22}] (z_{11,12}^{1,1}, x_{11,22}^{1,1}, z_{21,22}^{1,2} \mid y_{12}, y_{22})$$

The nodes N_{11}^{AI} ; $I_{1.5,1}^{AI}$; N_{21}^{AI} and $N_{12} + N_{22}$ of the third interim FN can be merged horizontally due to the parallel connection paths. This merging operation transforms the third interim FN into a final FN with a single equivalent node whereby the merging of nodes N_{11}^{AI} ; $I_{1.5,1}^{AI}$; N_{21}^{AI} and $N_{12} + N_{22}$ is reflected by their replacement with node $(N_{11}^{AI}; I_{1.5,1}^{AI}; N_{21}^{AI}) * (N_{12} + N_{22})$. This final FN can be described by the block-scheme in Fig.7.54 and the topological expression in Eq.(7.67) from where it can be seen that the two composite nodes are implicit in the single equivalent node and the nodes in the first composite node have augmented inputs.

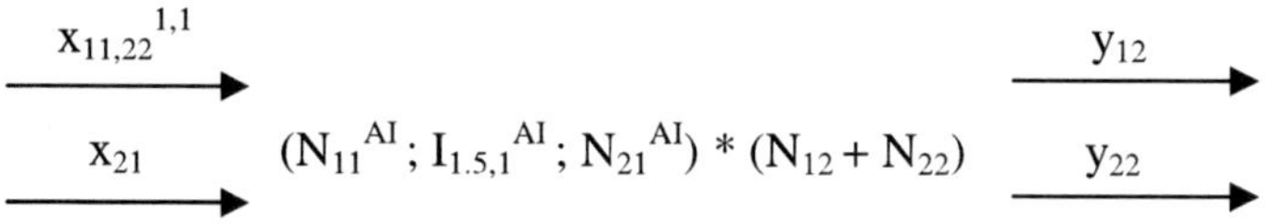

Fig. 7.54 Final FN for Example 7.15

$$[(N_{11}^{AI}; I_{1.5,1}^{AI}; N_{21}^{AI}) * (N_{12} + N_{22})] (x_{11,22}^{1,1}, x_{21} \mid y_{12}, y_{22}) \tag{7.67}$$

The considerations in Example 7.15 are concerned with network analysis when all network nodes are known. In the case of network design, at least one node is unknown. In this context, Algorithms 7.37-7.40 from Example 7.14 describe the process of deriving an unknown node in the initial FN from Fig.7.50 when the other nodes, the implicit identity node $I_{1.5,1}^{AI}$ and the single equivalent node N_E are known. In this case, node N_E is given by the Boolean matrix equation Eq.(7.62) from Example 7.14.

7.6 Summary on Feedforward Fuzzy Networks

The examples presented in this chapter illustrate the application of basic operations, their properties and advanced operations in feedforward FNs. These examples validate theoretically the linguistic composition approach used in the book. This applies particularly to FNs with multiple levels and multiple layers which are the most complex type of feedforward FNs. However, the other two less complex types of feedforward FNs, i.e. FNs with single level and multiple layers as well as FNs with multiple levels and single layer, are also quite useful in that they are usually part of FNs with multiple levels and multiple layers. In this context, the simplest type of feedforward FNs, i.e. FNs with single level and single layer, are discussed only briefly as they are identical with fuzzy systems.

The different types of feedforward FNs can accommodate different types of nodes, e.g. parallel nodes and sequential nodes. In this case, parallel nodes are used for tasks that can always be carried out at the same time whereas sequential nodes are used for tasks that can only be carried out one at a time. So, the type of tasks to be carried out in a FN determines the type of FN to be used.

The relationship between different types of FNs and nodes is described in Table 7.1.

Table 7.1 Relationship between types of FNs and nodes

Feedforward FN	Parallel nodes	Sequential nodes
Single level and single layer	No	No
Single level and multiple layers	No	Yes
Multiple levels and single layer	Yes	No
Multiple levels and multiple layers	Yes	Yes

The next chapter shows more applications of the theoretical results from Chapters 4-6 to FNs. In particular, several basic types of feedback FNs are considered.

Chapter 8
Feedback Fuzzy Networks

8.1 Preliminaries on Feedback Fuzzy Networks

The application of basic operations, their structural properties and advanced operations in Chapter 7 is illustrated only in feedforward FNs. The examples presented there show the application of the above operations and their properties to the overall structure of FNs with single or multiple levels and layers. Although these networks may have a fairly complex structure, they are assumed to have connections only in a forward direction. Therefore, it is necessary to extend these considerations to more complex networks in terms of the direction of their connections.

The current chapter describes the application of basic operations, their structural properties and advanced operations in feedback FNs. The latter are FNs some of whose connections are in a backward direction, i.e. from nodes residing in specific layers to nodes residing in the same or preceding layers. This feedback characteristic is reflected by left-sided arrows in the corresponding block scheme for the FN under consideration. In this context, a left-sided arrow represents an output from a node that is fed back as an input to the same or another node by means of a feedback node, i.e. a feedback rule base.

Four types of feedback FNs are considered in the context of both analysis and design. The analysis part is presented first and is then followed by the design part. In the case of analysis, all feedback and network nodes are known whereby the aim is to derive a formula for the single node representing the linguistically equivalent fuzzy system. In the case of design, each feedback node is unknown at a time with all other feedback nodes and all network nodes known whereby the aim is to derive an algorithm for the unknown feedback node from the known feedback nodes, the known network nodes and the single node representing the linguistically equivalent fuzzy system. The design task can be easily extended to cases with more than one unknown feedback node.

The four types of feedback FNs represent different network topologies with respect to single or multiple local and global feedback in the underlying grid structure for the FN under consideration. Each of these types is illustrated with several examples that are presented at a fairly high level of abstraction using mainly block schemes and topological expressions. These formal models for FNs are at network level and they both lend themselves easily to advanced manipulations in the context of the linguistic composition approach. Boolean matrices are used as formal models only implicitly in block schemes and topological expressions as well as in design tasks.

A. Gegov: Fuzzy Networks for Complex Systems, STUDFUZZ 259, pp. 201–246.
springerlink.com

All presented examples are for feedback FNs with a fairly small number of nodes but the extension of these examples to cases of higher dimension is straightforward. The only difference in this extension is the higher complexity of the formulas for the derivation of the single node representing the linguistically equivalent fuzzy system in the case of analysis and the algorithms for the derivation of the unknown feedback node in the case of design.

8.2 Networks with Single Local Feedback

The simplest type of FN is the one with single local feedback. This network has only one node embraced by a feedback connection with a feedback node in it. In this case, the feedback is single as it appears only once but it is also local as it embraces only one node. There may be an arbitrary number of feedforward connections between this node and any other nodes as well as between any pair of other nodes. However, the presence of any feedforward connections does not remove the feedback characteristics of this type of FN due to the presence of the feedback connection with the feedback node.

Example 8.1

This example considers a FN with network nodes N_{11}, N_{12} and feedback node F_{11} embracing N_{11} where x_{11} is an input for N_{11}, y_{12} is an output for N_{12}, $z_{11,12}^{1,1}$ is the feedforward connection from the first output for N_{11} to the first and only input for N_{12}, v_{11} is the part of the feedback connection to F_{11} and w_{11} is the part of the feedback connection from F_{11}. This initial FN represents a queue of two fuzzy systems that can be described by the block scheme in Fig.8.1 and the topological expression in Eq.(8.1) from where it can be seen that the node in the first layer is embraced by the feedback.

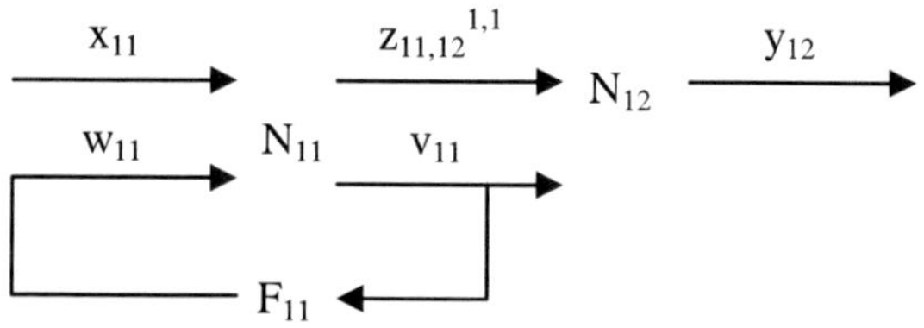

Fig. 8.1 Initial FN for Example 8.1

$$[N_{11}]\,(x_{11}, w_{11} \mid z_{11,12}^{2,1}, v_{11}) * [N_{12}]\,(z_{11,12}^{1,1} \mid y_{12}),\ [F_{11}]\,(v_{11} \mid w_{11}) \qquad (8.1)$$

The feedback node F_{11} represents a non-identity feedback connection. This is also implied by the use of different variable names for the input and the output for F_{11}, i.e. v_{11} and w_{11}. In order to apply the linguistic composition approach to the initial

FN, it is necessary to introduce a second level within the second layer of the underlying grid structure of the FN and to move F_{11} to this new grid cell.

The above movement transforms the initial FN into a first interim FN whereby the non-identity feedback connection embracing the network node N_{11} is represented as a feedforward connection v_{11} between N_{11} and F_{11} and an identity feedback connection w_{11} embracing both N_{11} and F_{11}. This first interim FN can be described by the block-scheme in Fig.8.2 and the topological expression in Eq.(8.2) from where it can be seen that F_{11} is already a feedforward node alongside the other two nodes.

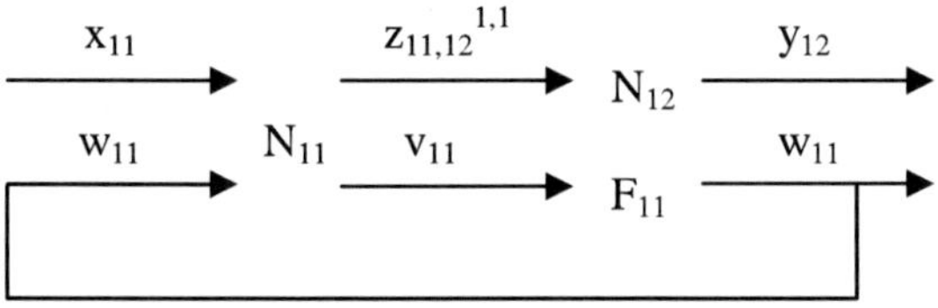

Fig. 8.2 First interim FN for Example 8.1

$$[N_{11}]\ (x_{11}, w_{11} \mid z_{11,12}^{2,1}, v_{11}) * \{[N_{12}]\ (z_{11,12}^{1,1} \mid y_{12}) + [F_{11}]\ (v_{11} \mid w_{11})\} \qquad (8.2)$$

Nodes N_{12} and F_{11} of the first interim FN can be merged vertically into a temporary node $N_{12} + F_{11}$. This temporary node can be further merged horizontally with node N_{11} on the left. These merging operations transform the first interim FN into a second interim FN with a single node whereby the horizontal merging of nodes N_{11} and $N_{12} + F_{11}$ is reflected by their replacement with node $N_{11} * (N_{12} + F_{11})$. This second interim FN can be described by the block-scheme in Fig.8.3 and the topological expression in Eq.(8.3) from where it can be seen that the single node is embraced by the identity feedback connection.

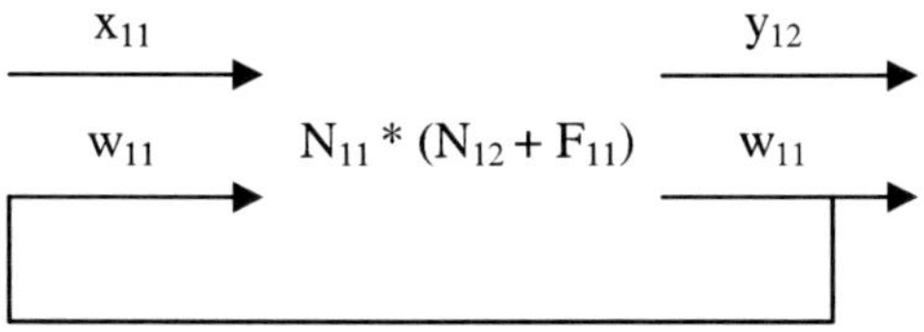

Fig. 8.3 Second interim FN for Example 8.1

$$[N_{11} * (N_{12} + F_{11})]\ (x_{11}, w_{11} \mid y_{12}, w_{11}) \qquad (8.3)$$

The single node $N_{11} * (N_{12} + F_{11})$ with input set $\{x_{11}, w_{11}\}$ and output set $\{y_{12}, w_{11}\}$ can be further transformed into a single node with equivalent feedback $(N_{11} * (N_{12} + F_{11}))^{EF}$ with input set $\{x_{11}, x^{EF}\}$ and output set $\{y_{12}, y^{EF}\}$.This transformation removes the identity feedback and makes the fuzzy system with feedback equivalent to a fuzzy system without feedback. As a result, the second interim FN is transformed into a final FN. This final FN can be described by the block-scheme in Fig.8.4 and the topological expression in Eq.(8.4) from where it can be seen that the single node is not embraced by the identity feedback connection anymore.

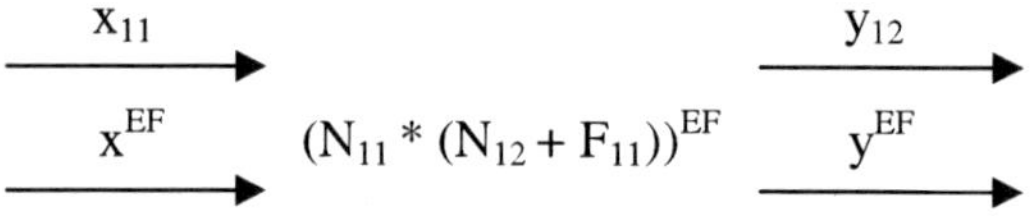

Fig. 8.4 Final FN for Example 8.1

$$[(N_{11} * (N_{12} + F_{11}))^{EF}] (x_{11}, x^{EF} \mid y_{12}, y^{EF}) \tag{8.4}$$

The considerations in Example 8.1 are concerned with network analysis when all network nodes and the feedback node are known. In the case of network design, the feedback node is unknown. In this context, Algorithm 8.1 describes the process of deriving the unknown feedback node in the initial FN from Fig.8.1 when the network nodes and the single equivalent node N_E are known. In this case, node N_E is given by the Boolean matrix equation in Eq.(8.5).

$$N_E = (N_{11} * (N_{12} + F_{11}))^{EF} \tag{8.5}$$

Algorithm 8.1

1. Define N_E, N_{11} and N_{12}.
2. Confirm that N_E satisfies the feedback constraints, if possible.
3. Make N_E equal to $N_{11} * (N_{12} + F_{11})$ in Eq.(8.5).
4. Derive $N_{12} + F_{11}$ from N_E in Eq.(8.5), if possible.
5. Derive F_{11} from $N_{12} + F_{11}$, if possible.

Example 8.2

This example considers a FN with network nodes N_{11}, N_{12} and feedback node F_{12} embracing N_{12} where x_{11} is an input for N_{11}, y_{12} is an output for N_{12}, $z_{11,12}^{1,1}$ is the feedforward connection from the first and only output for N_{11} to the first input for N_{12}, v_{12} is the part of the feedback connection to F_{12} and w_{12} is the part of the feedback connection from F_{12}. This initial FN represents a queue of two fuzzy systems that can be described by the block scheme in Fig.8.5 and the topological expression in Eq.(8.6) from where it can be seen that the node in the second layer is embraced by the feedback.

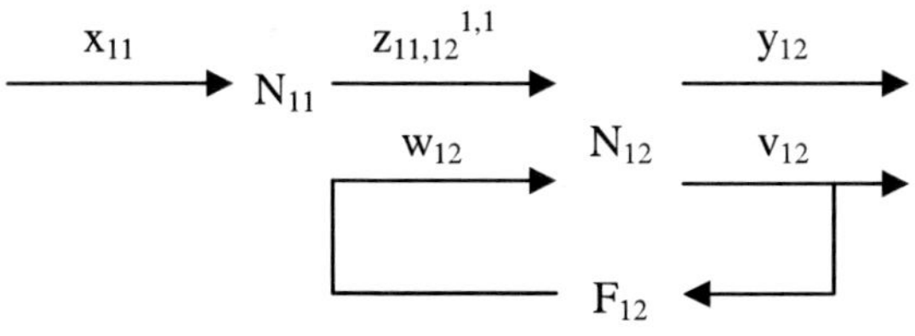

Fig. 8.5 Initial FN for Example 8.2

$$[N_{11}]\,(x_{11} \mid z_{11,12}^{2,1}) * [N_{12}]\,(z_{11,12}^{1,1}, w_{12} \mid y_{12}, v_{12}), [F_{12}]\,(v_{12} \mid w_{12}) \tag{8.6}$$

The feedback node F_{12} represents a non-identity feedback connection. This is also implied by the use of different variable names for the input and the output for F_{12}, i.e. v_{12} and w_{12}. In order to apply the linguistic composition approach to the initial FN, it is necessary to introduce a second level within a third layer of the underlying grid structure of the FN and to move F_{12} to this new grid cell. It is also necessary to propagate y_{12} forwards through the third layer and insert an implicit identity node I_{13} in level 1 of layer 3. Likewise, it is necessary to propagate w_{12} backwards through the first layer and insert an implicit identity node I_{21} in level 2 of layer 1.

The above movements and insertions transform the initial FN into a first interim FN whereby the non-identity feedback connection embracing the network node N_{12} is represented as a feedforward connection v_{12} between N_{12} and F_{12} and an identity feedback connection w_{12} embracing I_{21}, N_{12} and F_{12}. This first interim FN can be described by the block-scheme in Fig.8.6 and the topological expression in Eq.(8.7) from where it can be seen that F_{12} is already a feedforward node alongside the other four nodes.

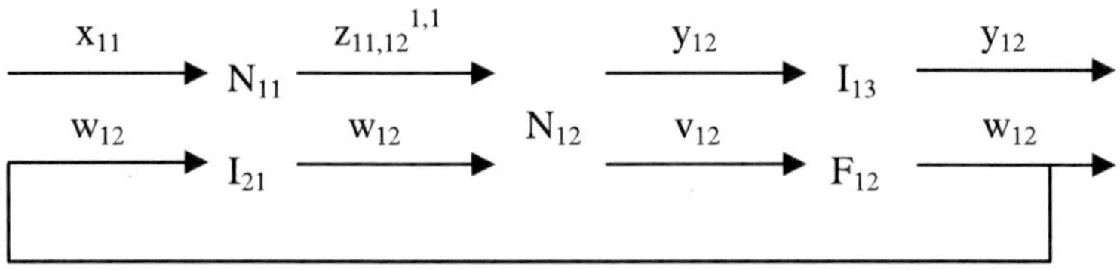

Fig. 8.6 First interim FN for Example 8.2

$$\{[N_{11}]\,(x_{11} \mid z_{11,12}^{2,1}) + [I_{21}]\,(w_{12} \mid w_{12})\} * [N_{12}]\,(z_{11,12}^{1,1}, w_{12} \mid y_{12}, v_{12}) * \tag{8.7}$$

$$\{[I_{13}]\,(y_{12} \mid y_{12}) + [F_{12}]\,(v_{12} \mid w_{12})\}$$

Nodes N_{11} and I_{21} of the first interim FN can be merged vertically into a temporary node $N_{11} + I_{21}$. Similarly, nodes I_{13} and F_{12} of the same interim FN can be merged vertically into another temporary node $I_{13} + F_{12}$. These two temporary nodes can be further merged horizontally with node N_{12} in the middle. These merging operations transform the first interim FN into a second interim FN with a single node whereby the horizontal merging of nodes $N_{11} + I_{21}$, N_{11} and $I_{13} + F_{12}$ is reflected by their replacement with node $(N_{11} + I_{21}) * N_{11} * (I_{13} + F_{12})$. This second interim FN can be described by the block-scheme in Fig.8.7 and the topological expression in Eq.(8.8) from where it can be seen that single node is embraced by the identity feedback connection.

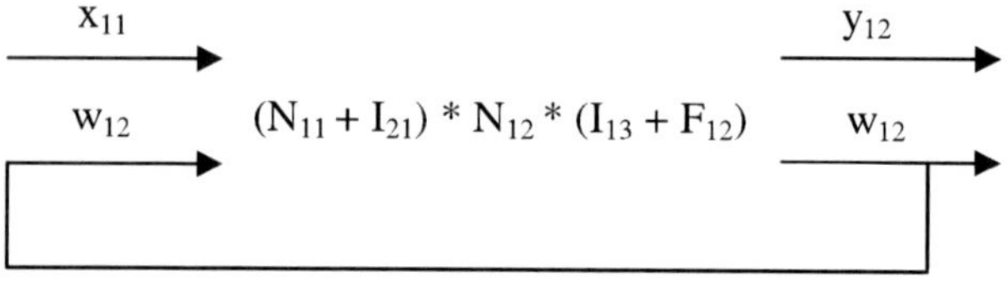

Fig. 8.7 Second interim FN for Example 8.2

$$[(N_{11} + I_{21}) * N_{12} * (I_{13} + F_{12})]\ (x_{11}, w_{12} \mid y_{12}, w_{12}) \tag{8.8}$$

The single node $(N_{11} + I_{21}) * N_{12} * (I_{13} + F_{12})$ with input set $\{x_{11}, w_{12}\}$ and output set $\{y_{12}, w_{12}\}$ can be further transformed into a single node with equivalent feedback $((N_{11} + I_{21}) * N_{12} * (I_{13} + F_{12}))^{EF}$ with input set $\{x_{11}, x^{EF}\}$ and output set $\{y_{12}, y^{EF}\}$. This transformation removes the identity feedback and makes the fuzzy system with feedback equivalent to a fuzzy system without feedback. As a result, the second interim FN is transformed into a final FN. This final FN can be described by the block-scheme in Fig.8.8 and the topological expression in Eq.(8.9) from where it can be seen that the single node is not embraced by the identity feedback connection anymore.

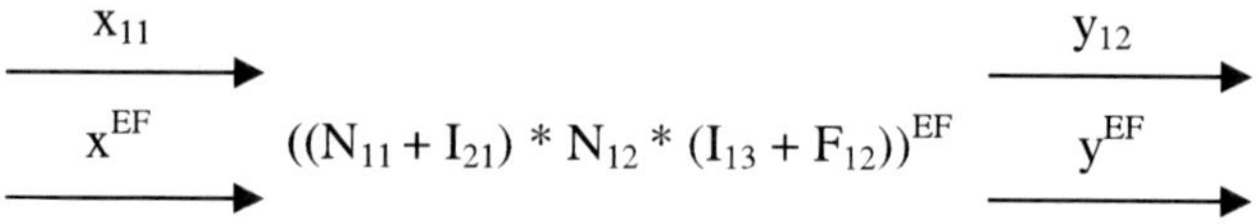

Fig. 8.8 Final FN for Example 8.2

$$[((N_{11} + I_{21}) * N_{12} * (I_{13} + F_{12}))^{EF}]\ (x_{11}, x^{EF} \mid y_{12}, y^{EF}) \tag{8.9}$$

The considerations in Example 8.2 are concerned with network analysis when all network nodes and the feedback nodes are known. In the case of network design, the feedback node is unknown. In this context, Algorithm 8.2 describes the process of deriving the unknown feedback node in the initial FN from Fig.8.5 when the network nodes and the single equivalent node N_E are known. In this case, node N_E is given by the Boolean matrix equation in Eq.(8.10).

$$N_E = ((N_{11} + I_{21}) * N_{12} * (I_{13} + F_{12}))^{EF} \tag{8.10}$$

Algorithm 8.2

1. Define N_E, N_{11}, N_{12}, I_{21} and I_{13}.
2. Confirm that N_E satisfies the feedback constraints, if possible.
3. Make N_E equal to $(N_{11} + I_{21}) * N_{12} * (I_{13} + F_{12})$ in Eq.(8.10).
4. Find $N_{11} + I_{21}$ by vertical merging of N_{11} and I_{21}.
5. Find $(N_{11} + I_{21}) * N_{12}$ by horizontal merging of $N_{11} + I_{21}$ and N_{12}.
6. Derive $I_{13} + F_{12}$ from N_E in Eq.(8.10), if possible.
7. Derive F_{12} from $I_{13} + F_{12}$, if possible.

Example 8.3

This example considers a FN with network nodes N_{11}, N_{21} and feedback node F_{11} embracing N_{11} where $x_{11,21}$ is a common input for N_{11} and N_{21}, y_{11} is an output for N_{11}, y_{21} is an output for N_{21}, v_{11} is the part of the feedback connection to F_{11} and w_{11} is the part of the feedback connection from F_{11}. This initial FN represents a stack of two fuzzy systems that can be described by the block scheme in Fig.8.9 and the topological expression in Eq.(8.11) from where it can be seen that the node in the first level is embraced by the feedback.

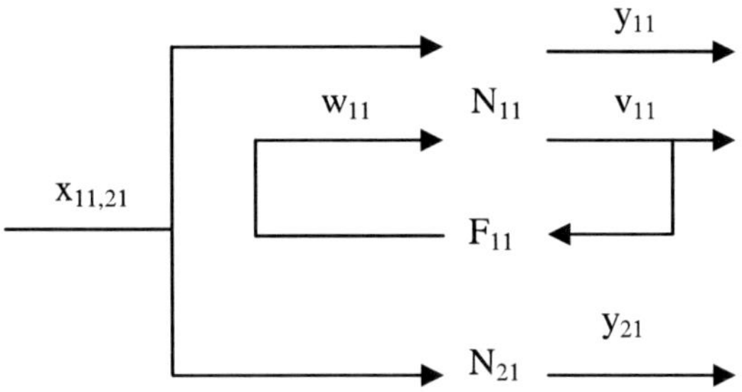

Fig. 8.9 Initial FN for Example 8.3

$$[N_{11}]\ (x_{11,21}, w_{11} \mid y_{11}, v_{11})\ ;\ [N_{21}]\ (x_{11,21} \mid y_{21}),\ [F_{11}]\ (v_{11} \mid w_{11}) \tag{8.11}$$

The feedback node F_{11} represents a non-identity feedback connection. This is also implied by the use of different variable names for the input and the output for F_{11}, i.e. v_{11} and w_{11}. In order to apply the linguistic composition approach to the initial FN, it is necessary to introduce a virtual intermediate level just under the first level within a second layer of the underlying grid structure of the FN and to move F_{11} to this new grid cell. It is also necessary to propagate y_{11} forwards through the second layer and insert an implicit identity node I_{12} in level 1 of layer 2. Likewise, it is necessary to propagate y_{21} forwards through the second layer and insert an implicit identity node I_{22} in level 2 of layer 2.

The above movements and insertions transform the initial FN into a first interim FN whereby the non-identity feedback connection embracing the network node N_{11} is represented as a feedforward connection v_{11} between N_{11} and F_{11} and an identity feedback connection w_{11} embracing both N_{11} and F_{11}. This first interim FN can be described by the block-scheme in Fig.8.10 and the topological expression in Eq.(8.12) from where it can be seen that F_{11} is already a feedforward node alongside the other four nodes.

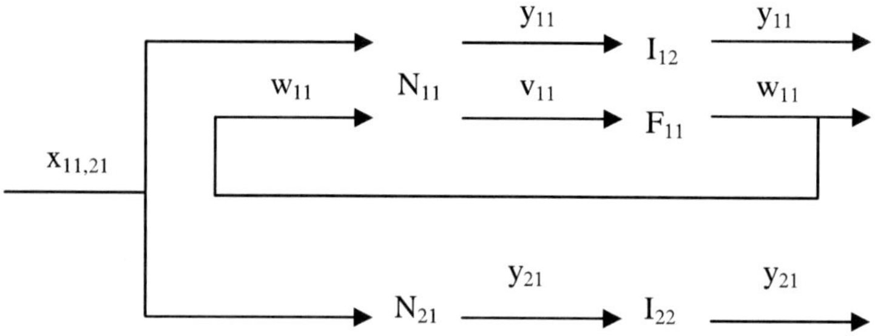

Fig. 8.10 First interim FN for Example 8.3

$$\{[N_{11}]\ (x_{11,21}, w_{11} \,|\, y_{11}, v_{11}) * \{[I_{12}]\ (y_{11} \,|\, y_{11}) + [F_{11}]\ (v_{11} \,|\, w_{11})\}\}\ ; \quad (8.12)$$

$$\{[N_{21}]\ (x_{11,21} \,|\, y_{21}) * [I_{22}]\ (y_{21} \,|\, y_{21})\}$$

Nodes I_{12} and F_{11} of the first interim FN can be merged vertically into a temporary node $I_{12} + F_{11}$. This temporary node can be further merged horizontally with node N_{11} on the left. Also, nodes N_{21} and I_{22} of the same FN can be merged horizontally into another temporary node $N_{21} * I_{22}$. These merging operations transform the first interim FN into a second interim FN with two nodes whereby the horizontal merging of nodes N_{11} and $I_{12} + F_{11}$ is reflected by their replacement with node $N_{11} * (I_{12} + F_{11})$ and the horizontal merging of nodes N_{21} and I_{22} is reflected by their replacement with node $N_{21} * I_{22}$. This second interim FN can be described by the block-scheme in Fig.8.11 and the topological expression in Eq.(8.13) from where it can be seen that the upper node is embraced by the identity feedback connection.

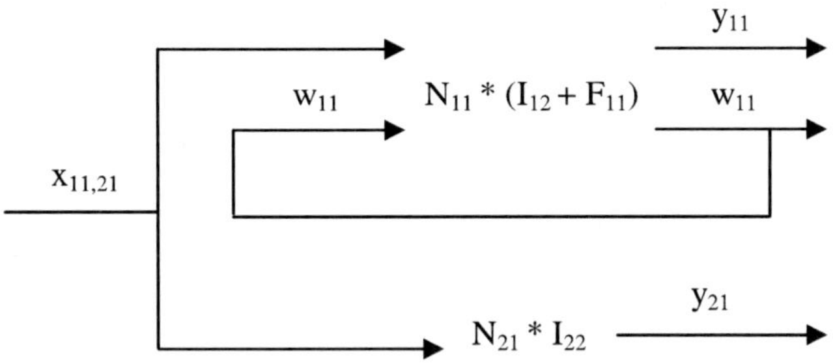

Fig. 8.11 Second interim FN for Example 8.3

$$[N_{11} * (I_{12} + F_{11})]\ (x_{11,21}, w_{11} \mid y_{11}, v_{11})\ ;\ [N_{21} * I_{22}]\ (x_{11,21} \mid y_{21}) \quad (8.13)$$

The upper node $N_{11} * (I_{12} + F_{11})$ with input set $\{x_{11,21}, w_{11}\}$ and output set $\{y_{11}, w_{11}\}$ can be further transformed into a node with equivalent feedback $(N_{11} * (I_{12} + F_{11}))^{EF}$ with input set $\{x_{11,21}, x^{EF}\}$ and output set $\{y_{11}, y^{EF}\}$. This transformation removes the identity feedback and makes the fuzzy subsystem with feedback equivalent to a fuzzy subsystem without feedback. As a result, the second interim FN is transformed into a third interim FN. This third interim FN can be described by the block-scheme in Fig.8.12 and the topological expression in Eq.(8.14) from where it can be seen that the upper node is not embraced by the identity feedback connection anymore.

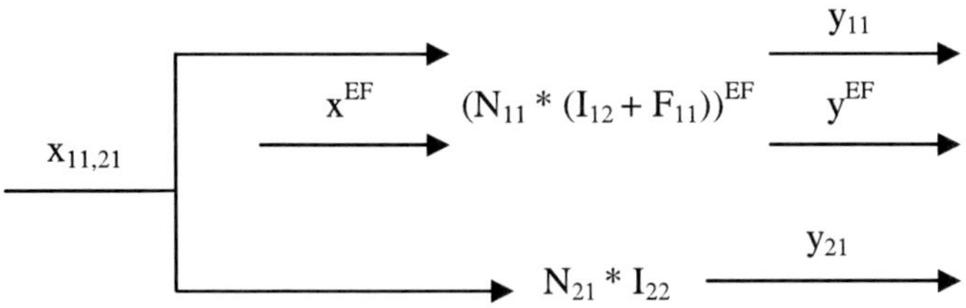

Fig. 8.12 Third interim FN for Example 8.3

$$[(N_{11} * (I_{12} + F_{11}))^{EF}]\ (x_{11,21}, x^{EF} \mid y_{11}, y^{EF})\ ;\ [N_{21} * I_{22}]\ (x_{11,21} \mid y_{21}) \quad (8.14)$$

In order to merge the outputs of nodes $(N_{11} * (I_{12} + F_{11}))^{EF}$ and $N_{21} * I_{22}$ of the third interim FN, it is necessary to augment the input $x_{11,21}$ for $N_{21} * I_{22}$ with the input x^{EF}. This augmentation operation transforms the third interim FN into a fourth interim FN with common inputs for the two nodes whereby the second node $N_{21} * I_{22}$ is transformed into a node $(N_{21} * I_{22})^{AI}$ with input set $\{x_{11,21}, x^{EF}\}$. This fourth interim FN can be described by the block-scheme in Fig.8.13 and the topological expression in Eq.(8.15) from where it can be seen that the lower node has augmented inputs.

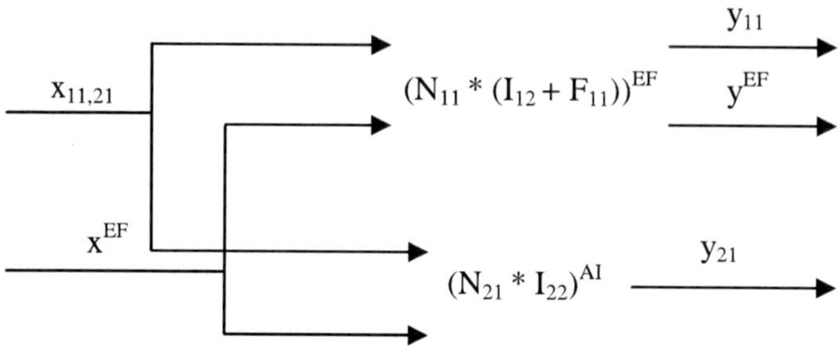

Fig. 8.13 Fourth interim FN for Example 8.3

$$[(N_{11} * (I_{12} + F_{11}))^{EF}] (x_{11,21}, x^{EF} \mid y_{11}, y^{EF}) ; [(N_{21} * I_{22})^{AI}] (x_{11,21}, x^{EF} \mid y_{21}) \tag{8.15}$$

The two composite nodes $(N_{11} * (I_{12} + F_{11}))^{EF}$ and $(N_{21} * I_{22})^{AI}$ of the fourth interim FN can be output merged into a single equivalent node $(N_{11} * (I_{12} + F_{11}))^{EF}$; $(N_{21} * I_{22})^{AI}$. As result of this merging operation, the fourth interim FN is transformed into a final FN. This final FN can be described by the block-scheme in Fig.8.14 and the topological expression in Eq.(8.16).

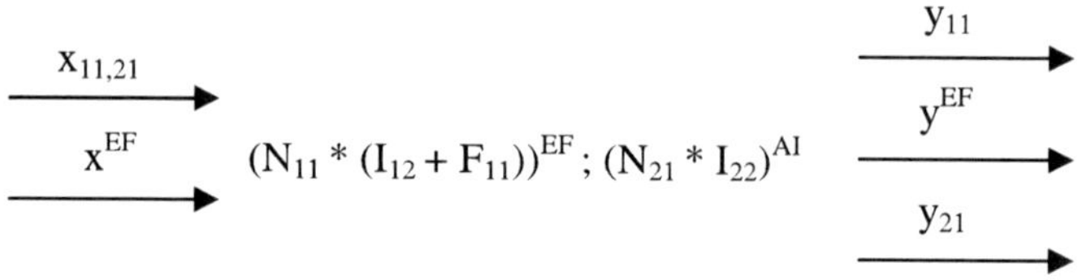

Fig. 8.14 Final FN for Example 8.3

$$[(N_{11} * (I_{12} + F_{11}))^{EF} ; (N_{21} * I_{22})^{AI}] (x_{11,21}, x^{EF} \mid y_{11}, y^{EF}, y_{12}) \tag{8.16}$$

The considerations in Example 8.3 are concerned with network analysis when all network nodes and the feedback node are known. In the case of network design, the feedback node is unknown. In this context, Algorithm 8.3 describes the process of deriving the unknown feedback node in the initial FN from Fig.8.9 when the network nodes and the single equivalent node in N_E are known. In this case, node N_E is given by the Boolean matrix equation in Eq.(8.17).

$$N_E = (N_{11} * (I_{12} + F_{11}))^{EF} ; (N_{21} * I_{22})^{AI} \tag{8.17}$$

Algorithm 8.3

1. Define N_E, N_{11}, N_{21}, I_{12} and I_{22}.
2. Find $N_{21} * I_{22}$ by horizontal merging of N_{21} and I_{22}.
3. Find $(N_{21} * I_{22})^{AI}$ by input augmentation of $N_{21} * I_{22}$.
4. Derive $N_{11} * (I_{12} + F_{11}))^{EF}$ from N_E in Eq.(8.17), if possible.
5. Confirm that $N_{11} * (I_{12} + F_{11}))^{EF}$ satisfies the feedback constraints, if possible.
6. Replace $(N_{11} * (I_{12} + F_{11}))^{EF}$ with $N_{11} * (I_{12} + F_{11})$ in Eq.(8.17).
7. Derive $I_{12} + F_{11}$ from $N_{11} * (I_{12} + F_{11})$, if possible.
8. Derive F_{11} from $I_{12} + F_{11}$, if possible.

Example 8.4

This example considers a FN with network nodes N_{11}, N_{21} and feedback node F_{21} embracing N_{21} where $x_{11,21}$ is a common input for N_{11} and N_{21}, y_{11} is an output for N_{11}, y_{21} is an output for N_{21}, v_{21} is the part of the feedback connection to F_{21} and w_{21} is the part of the feedback connection from F_{21}. This initial FN represents a stack of two fuzzy systems that can be described by the block scheme in Fig.8.15 and the topological expression in Eq.(8.18) from where it can be seen that the node in the second level is embraced by the feedback.

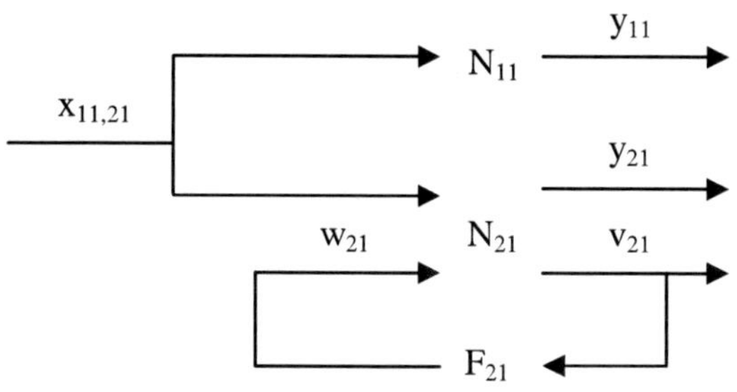

Fig. 8.15 Initial FN for Example 8.4

$$[N_{11}]\ (x_{11,21} \mid y_{11}) * [N_{21}]\ (x_{11,21}, w_{21} \mid y_{21}, v_{21}),\ [F_{21}]\ (v_{21} \mid w_{21}) \qquad (8.18)$$

The feedback node F_{21} represents a non-identity feedback connection. This is also implied by the use of different variable names for the input and the output for F_{21}, i.e. v_{21} and w_{21}. In order to apply the linguistic composition approach to the initial FN, it is necessary to introduce a virtual intermediate level just under the second level within a second layer of the underlying grid structure of the FN and to move F_{21} to this new grid cell. It is also necessary to propagate y_{11} forwards through the second layer and insert an implicit identity node I_{12} in level 1 of layer 2. Likewise, it is necessary to propagate y_{21} forwards through the second layer and insert an implicit identity node I_{22} in level 2 of layer 2.

The above movements and insertions transform the initial FN into a first interim FN whereby the non-identity feedback connection embracing the network node N_{21} is represented as a feedforward connection v_{21} between N_{21} and F_{21} and an identity feedback connection w_{21} embracing both N_{21} and F_{21}. This first interim FN can be described by the block-scheme in Fig.8.16 and the topological expression in Eq.(8.19) from where it can be seen that F_{21} is already a feedforward node alongside the other four nodes.

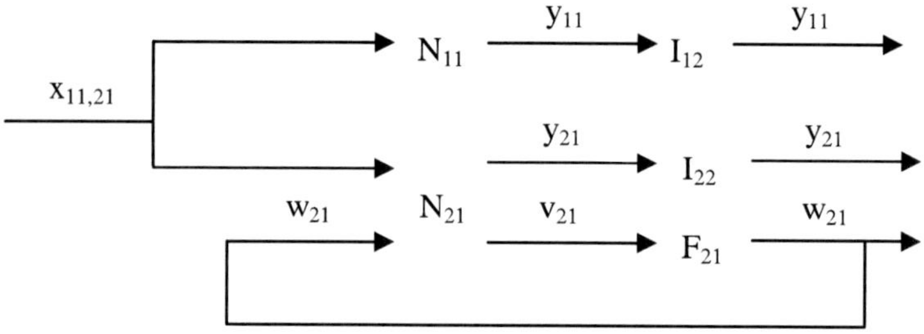

Fig. 8.16 First interim FN for Example 8.4

$$\{[N_{11}]\ (x_{11,21} \mid y_{11}) * [I_{12}]\ (y_{12} \mid y_{12})\}\ ; \qquad (8.19)$$

$$\{[N_{21}]\ (x_{11,21}, w_{21} \mid y_{21}, v_{21}) * \{[I_{22}]\ (y_{21} \mid y_{21}) + [F_{21}]\ (v_{21} \mid w_{21})\}\}$$

Nodes I_{22} and F_{21} of the first interim FN can be merged vertically into a temporary node $I_{22} + F_{21}$. This temporary node can be further merged horizontally with node N_{21} on the left. Also, nodes N_{11} and I_{12} of the same FN can be merged horizontally into another temporary node $N_{11} * I_{12}$. These merging operations transform the first interim FN into a second interim FN with two nodes whereby the horizontal merging of nodes N_{21} and $I_{22} + F_{21}$ is reflected by their replacement with node $N_{21} * (I_{22} + F_{21})$ and the horizontal merging of nodes N_{11} and I_{12} is reflected by their replacement with node $N_{11} * I_{12}$. This second interim FN can be described by the block-scheme in Fig.8.17 and the topological expression in Eq.(8.20) from where it can be seen that the lower node is embraced by the identity feedback connection.

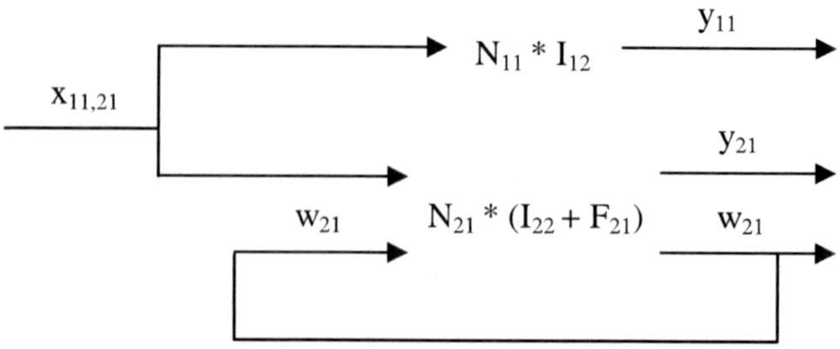

Fig. 8.17 Second interim FN for Example 8.4

$$[N_{11} * I_{12}]\ (x_{11,21} \mid y_{11})\ ;\ [N_{21} * (I_{22} + F_{21})]\ (x_{11,21}, w_{21} \mid y_{21}, w_{21}) \tag{8.20}$$

The lower node $N_{21} * (I_{22} + F_{21})$ with input set $\{x_{11,21}, w_{21}\}$ and output set $\{y_{21}, w_{21}\}$ can be further transformed into a node with equivalent feedback $(N_{21} * (I_{22} + F_{21}))^{EF}$ with input set $\{x_{11,21}, x^{EF}\}$ and output set $\{y_{21}, y^{EF}\}$. This transformation removes the identity feedback and makes the fuzzy subsystem with feedback equivalent to a fuzzy subsystem without feedback. As a result, the second interim FN is transformed into a third interim FN. This third interim FN can be described by the block-scheme in Fig.8.18 and the topological expression in Eq.(8.21) from where it can be seen that the lower node is not embraced by the identity feedback connection anymore.

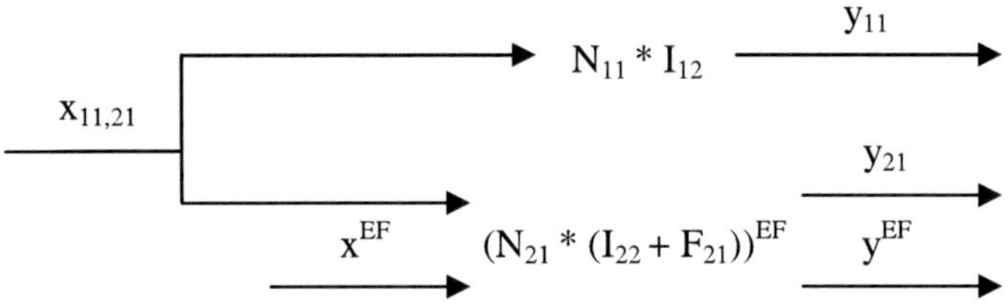

Fig. 8.18 Third interim FN for Example 8.4

$$[N_{11} * I_{12}]\ (x_{11,21} \mid y_{11})\ ;\ [(N_{21} * (I_{22} + F_{21}))^{EF}]\ (x_{11,21}, x^{EF} \mid y_{21}, y^{EF}) \tag{8.21}$$

In order to merge the outputs of nodes $N_{11} * I_{12}$ and $(N_{21} * (I_{22} + F_{21}))^{EF}$ of the third interim FN, it is necessary to augment the input $x_{11,21}$ for $N_{11} * I_{12}$ with the input x^{EF}. This augmentation operation transforms the third interim FN into a fourth interim FN with common inputs for the two nodes whereby the first node $N_{11} * I_{12}$ is transformed into a node $(N_{11} * I_{12})^{AI}$ with input set $\{x_{11,21}, x^{EF}\}$. This fourth interim FN can be described by the block-scheme in Fig.8.19 and the topological expression in Eq.(8.22) from where it can be seen that the upper node has augmented inputs.

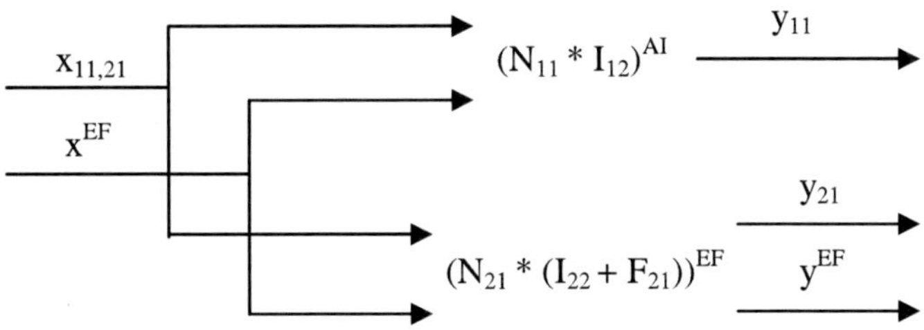

Fig. 8.19 Fourth interim FN for Example 8.4

$$[(N_{11} * I_{12})^{AI}] (x_{11,21}, x^{EF} \mid y_{11}) ; [(N_{21} * (I_{22} + F_{21}))^{EF}] (x_{11,21}, x^{EF} \mid y_{21}, y^{EF}) \quad (8.22)$$

The two nodes $(N_{11} * I_{12})^{AI}$ and $(N_{21} * (I_{22} + F_{21}))^{EF}$ of the fourth interim FN can be output merged into a single equivalent node $(N_{11} * I_{12})^{AI}$; $(N_{21} * (I_{22} + F_{21}))^{EF}$. As result of this merging operation, the fourth interim FN is transformed into a final FN. This final FN can be described by the block-scheme in Fig.8.20 and the topological expression in Eq.(8.23).

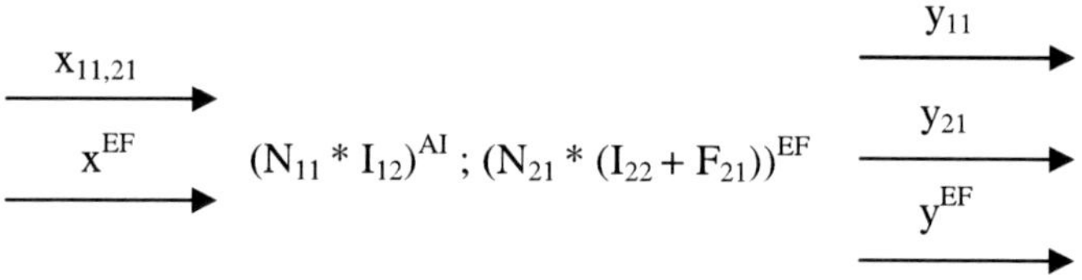

Fig. 8.20 Final FN for Example 8.4

$$[(N_{11} * I_{12})^{AI} ; (N_{21} * (I_{22} + F_{21}))^{EF}] (x_{11,21}, x^{EF} \mid y_{11}, y_{21}, y^{EF}) \quad (8.23)$$

The considerations in Example 8.4 are concerned with network analysis when all network nodes and the feedback node are known. In the case of network design, the feedback node is unknown. In this context, Algorithm 8.4 describes the process of deriving the unknown feedback node in the initial FN from Fig.8.15 when the network nodes and the single equivalent node N_E are known. In this case, node N_E is given by the Boolean matrix equation in Eq.(8.24).

$$N_E = (N_{11} * I_{12})^{AI} ; (N_{21} * (I_{22} + F_{21}))^{EF} \quad (8.24)$$

Algorithm 8.4

1. Define N_E, N_{11}, N_{21}, I_{12} and I_{22}.
2. Find $N_{11} * I_{12}$ by horizontal merging of N_{11} and I_{12}.
3. Find $(N_{11} * I_{12})^{AI}$ by input augmentation of $N_{11} * I_{12}$.
4. Derive $(N_{21} * (I_{22} + F_{21}))^{EF}$ from N_E in Eq.(8.24), if possible.
5. Confirm that $(N_{21} * (I_{22} + F_{21}))^{EF}$ satisfies the feedback constraints, if possible.
6. Replace $(N_{21} * (I_{22} + F_{21}))^{EF}$ with $N_{21} * (I_{22} + F_{21})$ in Eq.(8.24).
7. Derive $I_{22} + F_{21}$ from $N_{21} * (I_{22} + F_{21})$, if possible.
8. Derive F_{21} from $I_{22} + F_{21}$, if possible.

8.3 Networks with Multiple Local Feedback

A more complex type of FN is the one with multiple local feedback. This network has at least two nodes embraced by separate feedback connections with feedback nodes in each connection. In this case, the feedback is multiple as it appears more than once but it is also local as it embraces only one node. There may be an arbitrary number of feedforward connections between this node and any other nodes as well as between any pair of other nodes. However, the presence of any feedforward connections does not remove the feedback characteristics of this type of FN due to the presence of the feedback connections with the feedback nodes.

Example 8.5

This example considers a FN with network nodes N_{11}, N_{12}, feedback node F_{11} embracing N_{11} and feedback node F_{12} embracing N_{12} where x_{11} is an input for N_{11}, y_{12} is an output for N_{12}, $z_{11,12}^{1,1}$ is the feedforward connection from the first output for N_{11} to the first input for N_{12}, v_{11} is the part of the feedback connection to F_{11}, w_{11} is the part of the feedback connection from $F_{11,}$ v_{12} the is the part of the feedback connection to F_{12} and w_{12} the is the part of the feedback connection from F_{12}. This initial FN represents a queue of two fuzzy systems that can be described by the block scheme in Fig.8.21 and the topological expression in Eq.(8.25) from where it can be seen that the node in each of the two layers is embraced by a separate feedback.

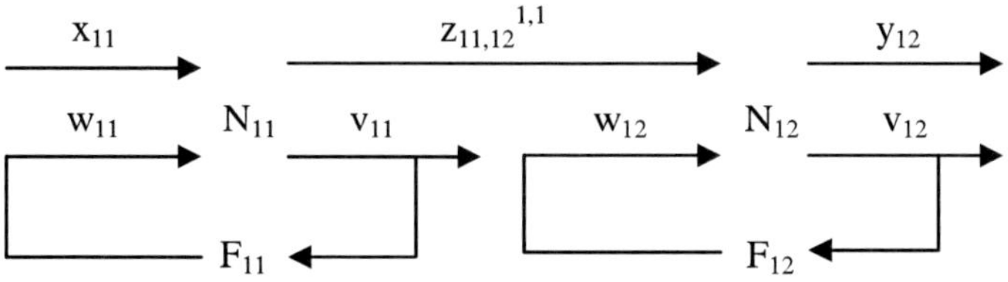

Fig. 8.21 Initial FN for Example 8.5

$$[N_{11}]\ (x_{11}, w_{11} \mid z_{11,12}^{1,1}, v_{11}) * [N_{12}]\ (z_{11,12}^{1,1}, w_{12} \mid y_{12}, v_{12}), \tag{8.25}$$

$$[F_{11}]\ (v_{11} \mid w_{11}),\ [F_{12}]\ (v_{12} \mid w_{12})$$

The feedback nodes F_{11} and F_{12} represent non-identity feedback connections. This is also implied by the use of different variable names for the inputs and the outputs for F_{11} and F_{12}, i.e. v_{11}, w_{11} and v_{12}, w_{12}. In order to apply the linguistic composition approach to the initial FN, it is necessary to introduce a second level within a third layer of the underlying grid structure of the FN and to move F_{12} to this new grid cell. It is also necessary to propagate y_{12} forwards through the third layer and insert an implicit identity node I_{13} in level 1 of layer 3. Further on, it is necessary to introduce a third level within the second layer of the underlying grid structure of the FN and to move F_{11} to this new grid cell. It is also necessary to propagate y_{12} forwards through the third layer and insert an implicit identity node I_{33} in level 3 of layer 3. Likewise, it is necessary to propagate w_{12} backwards through the first layer and insert an implicit identity node I_{31} in level 3 of layer 1.

The above movements and insertions transform the initial FN into a first interim FN whereby the non-identity feedback connection embracing the network node N_{11} is represented as a feedforward connection v_{11} between N_{11} and F_{11} and an identity feedback connection w_{11} embracing N_{11}, F_{11} and I_{33}. Also, the non-identity feedback connection embracing the network node N_{12} is represented as a feedforward connection v_{12} between N_{12} and F_{12} and an identity feedback connection w_{12} embracing I_{31}, N_{12} and F_{12}. This first interim FN can be described by the block-scheme in Fig.8.22 and the topological expression in Eq.(8.6) from where it can be seen that both F_{11} and F_{12} are already feedforward nodes alongside the other five nodes.

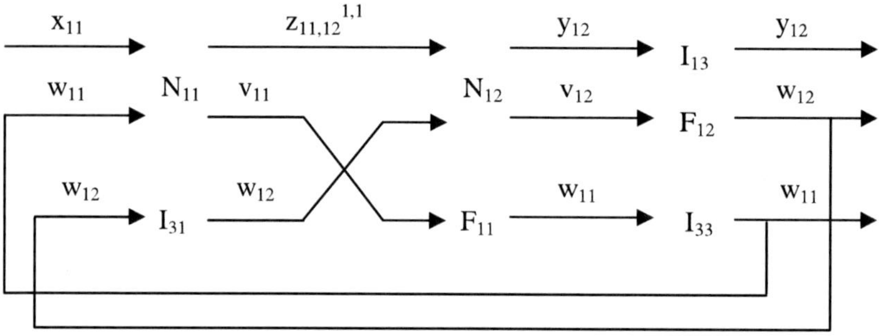

Fig. 8.22 First interim FN for Example 8.5

$$\{[N_{11}]\,(x_{11}, w_{11} \,|\, z_{11,12}^{1,1}, v_{11}) + [I_{31}]\,(w_{12} \,|\, w_{12})\} \;* \tag{8.26}$$

$$\{[N_{12}]\,(z_{11,12}^{1,1}, w_{12} \,|\, y_{12}, v_{12}) + [F_{11}]\,(v_{11} \,|\, w_{11})\} \;*$$

$$\{[I_{13}]\,(y_{12} \,|\, y_{12}) + [F_{12}]\,(v_{12} \,|\, w_{12}) + [I_{33}]\,(w_{11} \,|\, w_{11})\}$$

Nodes N_{11} and I_{31} of the first interim FN can be merged vertically into a temporary node $N_{11} + I_{31}$. Similarly, nodes N_{12} and F_{11} of the same interim FN can be merged vertically into a second temporary node $N_{12} + F_{11}$ whereas nodes I_{13}, F_{12} and I_{33} can be merged vertically into a third temporary node $I_{13} + F_{12} + I_{33}$. This second interim FN can be described by the block-scheme in Fig.8.23 and the topological expression in Eq.(8.27) from where it can be seen that the three nodes are embraced by the identity feedback connections and some of the feedforward connections between the first and the second node have crossing paths.

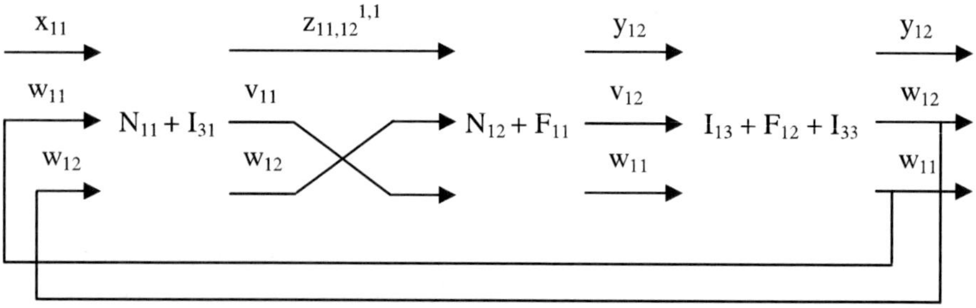

Fig. 8.23 Second interim FN for Example 8.5

$$[N_{11} + I_{31}]\ (x_{11}, w_{11}, w_{12} \mid z_{11,12}^{1,1}, v_{11}, w_{12})\ * \tag{8.27}$$

$$[N_{12} + F_{11}]\ (z_{11,12}^{1,1}, w_{12}, v_{11} \mid y_{12}, v_{12}, w_{11})\ *$$

$$[I_{13} + F_{12} + I_{33}]\ (y_{12}, v_{12}, w_{11} \mid y_{12}, w_{12}, w_{11})$$

In order to merge horizontally nodes $N_{11} + I_{31}$ and $N_{12} + F_{11}$ of the second interim FN, it is necessary to permute the outputs input v_{11} and w_{12} for $N_{11} + I_{31}$. This permutation operation transforms the second interim FN into a third interim FN whereby the first node $N_{11} + I_{31}$ is transformed into a node $(N_{11} + I_{31})^{PO}$. This third interim FN can be described by the block-scheme in Fig.8.24 and the topological expression in Eq.(8.28) from where it can be seen that all feedforward connections already have parallel paths.

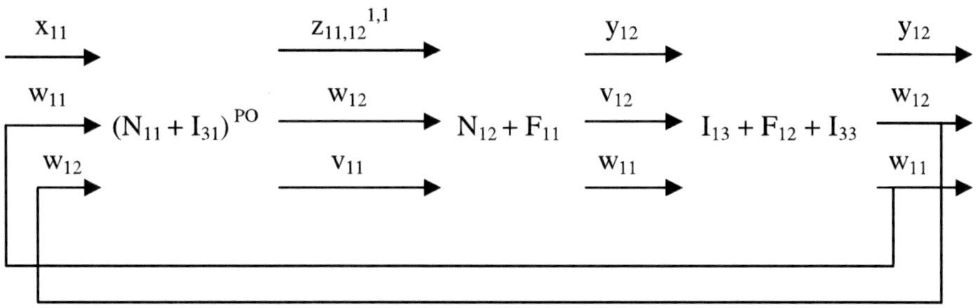

Fig. 8.24 Third interim FN for Example 8.5

$$[(N_{11} + I_{31})^{PO}]\ (x_{11}, w_{11}, w_{12} \mid z_{11,12}^{1,1}, w_{12}, v_{11})\ * \tag{8.28}$$

$$[N_{12} + F_{11}]\ (z_{11,12}^{1,1}, w_{12}, v_{11} \mid y_{12}, v_{12}, w_{11})\ *$$

$$[I_{13} + F_{12} + I_{33}]\ (y_{12}, v_{12}, w_{11} \mid y_{12}, w_{12}, w_{11})$$

The three composite nodes $(N_{11} + I_{31})^{PO}$, $N_{12} + F_{11}$ and $I_{13} + F_{12} + I_{33}$ of the third interim FN can be merged horizontally into a single equivalent node $(N_{11} + I_{31})^{PO} * (N_{12} + F_{11}) * (I_{13} + F_{12} + I_{33})$. As result of this merging operation, the third interim FN is transformed into a fourth interim FN. This fourth interim FN can be described by the block-scheme in Fig.8.25 and the topological expression in Eq.(8.29) from where it can be seen that the single equivalent node is embraced by the identity feedback connections.

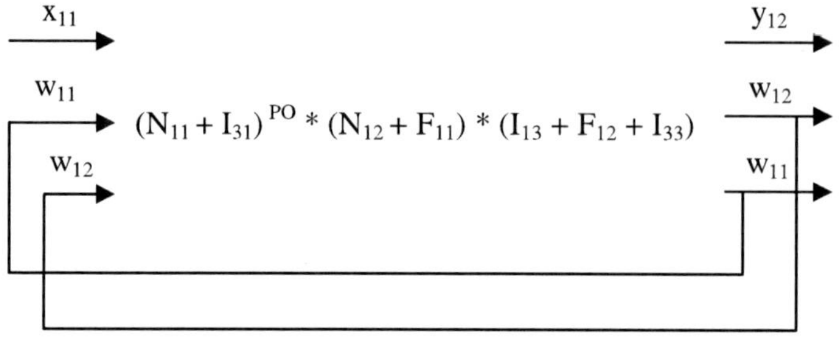

Fig. 8.25 Fourth interim FN for Example 8.5

$$[(N_{11} + I_{31})^{PO} * (N_{12} + F_{11}) * (I_{13} + F_{12} + I_{33})] \; (x_{11}, w_{11}, w_{12} \mid y_{12}, w_{12}, w_{11}) \tag{8.29}$$

The node $(N_{11} + I_{31})^{PO} * (N_{12} + F_{11}) * (I_{13} + F_{12} + I_{33})$ with input set $\{x_{11}, w_{11}, w_{12}\}$ and output set $\{y_{12}, w_{12}, w_{11}\}$ can be further transformed into a node with equivalent feedback $((N_{11} + I_{31})^{PO} * (N_{12} + F_{11}) * (I_{13} + F_{12} + I_{33}))^{FE}$ with input set $\{x_{11}, x_1^{FE}, x_2^{FE}\}$ and output set $\{y_{12}, y_2^{FE}, y_1^{FE}\}$.This transformation removes the two identity feedbacks and makes the fuzzy system with feedback equivalent to a fuzzy subsystem without feedback. As a result, the fourth interim FN is transformed into a final FN. This third interim FN can be described by the block-scheme in Fig.8.26 and the topological expression in Eq.(8.30) from where it can be seen that the single equivalent node is not embraced by the identity feedback connections anymore.

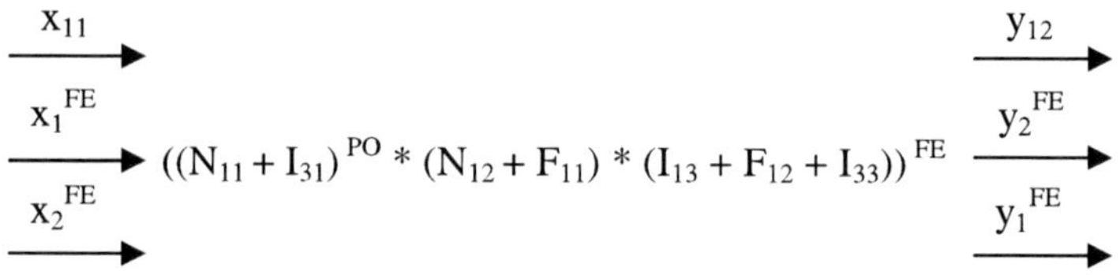

Fig. 8.26 Final FN for Example 8.5

$$[((N_{11} + I_{31})^{PO} * (N_{12} + F_{11}) * (I_{13} + F_{12} + I_{33}))^{FE}] \tag{8.30}$$

$$(x_{11}, x_1^{FE}, x_2^{FE} \mid y_{12}, y_2^{FE}, y_1^{FE})$$

The considerations in Example 8.5 are concerned with network analysis when all network and feedback nodes are known. In the case of network design, at least one feedback node is unknown. In this context, Algorithms 8.5-8.6 describe the process of deriving an unknown feedback node in the initial FN from Fig.8.21 when the network nodes, one of the feedback nodes and the single equivalent node N_E are known. In this case, node N_E is given by the Boolean matrix equation in Eq.(8.31).

$$N_E = ((N_{11} + I_{31})^{PO} * (N_{12} + F_{11}) * (I_{13} + F_{12} + I_{33}))^{FE} \tag{8.31}$$

Algorithm 8.5

1. Define N_E, N_{11}, N_{12}, I_{31}, I_{13}, I_{33} and F_{12}.
2. Confirm that N_E satisfies the feedback constraints, if possible.
3. Make N_E equal to $(N_{11} + I_{31})^{PO} * (N_{12} + F_{11}) * (I_{13} + F_{12} + I_{33})$ in Eq.(8.31).
4. Find $N_{11} + I_{31}$ by vertical merging of N_{11} and I_{31}.
5. Find $(N_{11} + I_{31})^{PO}$ by output permutation of $N_{11} + I_{31}$.
6. Find $I_{13} + F_{12} + I_{33}$ by vertical merging of I_{13}, F_{12} and I_{33}.
7. Derive $N_{12} + F_{11}$ from N_E in Eq.(8.31), if possible.
8. Derive F_{11} from $N_{12} + F_{11}$, if possible.

Algorithm 8.6

1. Define N_E, N_{11}, N_{12}, I_{31}, I_{13}, I_{33} and F_{11}.
2. Confirm that N_E satisfies the feedback constraints, if possible.
3. Make N_E equal to $(N_{11} + I_{31})^{PO} * (N_{12} + F_{11}) * (I_{13} + F_{12} + I_{33})$ in Eq.(8.31).
4. Find $N_{11} + I_{31}$ by vertical merging of N_{11} and I_{31}.
5. Find $(N_{11} + I_{31})^{PO}$ by output permutation of $N_{11} + I_{31}$.
6. Find $N_{12} + F_{11}$ by vertical merging of N_{12} and F_{11}.
7. Find $(N_{11} + I_{31})^{PO} * (N_{12} + F_{11})$ by horizontal merging of $(N_{11} + I_{31})^{PO}$ and $(N_{12} + F_{11})$.
8. Derive $I_{13} + F_{12} + I_{33}$ from N_E in Eq.(8.31), if possible.
9. Derive F_{12} from $I_{13} + F_{12} + I_{33}$, if possible.

Example 8.6

This example considers a FN with network nodes N_{11}, N_{21}, feedback node F_{11} embracing N_{11} and feedback node F_{21} embracing N_{21} where $x_{11,21}$ is a common input for N_{11} and N_{21}, y_{11} is an output for N_{11}, y_{21} is an output for N_{21}, v_{11} is the part of the feedback connection to F_{11}, w_{11} is the part of the feedback connection from F_{11}, v_{21} is the part of the feedback connection to F_{21} and w_{21} the is the part of the feedback connection from F_{21}. This initial FN represents a stack of two fuzzy systems that can be described by the block scheme in Fig.8.27 and the topological expression in Eq.(8.32) from where it can be seen that the node in each of the two levels is embraced by a separate feedback.

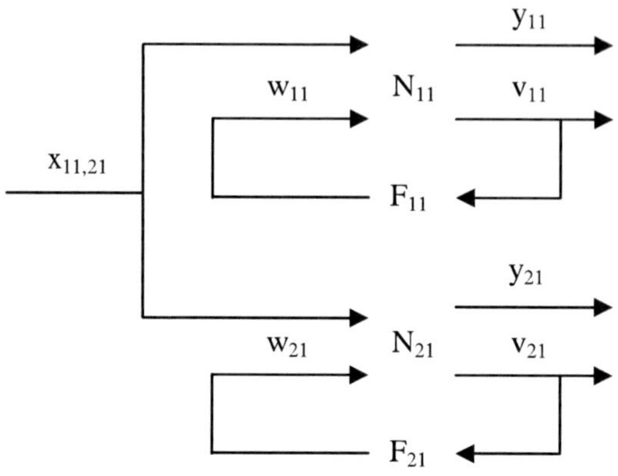

Fig. 8.27 Initial FN for Example 8.6

$$[N_{11}]\ (x_{11,21}, w_{11} \mid y_{11}, v_{11})\ ;\ [N_{21}]\ (x_{11,21}, w_{21} \mid y_{21}, v_{21}), \tag{8.32}$$

$$[F_{11}]\ (v_{11} \mid w_{11}),\ [F_{21}]\ (v_{21} \mid w_{21})$$

The feedback nodes F_{11} and F_{21} represent non-identity feedback connections. This is also implied by the use of different variable names for the inputs and the outputs for F_{11} and F_{21}, i.e. v_{11}, w_{11} and v_{21}, w_{21}. In order to apply the linguistic composition approach to the initial FN, it is necessary to introduce a virtual intermediate level just under the first level within a second layer of the underlying grid structure of the FN and to move F_{11} to this new grid cell as well as to introduce a virtual intermediate level just under the second level within the second layer and to move F_{21} to this new grid cell. Besides this, it is necessary to propagate y_{11} forwards through the second layer and insert an implicit identity node I_{12} in level 1 of layer 2 as well as to propagate y_{21} forwards through the second layer and insert an implicit identity node I_{22} in level 2 of layer 2.

The above movements and insertions transform the initial FN into a first interim FN. In this case, the non-identity feedback connection embracing the network node N_{11} is represented as a feedforward connection v_{11} between N_{11} and F_{11} and an identity feedback connection w_{11} embracing both N_{11} and F_{11} whereas the non-identity feedback connection embracing the network node N_{21} is represented as a feedforward connection v_{21} between N_{21} and F_{21} and an identity feedback connection w_{21} embracing both N_{21} and F_{21}. This first interim FN can be described by the block-scheme in Fig.8.28 and the topological expression in Eq.(8.33) from where it can be seen that both F_{11} and F_{21} are already feedforward nodes alongside the other four nodes.

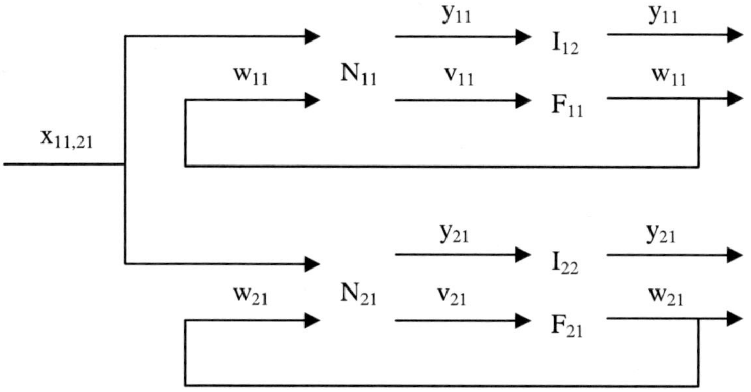

Fig. 8.28 First interim FN for Example 8.6

$$\{[N_{11}]\ (x_{11,21}, w_{11} \mid y_{11}, v_{11}) * \{[I_{12}]\ (y_{11} \mid y_{11}) + [F_{11}]\ (v_{11} \mid w_{11})\}\}\ ; \qquad (8.33)$$

$$\{[N_{21}]\ (x_{11,21}, w_{21} \mid y_{21}, v_{21}) * \{[I_{22}]\ (y_{21} \mid y_{21}) + [F_{21}]\ (v_{21} \mid w_{21})\}\}$$

Nodes I_{12} and F_{11} of the first interim FN can be merged vertically into a temporary node $I_{12} + F_{11}$. This temporary node can be further merged horizontally with node N_{11} on the left. Also, nodes I_{22} and F_{21} of the same FN can be merged vertically into another temporary node $I_{22} + F_{21}$. This temporary node can be further merged horizontally with node N_{21} on the left. These merging operations transform the first interim FN into a second interim FN with two nodes whereby the horizontal merging of nodes N_{11} and $I_{12} + F_{11}$ is reflected by their replacement with node $N_{11} * (I_{12} + F_{11})$ and the horizontal merging of nodes N_{21} and $I_{22} + F_{21}$ is reflected by their replacement with node $N_{21} * (I_{22} + F_{21})$. This second interim FN can be described by the block-scheme in Fig.8.29 and the topological expression in Eq.(8.34) from where it can be seen that each of the two nodes is embraced by separate identity feedback connections.

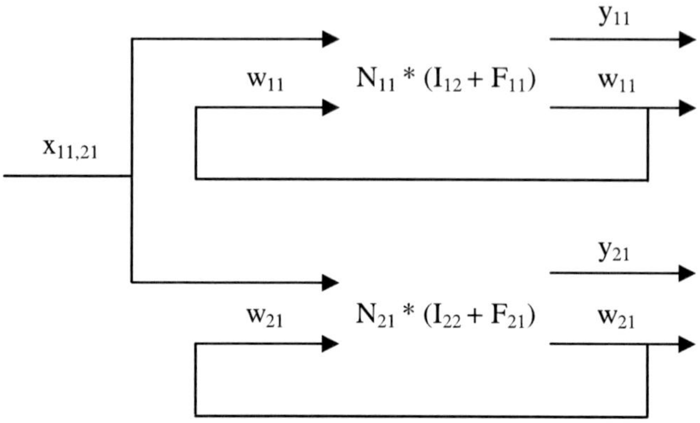

Fig. 8.29 Second interim FN for Example 8.6

$$[N_{11} * (I_{12} + F_{11})]\ (x_{11,21}, w_{11} \mid y_{11}, w_{11})\ ; \qquad (8.34)$$

$$[N_{21} * (I_{22} + F_{21})]\ (x_{11,21}, w_{21} \mid y_{21}, w_{21})$$

The upper node $N_{11} * (I_{12} + F_{11})$ with input set $\{x_{11,21}, w_{11}\}$ and output set $\{y_{11}, w_{11}\}$ can be further transformed into a node with equivalent feedback $(N_{11} * (I_{12} + F_{11}))^{EF}$ with input set $\{x_{11,21}, x_1^{EF}\}$ and output set $\{y_{11}, y_1^{EF}\}$. Similarly, the lower node $N_{21} * (I_{22} + F_{21})$ with input set $\{x_{11,21}, w_{21}\}$ and output set $\{y_{21}, w_{21}\}$ can be further transformed into a node with equivalent feedback $(N_{21} * (I_{22} + F_{21}))^{EF}$ with input set $\{x_{11,21}, x_2^{EF}\}$ and output set $\{y_{21}, y_2^{EF}\}$. These transformations remove the identity feedback and make the fuzzy subsystems with feedback equivalent to fuzzy subsystems without feedback. As a result, the second interim FN is transformed into a third interim FN. This third interim FN can be described by the block-scheme in Fig.8.30 and the topological expression in Eq.(8.35) from where it can be seen that the two nodes are not embraced by the identity feedback connections anymore.

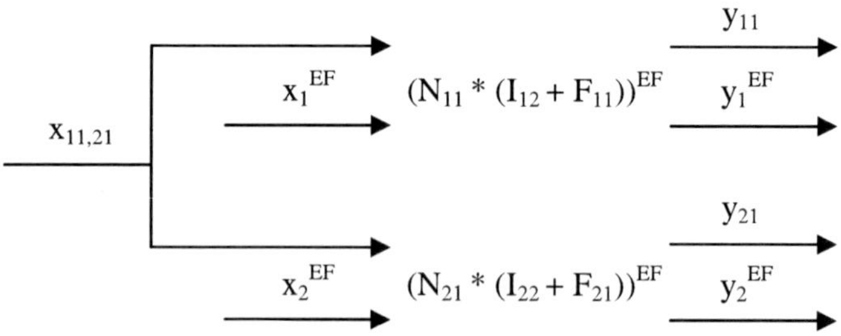

Fig. 8.30 Third interim FN for Example 8.6

$$[(N_{11} * (I_{12} + F_{11}))^{EF}]\ (x_{11,21}, x_1^{EF} \mid y_{11}, y_1^{EF})\ ; \qquad (8.35)$$

$$[(N_{21} * (I_{22} + F_{21}))^{EF}]\ (x_{11,21}, x_2^{EF} \mid y_{21}, y_2^{EF})$$

In order to merge the outputs of nodes $(N_{11} * (I_{12} + F_{11}))^{EF}$ and $(N_{21} * (I_{22} + F_{21}))^{EF}$ of the third interim FN, it is necessary to augment the input $x_{11,21}$ with the inputs x_1^{EF} and x_2^{EF}. This augmentation operation transforms the third interim FN into a fourth interim FN with common inputs for the two nodes whereby the upper node $(N_{11} * (I_{12} + F_{11}))^{EF}$ is transformed into a node $((N_{11} * (I_{12} + F_{11}))^{EF})^{AI}$ with input set $\{x_{11,21}, x_1^{EF}, x_2^{EF}\}$ and the lower node $(N_{21} * (I_{22} + F_{21}))^{EF}$ is transformed into a node $((N_{21} * (I_{22} + F_{21}))^{EF})^{AI}$ with the same input set. This fourth interim FN can be described by the block-scheme in Fig.8.31 and the topological expression in Eq.(8.36) from where it can be seen that the two nodes have augmented inputs.

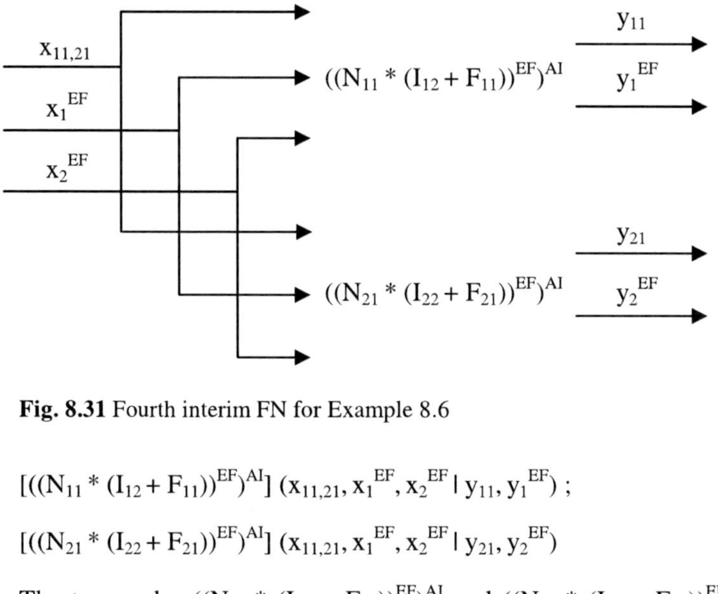

Fig. 8.31 Fourth interim FN for Example 8.6

$$[((N_{11} * (I_{12} + F_{11}))^{EF})^{AI}]\ (x_{11,21}, x_1^{EF}, x_2^{EF} \mid y_{11}, y_1^{EF})\ ; \qquad (8.36)$$

$$[((N_{21} * (I_{22} + F_{21}))^{EF})^{AI}]\ (x_{11,21}, x_1^{EF}, x_2^{EF} \mid y_{21}, y_2^{EF})$$

The two nodes $((N_{11} * (I_{12} + F_{11}))^{EF})^{AI}$ and $((N_{21} * (I_{22} + F_{21}))^{EF})^{AI}$ of the fourth interim FN can be output merged into a single equivalent node $((N_{11} * (I_{12} + F_{11}))^{EF})^{AI}$; $((N_{21} * (I_{22} + F_{21}))^{EF})^{AI}$. As result of this merging operation, the fourth interim FN is transformed into a final FN. This final FN can be described by the block-scheme in Fig.8.32 and the topological expression in Eq.(8.37).

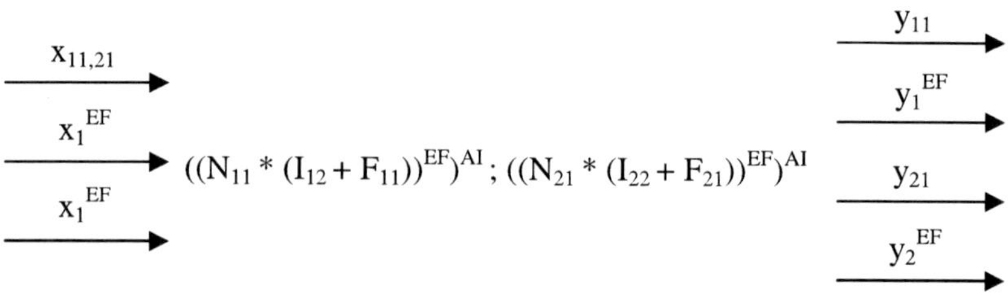

Fig. 8.32 Final FN for Example 8.6

$$[((N_{11} * (I_{12} + F_{11}))^{EF})^{AI}; ((N_{21} * (I_{22} + F_{21}))^{EF})^{AI}] \tag{8.37}$$

$$(x_{11,21}, x_1^{EF}, x_2^{EF} \mid y_{11}, y_1^{EF}, y_2^{EF})$$

The considerations in Example 8.6 are concerned with network analysis when all network and feedback nodes are known. In the case of network design, at least one feedback node is unknown. In this context, Algorithms 8.7-8.8 describe the process of deriving an unknown feedback node in the initial FN from Fig.8.27 when the network nodes, one of the feedback nodes and the single equivalent node N_E are known. In this case, node N_E is given by the Boolean matrix equation in Eq.(8.38).

$$N_E = ((N_{11} * (I_{12} + F_{11}))^{EF})^{AI}; ((N_{21} * (I_{22} + F_{21}))^{EF})^{AI} \tag{8.38}$$

Algorithm 8.7

1. Define N_E, N_{11}, N_{21}, I_{12}, I_{22} and F_{21}.
2. Find $I_{22} + F_{21}$ by vertical merging of I_{22} and F_{21}.
3. Find $N_{21} * (I_{22} + F_{21})$ by horizontal merging of N_{21} and $I_{22} + F_{21}$.
4. Confirm that $(N_{21} * (I_{22} + F_{21}))^{EF}$ satisfies the feedback constraints, if possible.
5. Replace $(N_{21} * (I_{22} + F_{21}))^{EF}$ with $N_{21} * (I_{22} + F_{21})$ in Eq.(8.38).
6. Find $(N_{21} * (I_{22} + F_{21}))^{AI}$ by input augmentation of $N_{21} * (I_{22} + F_{21})$.
7. Derive $((N_{11} * (I_{12} + F_{11}))^{EF})^{AI}$ from N_E in Eq.(8.38), if possible.
8. Find $(N_{11} * (I_{12} + F_{11}))^{EF}$ by inverse input augmentation of $((N_{11} * (I_{12} + F_{11}))^{EF})^{AI}$.
9. Confirm that $(N_{11} * (I_{12} + F_{11}))^{EF}$ satisfies the feedback constraints, if possible.
10. Replace $(N_{11} * (I_{12} + F_{11}))^{EF}$ with $N_{11} * (I_{12} + F_{11})$ in Eq.(8.38).
11. Derive $I_{12} + F_{11}$ from $N_{11} * (I_{12} + F_{11})$, if possible.
12. Derive F_{11} from $I_{12} + F_{11}$, if possible.

Algorithm 8.8

1. Define N_E, N_{11}, N_{21}, I_{12}, I_{22} and F_{11}.
2. Find $I_{12} + F_{11}$ by vertical merging of I_{12} and F_{11}.
3. Find $N_{11} * (I_{12} + F_{11})$ by horizontal merging of N_{11} and $I_{12} + F_{11}$.
4. Confirm that $(N_{11} * (I_{12} + F_{11}))^{EF}$ satisfies the feedback constraints, if possible.
5. Replace $(N_{11} * (I_{12} + F_{11}))^{EF}$ with $N_{11} * (I_{12} + F_{11})$ in Eq.(8.38).
6. Find $(N_{11} * (I_{12} + F_{11}))^{AI}$ by input augmentation of $N_{11} * (I_{12} + F_{11})$.
7. Derive $((N_{21} * (I_{22} + F_{21}))^{EF})^{AI}$ from N_E in Eq.(8.38), if possible.
8. Find $(N_{21} * (I_{22} + F_{21}))^{EF}$ by inverse input augmentation of $((N_{21} * (I_{22} + F_{21}))^{EF})^{AI}$.
9. Confirm that $(N_{21} * (I_{22} + F_{21}))^{EF}$ satisfies the feedback constraints, if possible.
10. Replace $(N_{21} * (I_{22} + F_{21}))^{EF}$ with $N_{21} * (I_{22} + F_{21})$ in Eq.(8.38).
11. Derive $I_{22} + F_{21}$ from $N_{21} * (I_{22} + F_{21})$, if possible.
12. Derive F_{21} from $I_{22} + F_{21}$, if possible.

8.4 Networks with Single Global Feedback

Another fairly simple type of FN is the one with single global feedback. This network has at least two nodes embraced by a feedback connection with a single feedback node in this connection. In this case, the feedback is single as it appears only once but it is also global as it embraces more than one node. There may be an arbitrary number of feedforward connections between these nodes and any other nodes as well as between any pair of other nodes. However, the presence of any feedforward connections does not remove the feedback characteristics of this type of FN due to the presence of the feedback connections with the feedback node.

Example 8.7

This example considers a FN with network nodes N_{11}, N_{12} and feedback node $F_{12,11}$ embracing N_{11} and N_{12} where x_{11} is an input for N_{11}, y_{12} is an output for N_{12}, $z_{11,12}^{1,1}$ is the feedforward connection from the first and only output for N_{11} to the first and only input for N_{12}, $v_{12,11}$ is the part of the feedback connection to $F_{12,11}$ and $w_{12,11}$ is the part of the feedback connection from $F_{12,11}$. This initial FN represents a queue of two fuzzy systems that can be described by the block scheme in Fig.8.33 and the topological expression in Eq.(8.39) from where it can be seen that the two nodes in the first and only level are embraced by the feedback.

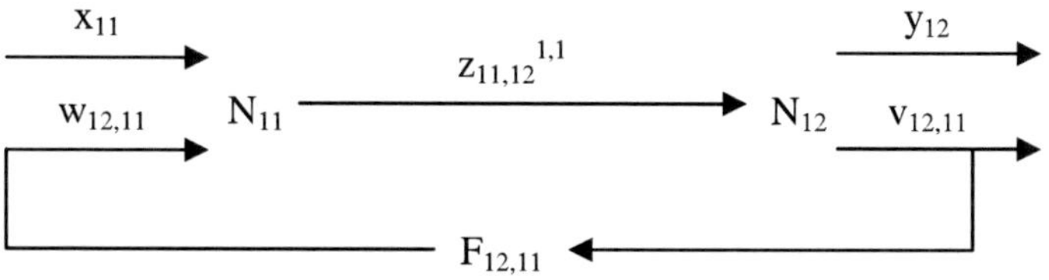

Fig. 8.33 Initial FN for Example 8.7

$$[N_{11}]\,(x_{11}, w_{12,11} \mid z_{11,12}^{\;1,1}) * [N_{12}]\,(z_{11,12}^{\;1,1} \mid y_{12}, v_{12,11}), [F_{12,11}]\,(v_{12,11} \mid w_{12,11}) \quad (8.39)$$

The feedback node $F_{12,11}$ represents a non-identity feedback connection. This is also implied by the use of different variable names for the input and the output for $F_{12,11}$, i.e. $v_{12,11}$ and $w_{12,11}$. In order to apply the linguistic composition approach to the initial FN, it is necessary to introduce a second level within a third layer of the underlying grid structure of the FN and to move $F_{12,11}$ to this new grid cell. It is also necessary to propagate y_{12} forwards through the third layer and insert an implicit identity node I_{13} in level 1 of layer 3.

The above movements and insertions transform the initial FN into a first interim FN whereby the non-identity feedback connection embracing the network nodes N_{11} and N_{12} is represented as a feedforward connection $v_{12,11}$ between N_{12} and $F_{12,11}$ and an identity feedback connection $w_{12,11}$ embracing N_{11}, N_{12} and $F_{12,11}$. This first interim FN can be described by the block-scheme in Fig.8.34 and the topological expression in Eq.(8.40) from where it can be seen that $F_{12,11}$ is already a feedforward node alongside the other three nodes.

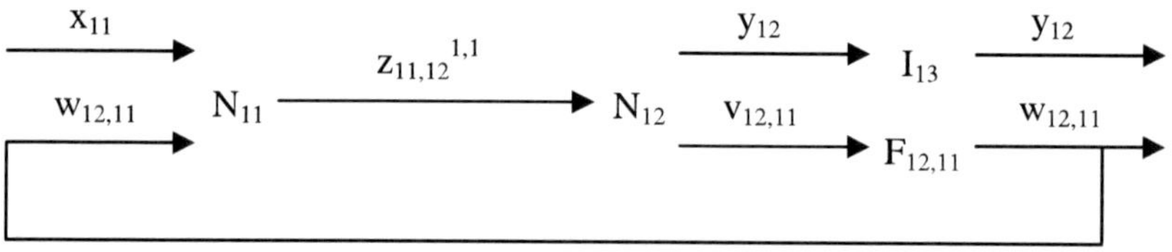

Fig. 8.34 First interim FN for Example 8.7

$$[N_{11}]\,(x_{11}, w_{12,11} \mid z_{11,12}^{\;1,1}) * [N_{12}]\,(z_{11,12}^{\;1,1} \mid y_{12}, v_{12,11}) * \quad (8.40)$$
$$\{[I_{13}]\,(y_{12} \mid y_{12}) + [F_{12,11}]\,(v_{12,11} \mid w_{12,11})\}$$

Nodes I_{13} and $F_{12,11}$ of the first interim FN can be merged vertically into a temporary node $I_{13} + F_{12,11}$. This temporary node can be further merged horizontally with nodes N_{11} and N_{12} on the left. These merging operations transform the first interim FN into a second interim FN with a single equivalent node whereby the horizontal merging of nodes N_{11}, N_{12} and $I_{13} + F_{12,11}$ is reflected by their replacement with node $N_{11} * N_{12} * (I_{13} + F_{12,11})$. This second interim FN can be described by the block-scheme in Fig.8.35 and the topological expression in Eq.(8.41) from where it can be seen that the single equivalent node is embraced by the identity feedback connection.

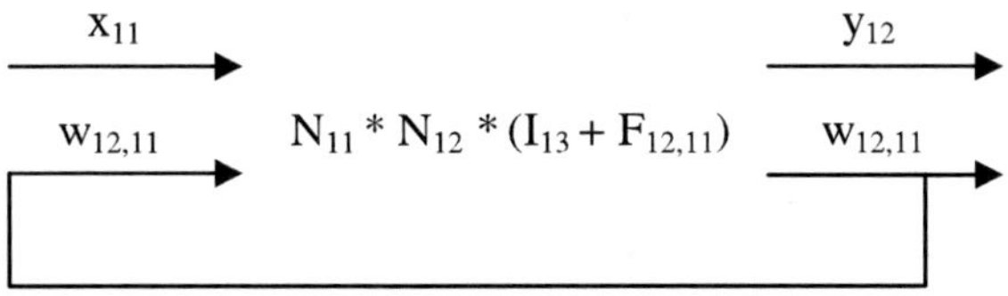

Fig. 8.35 Second interim FN for Example 8.7

$$[N_{11} * N_{12} * (I_{13} + F_{12,11})]\ (x_{11}, w_{12,11} \mid y_{12}, w_{12,11}) \qquad (8.41)$$

The node $N_{11} * N_{12} * (I_{13} + F_{12,11})$ with input set $\{x_{11}, w_{12,11}\}$ and output set $\{y_{12}, w_{12,11}\}$ can be further transformed into a node with equivalent feedback $(N_{11} * N_{12} * (I_{13} + F_{12,11}))^{EF}$ with input set $\{x_{11}, x^{EF}\}$ and output set $\{y_{12}, y^{EF}\}$. This transformation removes the identity feedback and makes the fuzzy system with feedback equivalent to a fuzzy system without feedback. As a result, the second interim FN is transformed into a final FN. This final interim FN can be described by the block-scheme in Fig.8.24 and the topological expression in Eq.(8.42), from where it can be seen that the single equivalent node is not embraced by the identity feedback connection anymore.

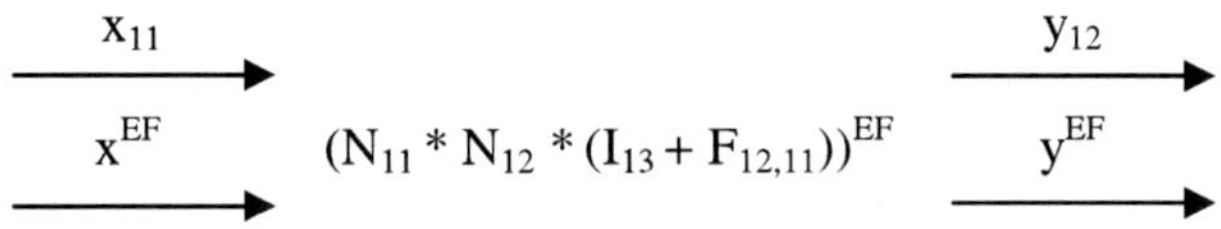

Fig. 8.36 Final FN for Example 8.7

$$[(N_{11} * N_{12} * (I_{13} + F_{12,11}))^{EF}]\ (x_{11}, x^{EF} \mid y_{12}, y^{EF}) \qquad (8.42)$$

The considerations in Example 8.7 are concerned with network analysis when all network nodes and the feedback node are known. In the case of network design, the feedback node is unknown. In this context, Algorithm 8.9 describes the process of deriving the unknown feedback node in the initial FN from Fig.8.33 when the network nodes and the single equivalent node N_E are known. In this case, node N_E is given by the Boolean matrix equation in Eq.(8.43).

$$N_E = (N_{11} * N_{12} * (I_{13} + F_{12,11}))^{EF} \qquad (8.43)$$

Algorithm 8.9

1. Define N_E, N_{11}, N_{12}, I_{13} and $F_{12,11}$.
2. Confirm that N_E satisfies the feedback constraints, if possible.
3. Make N_E equal to $N_{11} * N_{12} * (I_{13} + F_{12,11})$ in Eq.(8.43).
4. Find $N_{11} * N_{12}$ by horizontal merging of N_{11} and N_{12}.
5. Derive $I_{13} + F_{12,11}$ from N_E in Eq.(8.43), if possible.
6. Derive $F_{12,11}$ from $I_{13} + F_{12,11}$, if possible.

Example 8.8

This example considers a FN with network nodes N_{11}, N_{21} and feedback node $F_{11,21}$ embracing N_{11} and N_{21} where $x_{11,21}$ is a common input for N_{11} and N_{21}, y_{11} is an output for N_{11}, y_{21} is an output for N_{21}, $v_{11,21}$ is the part of the feedback connection to $F_{11,21}$ and $w_{11,21}$ is the part of the feedback connection from $F_{11,21}$. This initial FN represents a stack of two fuzzy systems that can be described by the block scheme in Fig.8.37 and the topological expression in Eq.(8.44) from where it can be seen that the feedback has a downward direction from the first to the second level.

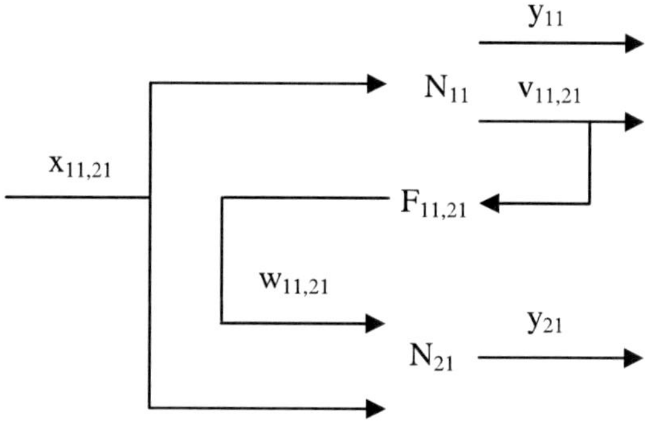

Fig. 8.37 Initial FN for Example 8.8

$$[N_{11}]\ (x_{11,21} \mid y_{11}, v_{11,21})\ ;\ [N_{21}]\ (w_{11,21}, x_{11,21} \mid y_{21}),\ [F_{11,21}]\ (v_{11,21} \mid w_{11,21}) \qquad (8.44)$$

The feedback node $F_{11,21}$ represents a non-identity feedback connection. This is also implied by the use of different variable names for the input and the output for $F_{11,21}$, i.e. $v_{11,21}$ and $w_{11,21}$. In order to apply the linguistic composition approach to the initial FN, it is necessary to introduce a virtual intermediate level just under the first level within a second layer of the underlying grid structure of the FN and to move $F_{11,21}$ to this new grid cell. Besides this, it is necessary to propagate y_{11} forwards through the second layer and insert an implicit identity node I_{12} in level 1 of layer 2 as well as to propagate y_{21} forwards through the second layer and insert an implicit identity node I_{22} in level 2 of layer 2.

The above movements and insertions transform the initial FN into a first interim FN. In this case, the non-identity feedback connection embracing the network nodes N_{11} and N_{21} is represented as a feedforward connection $v_{11,21}$ between N_{11} and $F_{11,21}$ and an identity feedback connection $w_{11,21}$ from $F_{11,21}$ to N_{21}. This first interim FN can be described by the block-scheme in Fig.8.38 and the topological expression in Eq.(8.45) from where it can be seen that $F_{11,21}$ is already a feedforward node alongside the other four nodes.

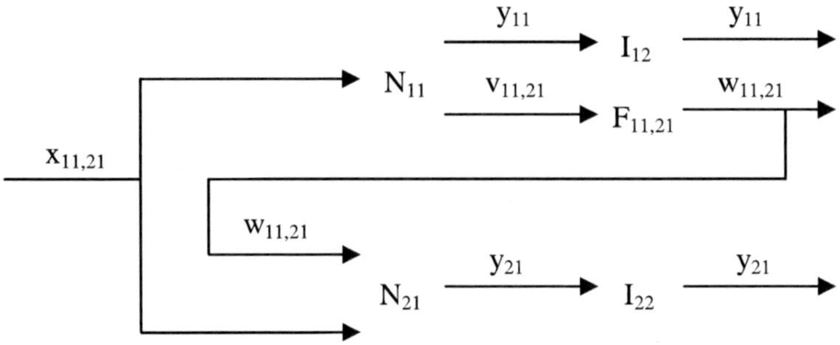

Fig. 8.38 First interim FN for Example 8.8

$$\{[N_{11}]\ (x_{11,21} \mid y_{11}, v_{11,21}) * \{[I_{12}]\ (y_{11} \mid y_{11}) + [F_{11,21}]\ (v_{11,21} \mid w_{11,21})\}\}\ ; \tag{8.45}$$

$$\{[N_{21}]\ (w_{11,21}, x_{11,21} \mid y_{21}) * [I_{22}]\ (y_{21} \mid y_{21})\}$$

Nodes I_{12} and $F_{11,21}$ of the first interim FN can be merged vertically into a temporary node $I_{12} + F_{11,21}$. This temporary node can be further merged horizontally with node N_{11} on the left. Also, nodes N_{21} and I_{22} of the same FN can be merged horizontally into another temporary node $N_{21} + I_{22}$. These merging operations transform the first interim FN into a second interim FN with two nodes whereby the horizontal merging of nodes N_{11} and $I_{12} + F_{11,21}$ is reflected by their replacement with node $N_{11} * (I_{12} + F_{11,21})$ and the horizontal merging of nodes N_{21} and I_{22} is reflected by their replacement with node $N_{21} * I_{22}$. This second interim FN can be described by the block-scheme in Fig.8.39 and the topological expression in Eq.(8.46) from where it can be seen that the lower node has an uncommon input.

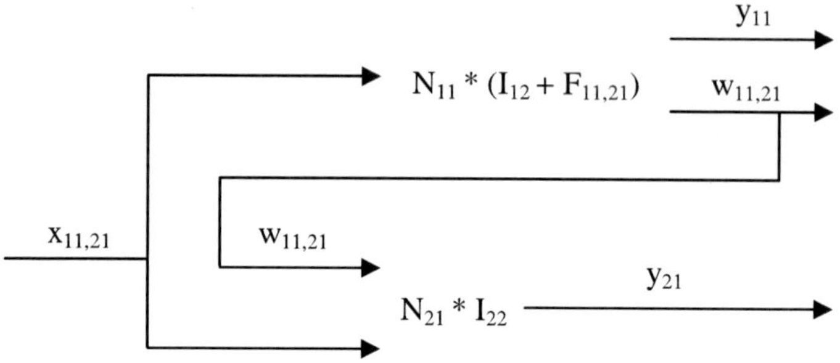

Fig. 8.39 Second interim FN for Example 8.8

$$[N_{11} * (I_{12} + F_{11,21})]\ (x_{11,21} \mid y_{11}, w_{11,21})\ ;\ [N_{21} * I_{22}]\ (w_{11,21}, x_{11,21} \mid y_{21}) \tag{8.46}$$

In order to merge the outputs of nodes $N_{11} * (I_{12} + F_{11,21})$ and $N_{21} * I_{22}$ of the second interim FN, it is necessary to augment the input $x_{11,21}$ for $N_{11} * (I_{12} + F_{11,21})$ with the input $w_{11,21}$. This augmentation operation transforms the second interim FN into a third interim FN with common inputs for the two nodes whereby the first node $N_{11} * (I_{12} + F_{11,21})$ is transformed into a node $(N_{11} * (I_{12} + F_{11,21}))^{AI}$ with input set $\{w_{11,21}, x_{11,21}\}$. This third interim FN can be described by the block-scheme in Fig.8.40 and the topological expression in Eq.(8.47) from where it can be seen that the upper node has augmented inputs.

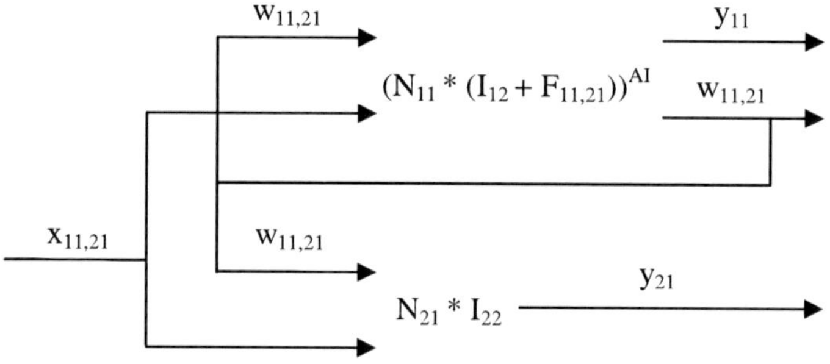

Fig. 8.40 Third interim FN for Example 8.8

$$[(N_{11} * (I_{12} + F_{11,21}))^{AI}]\ (w_{11,21}, x_{11,21} \mid y_{11}, w_{11,21})\ ; \tag{8.47}$$

$$[N_{21} * I_{22}]\ (w_{11,21}, x_{11,21} \mid y_{21})$$

The two nodes $(N_{11} * (I_{12} + F_{11,21}))^{AI}$ and $N_{21} * I_{22}$ of the third interim FN can be output merged into a single equivalent node $(N_{11} * (I_{12} + F_{11,21}))^{AI}\ ;\ (N_{21} * I_{22})$. As a result of this merging operation, the third interim FN is transformed into a fourth interim FN. This fourth interim FN can be described by the block-scheme in Fig.8.41 and the topological expression in Eq.(8.48) from where it can be seen that the single equivalent node is embraced by the identity feedback connection.

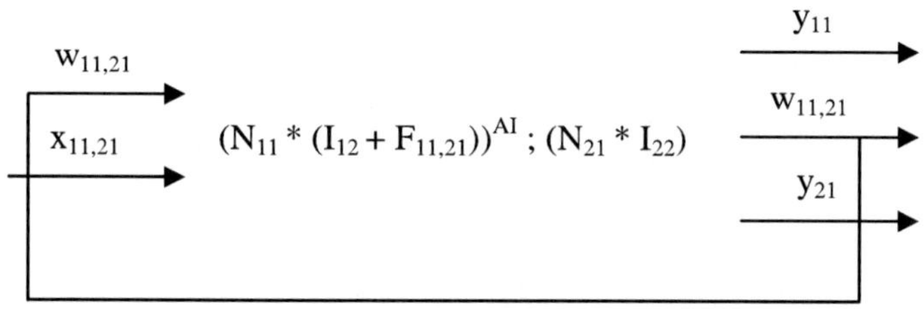

Fig. 8.41 Fourth interim FN for Example 8.8

$$[(N_{11} * (I_{12} + F_{11,21}))^{AI} ; (N_{21} * I_{22})] \; (w_{11,21}, x_{11,21} \mid y_{11}, w_{11,21}, y_{21}) \tag{8.48}$$

The node $(N_{11} * (I_{12} + F_{11,21}))^{AI}$; $(N_{21} * I_{22})$ with input set $\{w_{11,21}, x_{11,21}\}$ and output set $\{y_{11}, w_{11,21}, y_{21}\}$ can be further transformed into a node with equivalent feedback $((N_{11} * (I_{12} + F_{11,21}))^{AI} ; (N_{21} * I_{22}))^{EF}$ with input set $\{x^{EF}, x_{11,21}\}$ and output set $\{y_{11}, y^{EF}, y_{21}\}$.This transformation removes the identity feedback and makes the fuzzy system with feedback equivalent to a fuzzy system without feedback. As a result, the fourth interim FN is transformed into a final FN. This final interim FN can be described by the block-scheme in Fig.8.42 and the topological expression in Eq.(8.49) from where it can be seen that the single equivalent node is not embraced by the identity feedback connection anymore.

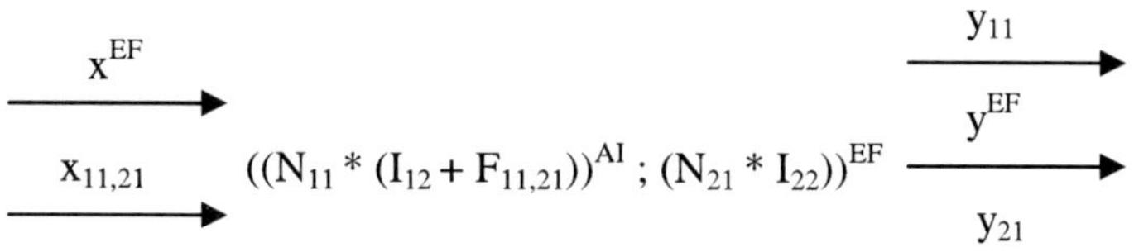

Fig. 8.42 Final interim FN for Example 8.8

$$[((N_{11} * (I_{12} + F_{11,21}))^{AI} ; (N_{21} * I_{22}))^{EF}] \; (x^{EF}, x_{11,21} \mid y_{11}, y^{EF}, y_{21}) \tag{8.49}$$

The considerations in Example 8.8 are concerned with network analysis when all network nodes and the feedback node are known. In the case of network design, the feedback node is unknown. In this context, Algorithm 8.10 describes the process of deriving the unknown feedback node in the initial FN from Fig.8.37 when the network nodes and the single equivalent node N_E are known. In this case, node N_E is given by the Boolean matrix equation in Eq.(8.50).

$$N_E = ((N_{11} * (I_{12} + F_{11,21}))^{AI} ; (N_{21} * I_{22}))^{EF} \quad (8.50)$$

Algorithm 8.10

1. Define N_E, N_{11}, N_{21}, I_{12}, I_{22} and $F_{11,21}$.
2. Confirm that N_E satisfies the feedback constraints, if possible.
3. Make N_E equal to $(N_{11} * (I_{12} + F_{11,21}))^{AI} ; (N_{21} * I_{22})$ in Eq.(8.50).
4. Find $N_{21} * I_{22}$ by vertical merging of N_{21} and I_{22}.
5. Derive $(N_{11} * (I_{12} + F_{11,21}))^{AI}$ from N_E in Eq.(8.50), if possible.
6. Find $N_{11} * (I_{12} + F_{11,21})$ by inverse input augmentation of $(N_{11} * (I_{12} + F_{11,21}))^{AI}$.
7. Derive $I_{12} + F_{11,21}$ from $N_{11} * (I_{12} + F_{11,21})$, if possible.
8. Derive $F_{11,21}$ from $I_{12} + F_{11,21}$, if possible.

Example 8.9

This example considers a FN with network nodes N_{11}, N_{21} and feedback node $F_{21,11}$ embracing N_{21} and N_{11} where $x_{11,21}$ is a common input for N_{11} and N_{21}, y_{11} is an output for N_{11}, y_{21} is an output for N_{21}, $v_{21,11}$ is the part of the feedback connection to $F_{21,11}$ and $w_{21,11}$ is the part of the feedback connection from $F_{21,11}$. This initial FN represents a stack of two fuzzy systems that can be described by the block scheme in Fig.8.43 and the topological expression in Eq.(8.51) from where it can be seen that the feedback has an upward direction from the first to the second level.

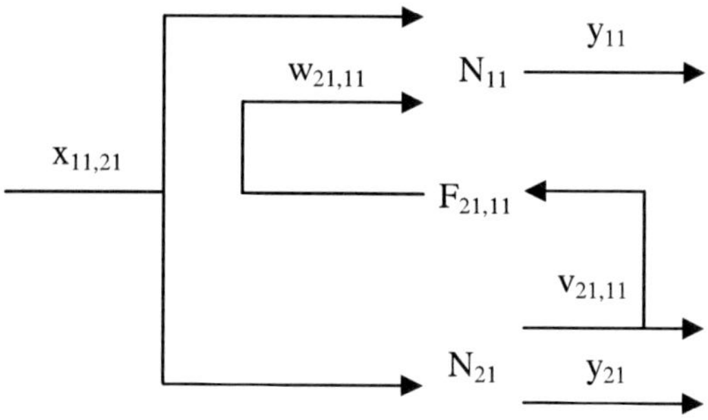

Fig. 8.43 Initial FN for Example 8.9

$$[N_{11}] (x_{11,21}, w_{21,11} \mid y_{11}) ; [N_{21}] (x_{11,21} \mid v_{21,11}, y_{21}), [F_{21,11}] (v_{21,11} \mid w_{21,11}) \quad (8.51)$$

The feedback node $F_{21,11}$ represents a non-identity feedback connection. This is also implied by the use of different variable names for the input and the output for $F_{21,11}$, i.e. $v_{21,11}$ and $w_{21,11}$. In order to apply the linguistic composition approach to the initial FN, it is necessary to introduce a virtual intermediate level just above the second level within a second layer of the underlying grid structure of the FN and to move $F_{21,11}$ to this new grid cell. Besides this, it is necessary to propagate y_{11} forwards through the second layer and insert an implicit identity node I_{12} in level 1 of layer 2 as well as to propagate y_{21} forwards through the second layer and insert an implicit identity node I_{22} in level 2 of layer 2.

The above movements and insertions transform the initial FN into a first interim FN. In this case, the non-identity feedback connection embracing the network nodes N_{11} and N_{21} is represented as a feedforward connection $v_{21,11}$ between N_{21} and $F_{21,11}$ and an identity feedback connection $w_{21,11}$ from $F_{21,11}$ to N_{11}. This first interim FN can be described by the block-scheme in Fig.8.44 and the topological expression in Eq.(8.52), from where it can be seen that $F_{21,11}$ is already a feedforward node alongside the other four nodes.

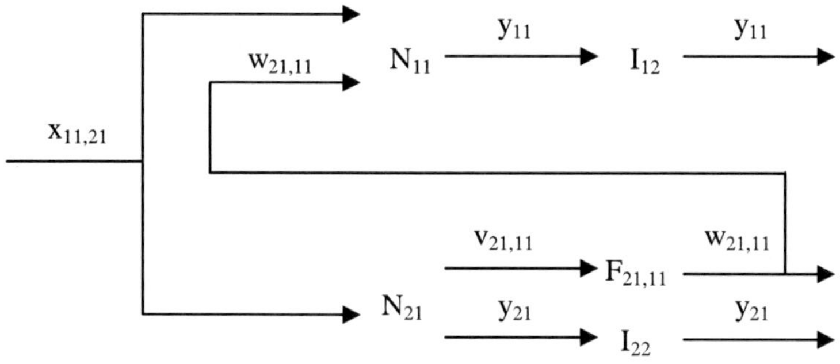

Fig. 8.44 First interim FN for Example 8.9

$$\{[N_{11}]\,(x_{11,21}, w_{21,11} \mid y_{11}) * [I_{12}]\,(y_{11} \mid y_{11})\}\ ;$$

$$\{[N_{21}]\,(x_{11,21} \mid v_{21,11}, y_{21}) * \{[F_{21,11}]\,(v_{21,11} \mid w_{21,11}) + [I_{22}]\,(y_{21} \mid y_{21})\}\} \qquad (8.52)$$

Nodes $F_{21,11}$ and I_{22} of the first interim FN can be merged vertically into a temporary node $F_{21,11} + I_{22}$. This temporary node can be further merged horizontally with node N_{21} on the left. Also, nodes N_{11} and I_{12} of the same FN can be merged horizontally into another temporary node $N_{11} + I_{12}$. These merging operations transform the first interim FN into a second interim FN with two nodes whereby the horizontal merging of nodes N_{11} and I_{12} is reflected by their replacement with node $N_{11} * I_{12}$ and the horizontal merging of nodes N_{21} and $F_{21,11} + I_{22}$ is reflected by their replacement with node $N_{21} * (F_{21,11} + I_{22})$. This second interim FN can be described by the block-scheme in Fig.8.45 and the topological expression in Eq.(8.53) from where it can be seen that the upper node has an uncommon input.

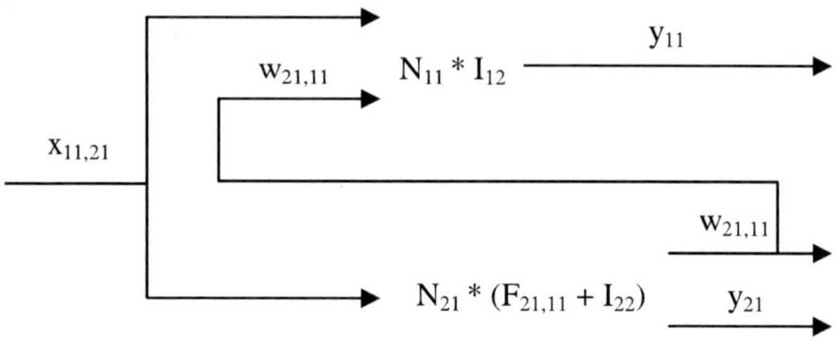

Fig. 8.45 Second interim FN for Example 8.9

$$[N_{11} * I_{12}] (x_{11,21}, w_{21,11} \mid y_{11}) ; [N_{21} * (F_{21,11} + I_{22})] (x_{11,21} \mid w_{21,11}, y_{21}) \quad (8.53)$$

In order to merge the outputs of nodes $N_{11} * I_{12}$ and $N_{21} * (F_{21,11} + I_{22})$ of the second interim FN, it is necessary to augment the input $x_{11,21}$ for $N_{21} * (F_{21,11} + I_{22})$ with the input $w_{21,11}$. This augmentation operation transforms the second interim FN into a third interim FN with common inputs for the two nodes whereby the second node $N_{21} * (F_{21,11} + I_{22})$ is transformed into a node $(N_{21} * (F_{21,11} + I_{22}))^{AI}$ with input set $\{x_{11,21}, w_{21,11}\}$. This third interim FN can be described by the block-scheme in Fig.8.46 and the topological expression in Eq.(8.54) from where it can be seen that the lower node has augmented inputs.

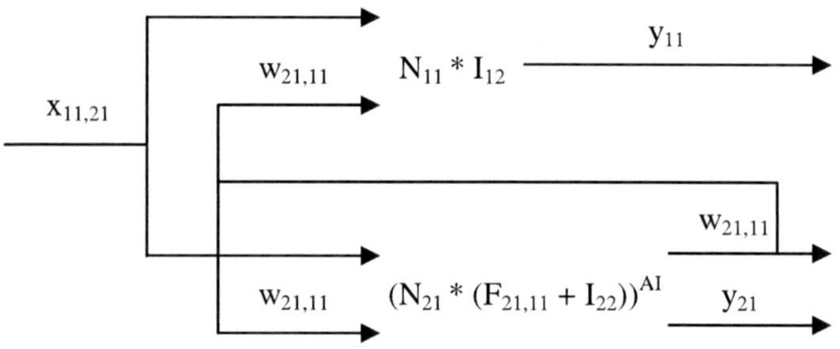

Fig. 8.46 Third interim FN for Example 8.9

$$[N_{11} * I_{12}] (x_{11,21}, w_{21,11} \mid y_{11}) ; \quad (8.54)$$

$$[(N_{21} * (F_{21,11} + I_{22}))^{AI}] (x_{11,21}, w_{21,11} \mid w_{21,11}, y_{21})$$

The two nodes $N_{11} * I_{12}$ and $(N_{21} * (F_{21,11} + I_{22}))^{AI}$ of the third interim FN can be output merged into a single equivalent node $(N_{11} * I_{12}) ; (N_{21} * (F_{21,11} + I_{22}))^{AI}$. As result of this merging operation, the third interim FN is transformed into a fourth interim FN. This fourth interim FN can be described by the block-scheme in

Fig.8.47 and the topological expression in Eq.(8.55) from where it can be seen that the single equivalent node is embraced by the identity feedback connection.

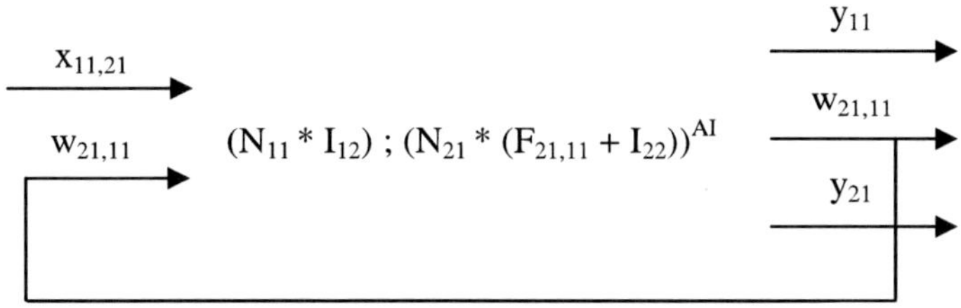

Fig. 8.47 Fourth interim FN for Example 8.9

$$[(N_{11} * I_{12}) ; (N_{21} * (F_{21,11} + I_{22}))^{AI}] (x_{11,21}, w_{21,11} \mid y_{11}, w_{21,11}, y_{21}) \quad (8.55)$$

The node $(N_{11} * I_{12}) ; (N_{21} * (F_{21,11} + I_{22}))^{AI}$ with input set $\{x_{11,21}, w_{21,11}\}$ and output set $\{y_{11}, w_{21,11}, y_{21}\}$ can be further transformed into a node with equivalent feedback $((N_{11} * I_{12}) ; (N_{21} * (F_{21,11} + I_{22}))^{AI})^{EF}$ with input set $\{x_{11,21}, x^{EF}\}$ and output set $\{y_{11}, y^{EF}, y_{21}\}$.This transformation removes the identity feedback and makes the fuzzy system with feedback equivalent to a fuzzy system without feedback. As a result, the fourth interim FN is transformed into a final FN. This final interim FN can be described by the block-scheme in Fig.8.48 and the topological expression in Eq.(8.56), from where it can be seen that the single equivalent node is not embraced by the identity feedback connection anymore.

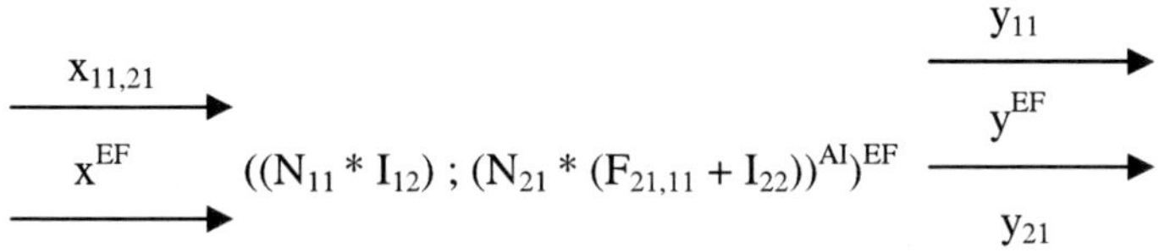

Fig. 8.48 Final interim FN for Example 8.9

$$[((N_{11} * I_{12}) ; (N_{21} * (F_{21,11} + I_{22}))^{AI})^{EF}] (x_{11,21}, x^{EF} \mid y_{11}, y^{EF}, y_{21}) \quad (8.56)$$

The considerations in Example 8.9 are concerned with network analysis when all network nodes and the feedback node are known. In the case of network design, the feedback node is unknown. In this context, Algorithm 8.11 describes the process of deriving the unknown feedback node in the initial FN from Fig.8.43 when the network nodes and the single equivalent node N_E are known. In this case, node N_E is given by the Boolean matrix equation in Eq.(8.57).

$$N_E = ((N_{11} * I_{12}) ; (N_{21} * (F_{21,11} + I_{22}))^{AI})^{EF} \quad (8.57)$$

Algorithm 8.11

1. Define N_E, N_{11}, N_{21}, I_{12}, I_{22} and $F_{21,11}$.
2. Confirm that N_E satisfies the feedback constraints, if possible.
3. Make N_E equal to $(N_{11} * I_{12})$; $(N_{21} * (F_{21,11} + I_{22}))^{AI}$ in Eq.(8.57).
4. Find $N_{11} * I_{12}$ by vertical merging of N_{11} and I_{12}.
5. Derive $(N_{21} * (F_{21,11} + I_{22}))^{AI}$ from N_E in Eq.(8.57), if possible.
6. Find $N_{21} * (F_{21,11} + I_{22})$ by inverse input augmentation of $(N_{21} * (F_{21,11} + I_{22}))^{AI}$.
7. Derive $F_{21,11} + I_{22}$ from $N_{21} * (F_{21,11} + I_{22})$, if possible.
8. Derive $F_{21,11}$ from $F_{21,11} + I_{22}$, if possible.

8.5 Networks with Multiple Global Feedback

The most complex type of FN is the one with multiple global feedback. This network has at least two sequences of nodes with at least two nodes in each sequence such that all nodes in a sequence are embraced by a separate feedback connection with a feedback node in it. In this case, the feedback is multiple as it appears more than once but it is also global as it embraces more than one node within a sequence. There may be an arbitrary number of feedforward connections between these nodes and any other nodes as well as between any pair of other nodes. However, the presence of any feedforward connections does not remove the feedback characteristics of this type of FN due to the presence of the feedback connections with the feedback nodes.

Example 8.10

This example considers a FN with network nodes N_{11}, N_{12} and N_{13}, feedback node $F_{12,11}$ embracing N_{11} and N_{12} and feedback node $F_{13,12}$ embracing N_{12} and N_{13} where x_{11} is an input for N_{11}, y_{13} is an output for N_{13}, $z_{11,12}^{1,1}$ is the feedforward connection from the first and only output for N_{11} to the first input for N_{12}, $z_{12,13}^{1,1}$ is the feedforward connection from the first output for N_{12} to the first and only input for N_{13}, $v_{12,11}$ is the part of the feedback connection to $F_{12,11}$, $w_{12,11}$ is the part of the feedback connection from $F_{12,11}$, $v_{13,12}$ is the part of the feedback connection to $F_{13,12}$ and $w_{13,12}$ is the part of the feedback connection from $F_{13,12}$. This initial FN represents a queue of three fuzzy systems that can be described by the block scheme in Fig.8.49 and the topological expression in Eq.(8.58) from where it can be seen that the three nodes in the first and only level are embraced by two partially overlapping feedbacks.

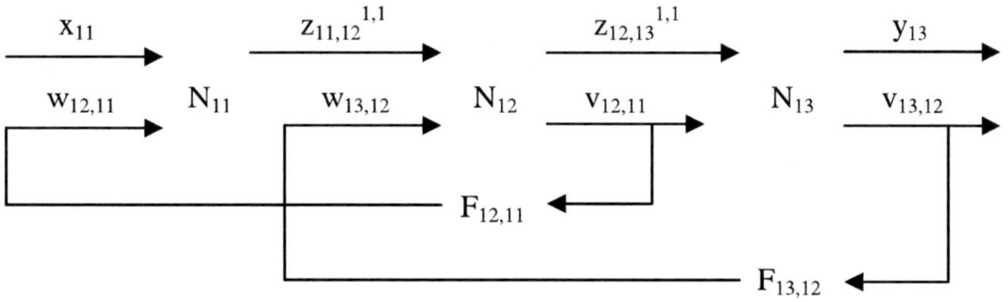

Fig. 8.49 Initial FN for Example 8.10

$$[N_{11}]\,(x_{11}, w_{12,11} \mid z_{11,12}^{1,1}) * [N_{12}]\,(z_{11,12}^{1,1}, w_{13,12} \mid z_{12,13}^{1,1}, v_{12,11}) * \quad (8.58)$$

$$[N_{13}]\,(z_{12,13}^{1,1} \mid y_{13}, v_{13,12}),\ [F_{12,11}]\,(v_{12,11} \mid w_{12,11}),\ [F_{13,12}]\,(v_{13,12} \mid w_{13,12})$$

The feedback nodes $F_{12,11}$ and $F_{13,12}$ represent non-identity feedback connections. This is also implied by the use of different variable names for the inputs and the outputs for $F_{12,11}$ and $F_{13,12}$, i.e. $v_{12,11}$, $w_{12,11}$ and $v_{13,12}$, $w_{13,12}$. In order to apply the linguistic composition approach to the initial FN, it is necessary to introduce a second level within a fourth layer of the underlying grid structure of the FN and to move $F_{13,12}$ to this new grid cell. It is also necessary to propagate y_{13} forwards through the fourth layer and insert an implicit identity node I_{14} in level 1 of layer 4. Further on, it is necessary to introduce a third level within the fourth layer of the underlying grid structure of the FN and to move $F_{12,11}$ to this new grid cell. It is also necessary to propagate $v_{12,11}$ forwards through the third layer and insert an implicit identity node I_{33} in level 3 of layer 3. Likewise, it is necessary to propagate $w_{13,12}$ backwards through the first layer and insert an implicit identity node I_{31} in level 3 of layer 1.

The above movements and insertions transform the initial FN into a first interim FN whereby the non-identity feedback connection embracing the network nodes N_{11} and N_{12} is represented as a feedforward connection $v_{12,11}$ between N_{12} and $F_{12,11}$ and an identity feedback connection $w_{12,11}$ embracing N_{11}, N_{12}, I_{33} and $F_{12,11}$. Also, the non-identity feedback connection embracing the network node N_{12} and N_{13} is represented as a feedforward connection $v_{13,12}$ between N_{13} and $F_{13,12}$ and an identity feedback connection $w_{13,12}$ embracing I_{31}, N_{12}, N_{13} and $F_{13,12}$. This first interim FN can be described by the block-scheme in Fig.8.50 and the topological expression in Eq.(8.59) from where it can be seen that both $F_{12,11}$ and $F_{13,12}$ are already feedforward nodes alongside the other six nodes.

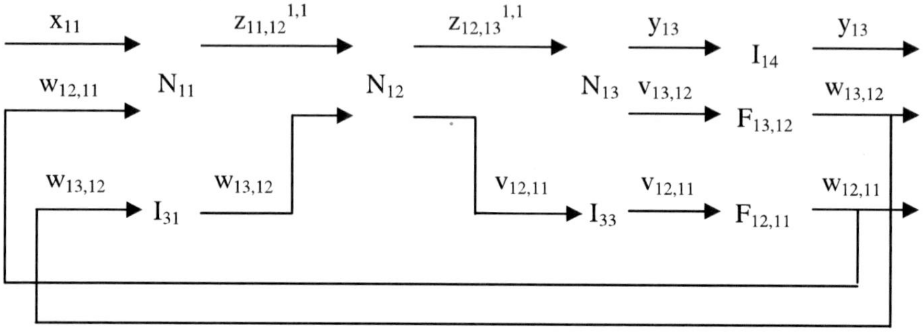

Fig. 8.50 First interim FN for Example 8.10

$$\{[N_{11}]\,(x_{11}, w_{12,11} \mid z_{11,12}^{1,1}) + [I_{31}]\,(w_{13,12} \mid w_{13,12})\} \; * \tag{8.59}$$
$$[N_{12}]\,(z_{11,12}^{1,1}, w_{13,12} \mid z_{12,13}^{1,1}, v_{12,11}) \; *$$
$$\{[N_{13}]\,(z_{12,13}^{1,1} \mid y_{13}, v_{13,12}) + [I_{33}]\,(v_{12,11} \mid v_{12,11})\} \; *$$
$$\{[I_{14}]\,(y_{13} \mid y_{13}) + [F_{13,12}]\,(v_{13,12} \mid w_{13,12}) + [F_{12,11}]\,(v_{12,11} \mid w_{12,11})\}$$

Nodes N_{11} and I_{31} of the first interim FN can be merged vertically into a temporary node $N_{11} + I_{31}$. Similarly, nodes N_{13} and I_{33} of the same interim FN can be merged vertically into a second temporary node $N_{13} + I_{33}$. Also, nodes I_{13}, $F_{13,12}$ and $F_{12,11}$ can be merged vertically into a third temporary node $I_{14} + F_{13,12} + F_{12,11}$. These three temporary nodes can be further merged horizontally with node N_{12} between the first and the second temporary node into a single equivalent node $(N_{11} + I_{31}) * N_{12} * (N_{13} + I_{33}) * (I_{14} + F_{13,12} + F_{12,11})$. These merging operations transform the first interim FN into a second interim FN with a single node. This second FN can be described by the block-scheme in Fig.8.51 and the topological expression in Eq.(8.60) from where it can be seen that the single equivalent node is embraced by the identity feedback connections.

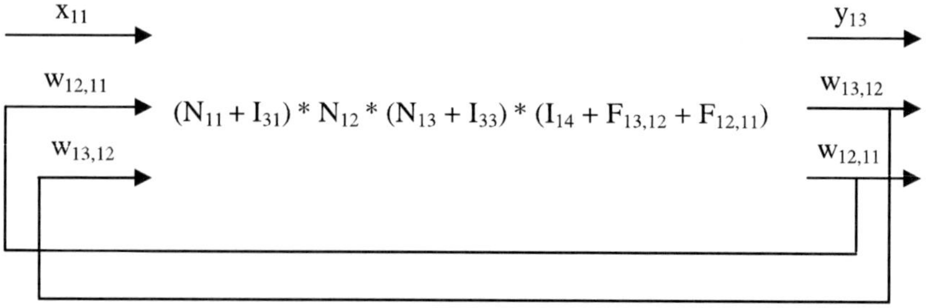

Fig. 8.51 Second interim FN for Example 8.10

$$[(N_{11} + I_{31}) * (N_{12} + F_{11}) * (I_{14} + F_{12} + I_{33})] \tag{8.60}$$

$$(x_{11}, w_{12,11}, w_{13,12} \mid y_{12}, w_{13,12}, w_{12,11})$$

The node $(N_{11} + I_{31}) * N_{12} * (N_{13} + I_{33}) * (I_{14} + F_{13,12} + F_{12,11})$ with input set $\{x_{11}, w_{12,11}, w_{13,12}\}$ and output set $\{y_{13}, w_{13,12}, w_{12,11}\}$ can be further transformed into a node with equivalent feedback $((N_{11} + I_{31}) * N_{12} * (N_{13} + I_{33}) * (I_{14} + F_{13,12} + F_{12,11}))^{FE}$ with input set $\{x_{11}, x_1^{FE}, x_2^{FE}\}$ and output set $\{y_{12}, y_2^{FE}, y_1^{FE}\}$.This transformation removes the two identity feedbacks and makes the fuzzy system with feedback equivalent to a fuzzy subsystem without feedback. As a result, the second interim FN is transformed into a final FN. This final FN can be described by the block-scheme in Fig.8.52 and the topological expression in Eq.(8.61) from where it can be seen that the single equivalent node is not embraced by the identity feedback connections.

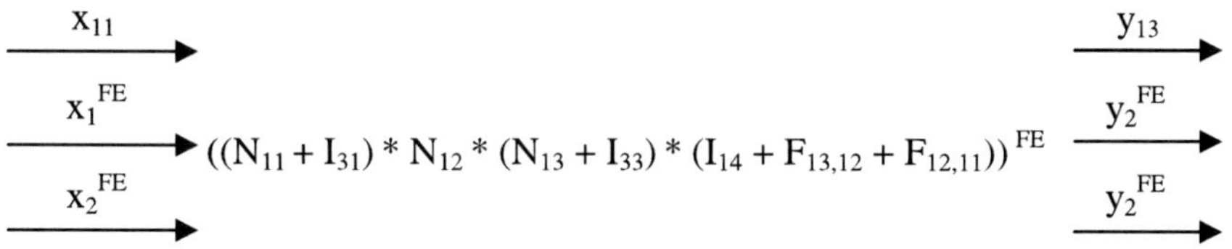

Fig. 8.52 Final FN for Example 8.10

$$[((N_{11} + I_{31}) * N_{12} * (N_{13} + I_{33}) * (I_{14} + F_{13,12} + F_{12,11}))^{FE}] \tag{8.61}$$

$$(x_{11}, x_1^{FE}, x_2^{FE} \mid y_{12}, y_2^{FE}, y_1^{FE})$$

The considerations in Example 8.10 are concerned with network analysis when all network and feedback nodes are known. In the case of network design, at least one feedback node is unknown. In this context, Algorithms 8.12-8.13 describe the process of deriving an unknown feedback node in the initial FN from Fig.8.49 when the network nodes, one of the feedback nodes and the single equivalent node N_E are known. In this case, node N_E is given by the Boolean matrix equation in Eq.(8.62).

$$N_E = ((N_{11} + I_{31}) * N_{12} * (N_{13} + I_{33}) * (I_{14} + F_{13,12} + F_{12,11}))^{FE} \tag{8.62}$$

Algorithm 8.12

1. Define N_E, N_{11}, N_{12}, N_{13}, I_{14}, I_{31}, I_{33} and $F_{13,12}$.
2. Confirm that N_E satisfies the feedback constraints, if possible.
3. Make N_E equal to $(N_{11} + I_{31}) * N_{12} * (N_{13} + I_{33}) * (I_{14} + F_{13,12} + F_{12,11})$ in Eq.(8.62).
4. Find $N_{11} + I_{31}$ by vertical merging of N_{11} and I_{31}.
5. Find $(N_{11} + I_{31}) * N_{12}$ by horizontal merging of $(N_{11} + I_{31})$ and N_{12}.
6. Find $N_{13} + I_{33}$ by vertical merging of N_{13} and I_{33}.
7. Find $(N_{11} + I_{31}) * N_{12} * (N_{13} + I_{33})$ by horizontal merging of $(N_{11} + I_{31}) * N_{12}$ and $(N_{13} + I_{33})$.
8. Derive $I_{14} + F_{13,12} + F_{12,11}$ from N_E in Eq.(8.62), if possible.
9. Find $I_{14} + F_{13,12}$ by vertical merging of I_{14} and $F_{13,12}$.
10. Derive $F_{12,11}$ from $I_{14} + F_{13,12} + F_{12,11}$, if possible.

Algorithm 8.13

1. Define N_E, N_{11}, N_{12}, N_{13}, I_{14}, I_{31}, I_{33} and $F_{11,12}$.
2. Confirm that N_E satisfies the feedback constraints, if possible.
3. Make N_E equal to $(N_{11} + I_{31}) * N_{12} * (N_{13} + I_{33}) * (I_{14} + F_{13,12} + F_{12,11})$ in Eq.(8.62).
4. Find $N_{11} + I_{31}$ by vertical merging of N_{11} and I_{31}.
5. Find $(N_{11} + I_{31}) * N_{12}$ by horizontal merging of $(N_{11} + I_{31})$ and N_{12}.
6. Find $N_{13} + I_{33}$ by vertical merging of N_{13} and I_{33}.
7. Find $(N_{11} + I_{31}) * N_{12} * (N_{13} + I_{33})$ by horizontal merging of $(N_{11} + I_{31}) * N_{12}$ and $(N_{13} + I_{33})$.
8. Derive $I_{14} + F_{13,12} + F_{12,11}$ from N_E in Eq.(8.62), if possible.
9. Derive $F_{13,12}$ from $I_{14} + F_{13,12} + F_{12,11}$, if possible.

Example 8.11

This example considers a FN with network nodes N_{11} and N_{21}, feedback node $F_{11,21}$ embracing N_{11} and N_{21} and feedback node $F_{21,11}$ embracing N_{21} and N_{11} where $x_{11,21}$ is a common input for N_{11} and N_{21}, y_{11} is an output for N_{11}, y_{21} is an output for N_{21}, $v_{11,21}$ is the part of the feedback connection to $F_{11,21}$, $w_{11,21}$ is the part of the feedback connection from $F_{11,21}$, $v_{21,11}$ is the part of the feedback connection to $F_{21,11}$ and $w_{21,11}$ is the part of the feedback connection from $F_{21,11}$. This initial FN represents a stack of two fuzzy systems that can be described by the block scheme in Fig.8.53 and the topological expression in Eq.(8.63) from where it can be seen that the feedback is bidirectional from the first to the second level and vice versa.

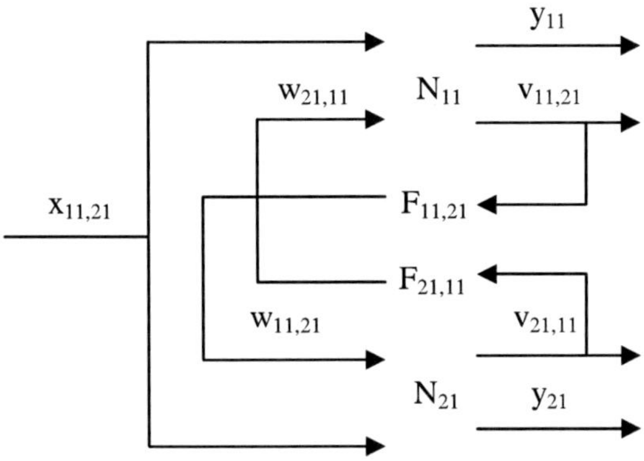

Fig. 8.53 Initial FN for Example 8.11

$$[N_{11}]\ (x_{11,21}, w_{21,11} \mid y_{11}, v_{11,21})\ ;\ [N_{21}]\ (w_{11,21}, x_{11,21} \mid v_{21,11}, y_{21}), \quad (8.63)$$

$$[F_{11,21}]\ (v_{11,21} \mid w_{11,21}),\ [F_{21,11}]\ (v_{21,11} \mid w_{21,11})$$

The feedback nodes $F_{11,12}$ and $F_{12,11}$ represent non-identity feedback connections. This is also implied by the use of different variable names for the inputs and the outputs for $F_{11,12}$ and $F_{12,11}$, i.e. $v_{11,12}$, $w_{11,12}$ and $v_{12,11}$, $w_{12,11}$. In order to apply the linguistic composition approach to the initial FN, it is necessary to introduce a virtual intermediate level just under the first level within a second layer of the underlying grid structure of the FN and to move $F_{11,21}$ to this new grid cell as well as to propagate y_{11} forwards through the second layer and insert an implicit identity node I_{12} in level 1 of layer 2. It is also necessary to introduce a virtual intermediate level just above the second level within the second layer and to move $F_{21,11}$ to this new grid cell of the underlying grid structure of the FN as well as to propagate y_{21} forwards through the second layer and insert an implicit identity node I_{22} in level 2 of layer 2.

The above movements and insertions transform the initial FN into a first interim FN. In this case, the non-identity feedback connection embracing the network nodes N_{11} and N_{21} is represented as a feedforward connection $v_{11,21}$ between N_{11} and $F_{11,21}$ and an identity feedback connection $w_{11,21}$ from $F_{11,21}$ to N_{21}. Also, the non-identity feedback connection embracing the network nodes N_{21} and N_{11} is represented as a feedforward connection $v_{21,11}$ between N_{21} and $F_{21,11}$ and an identity feedback connection $w_{21,11}$ from $F_{21,11}$ to N_{11}. This first interim FN can be described by the block-scheme in Fig.8.54 and the topological expression in Eq.(8.64) from where it can be seen that $F_{11,21}$ and $F_{21,11}$ are already feedforward nodes alongside the other four nodes.

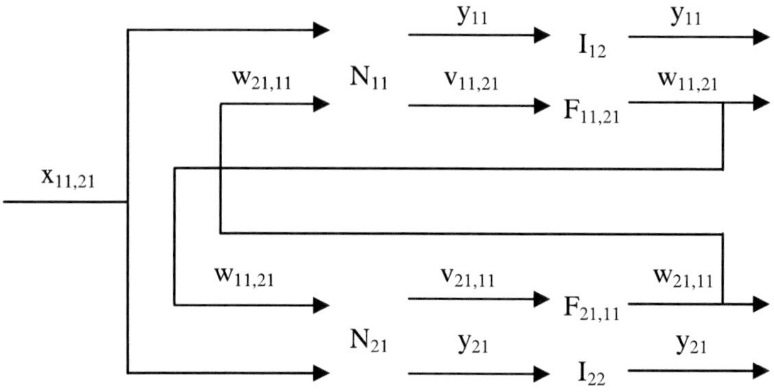

Fig. 8.54 First interim FN for Example 8.11

$$\{[N_{11}]\ (x_{11,21}, w_{21,11} \mid y_{11}, v_{11,21}) * \{[I_{12}]\ (y_{11} \mid y_{11}) + [F_{11,21}]\ (v_{11,21} \mid w_{11,21})\}\}\ ; \quad (8.64)$$

$$\{[N_{21}]\ (w_{11,21}, x_{11,21} \mid v_{21,11}, y_{21}) * \{[F_{21,11}]\ (v_{21,11} \mid w_{21,11}) + [I_{22}]\ (y_{21} \mid y_{21})\}\}$$

Nodes I_{12} and $F_{11,21}$ of the second interim FN can be merged vertically into a temporary node $I_{12} + F_{11,21}$ which can be further merged horizontally with node N_{11} on the left. Also, nodes $F_{21,11}$ and I_{22} can be merged vertically into another temporary node $F_{21,11} + I_{22}$ which can be further merged horizontally with node N_{21} on the left. These merging operations transform the first interim FN into a second interim FN with two nodes whereby the horizontal merging of nodes N_{11} and $I_{12} + F_{11,21}$ is reflected by their replacement with node $N_{11} * (I_{12} + F_{11,21})$ and the horizontal merging of nodes N_{21} and $(F_{21,11} + I_{22})$ is reflected by their replacement with node N_{21} and $(F_{21,11} + I_{22})$. This second interim FN can be described by the block-scheme in Fig.8.55 and the topological expression in Eq.(8.65) from where it can be seen that each of the two nodes has an uncommon input.

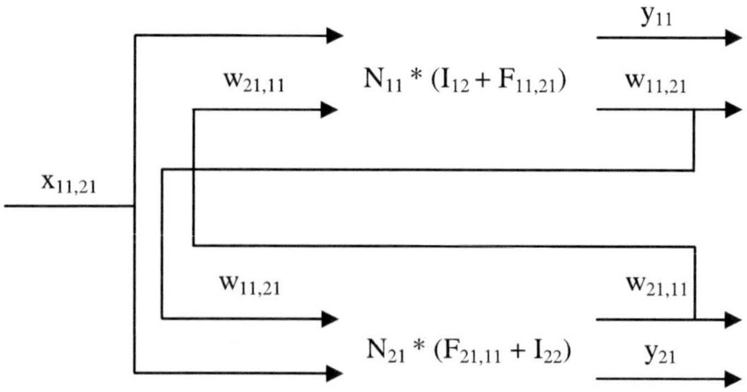

Fig. 8.55 Second interim FN for Example 8.11

$$[N_{11} * (I_{12} + F_{11,21})]\ (x_{11,21}, w_{21,11} \mid y_{11}, w_{11,21})\ ; \qquad (8.65)$$

$$[N_{21} * (F_{21,11} + I_{22})]\ (w_{11,21}, x_{11,21} \mid w_{21,11}, y_{21})$$

In order to merge the outputs of nodes $N_{11} * (I_{12} + F_{11,21})$ and $N_{21} * (F_{21,11} + I_{22})$ of the second interim FN, it is necessary to augment the common input $x_{11,21}$ for $N_{11} * (I_{12} + F_{11,21})$ with the input $w_{11,21}$. It is also necessary to augment this common input for $N_{21} * (F_{21,11} + I_{22})$ with the input $w_{21,11}$. These augmentation operations transform the second interim FN into a third interim FN with common inputs for the two nodes whereby the first node $N_{11} * (I_{12} + F_{11,21})$ is transformed into a node $(N_{11} * (I_{12} + F_{11,21}))^{AI}$ with input set $\{w_{11,21}, x_{11,21}, w_{21,11}\}$and the second node $N_{21} * (F_{21,11} + I_{22})$ is transformed into a node $(N_{21} * (F_{21,11} + I_{22}))^{AI}$ with the same input set. This third interim FN can be described by the block-scheme in Fig.8.56 and the topological expression in Eq.(8.66), from where it can be seen that the both nodes already have augmented inputs.

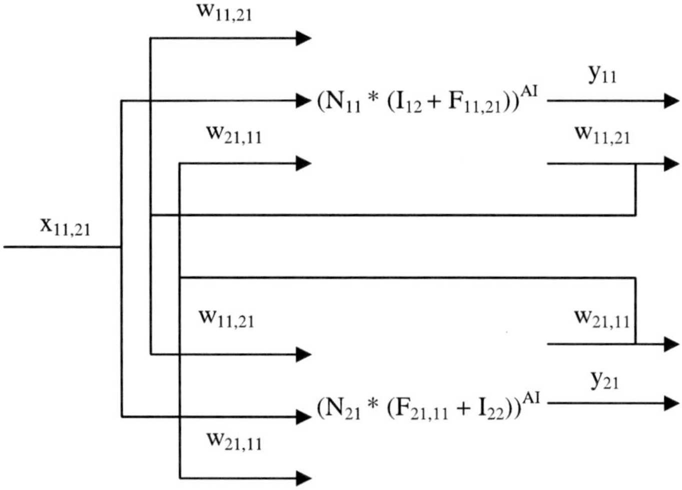

Fig. 8.56 Third interim FN for Example 8.11

$$[N_{11} * (I_{12} + F_{11,21})]\ (w_{11,21}, x_{11,21}, w_{21,11} \mid y_{11}, w_{11,21})\ ; \qquad (8.66)$$

$$[N_{21} * (F_{21,11} + I_{22})]\ (w_{11,21}, x_{11,21}, w_{21,11} \mid w_{21,11}, y_{21})$$

The two nodes $(N_{11} * (I_{12} + F_{11,21}))^{AI}$ and $(N_{21} * (F_{21,11} + I_{22}))^{AI}$ of the third interim FN can be output merged into a single equivalent node $(N_{11} * (I_{12} + F_{11,21}))^{AI}$; $(N_{21} * (F_{21,11} + I_{22}))^{AI}$. As result of this merging operation, the third interim FN is transformed into a fourth interim FN. This fourth interim FN can be described by the block-scheme in Fig.8.57 and the topological expression in Eq.(8.67) from where it can be seen that the single equivalent node is embraced by the identity feedback connections.

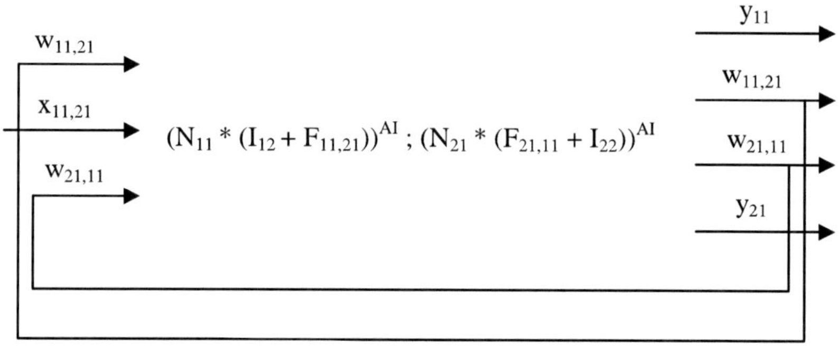

Fig. 8.57 Fourth interim FN for Example 8.11

$$[(N_{11} * (I_{12} + F_{11,21}))^{AI} ; (N_{21} * (F_{21,11} + I_{22}))^{AI}] \quad (8.67)$$

$$(w_{11,21}, x_{11,21}, w_{21,11} \mid y_{11}, w_{11,21}, w_{21,11}, y_{21})$$

The node $(N_{11} * (I_{12} + F_{11,21}))^{AI}$; $(N_{21} * (F_{21,11} + I_{22}))^{AI}$ with input set $\{w_{11,21}, x_{11,21}, w_{21,11}\}$ and output set $\{y_{11}, w_{11,21}, w_{21,11}, y_{21}\}$ can be further transformed into a node with equivalent feedback $((N_{11} * (I_{12} + F_{11,21}))^{AI} ; (N_{21} * (F_{21,11} + I_{22}))^{AI})^{FE}$ with input set $\{x_1^{FE}, x_{11,21}, x_2^{FE}\}$ and output set $\{y_{11}, y_1^{FE}, y_2^{FE}, y_{21}\}$.This transformation removes the two identity feedbacks and makes the fuzzy system with feedback equivalent to a fuzzy subsystem without feedback. As a result, the fourth interim FN is transformed into a final FN. This final FN can be described by the block-scheme in Fig.8.58 and the topological expression in Eq.(8.68) from where it can be seen that the single equivalent node is not embraced by the identity feedback connections anymore.

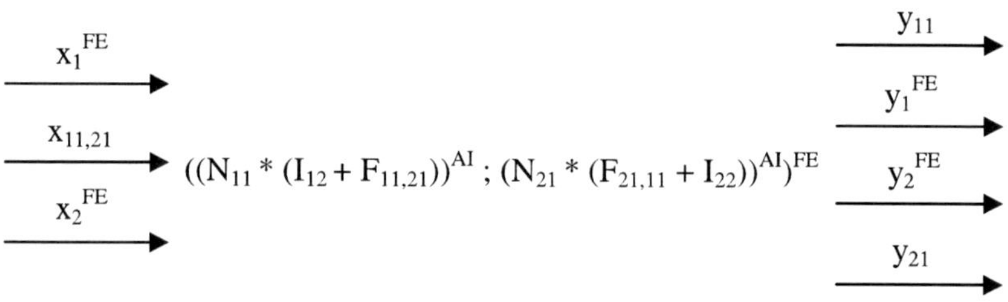

Fig. 8.58 Final FN for Example 8.11

$$[((N_{11} * (I_{12} + F_{11,21}))^{AI} ; (N_{21} * (F_{21,11} + I_{22}))^{AI})^{FE}] \quad (8.68)$$

$$(x_1^{FE}, x_{11,21}, x_2^{FE} \mid y_{11}, y_1^{FE}, y_2^{FE}, y_{21})$$

The considerations in Example 8.11 are concerned with network analysis when all network and feedback nodes are known. In the case of network design, at least one feedback node is unknown. In this context, Algorithms 8.14-8.15 describe the process of deriving an unknown feedback node in the initial FN from Fig.8.53 when the network nodes, one of the feedback nodes and the single equivalent node N_E are known. In this case, node N_E is given by the Boolean matrix equation in Eq.(8.69).

$$N_E = ((N_{11} * (I_{12} + F_{11,21}))^{AI} ; (N_{21} * (F_{21,11} + I_{22}))^{AI})^{FE} \tag{8.69}$$

Algorithm 8.14

1. Define N_E, N_{11}, N_{21}, I_{12}, I_{22} and $F_{21,11}$.
2. Confirm that N_E satisfies the feedback constraints, if possible.
3. Make N_E equal to $(N_{11} * (I_{12} + F_{11,21}))^{AI}$; $(N_{21} * (F_{21,11} + I_{22}))^{AI}$ in Eq.(8.69).
4. Find $F_{21,11} + I_{22}$ by vertical merging of $F_{21,11}$ and I_{22}.
5. Find $N_{21} * (F_{21,11} + I_{22})$ by horizontal merging of N_{21} and $(F_{21,11} + I_{22})$.
6. Find $(N_{21} * (F_{21,11} + I_{22}))^{AI}$ by input augmentation of $N_{21} * (F_{21,11} + I_{22})$.
7. Derive $(N_{11} * (I_{12} + F_{11,21}))^{AI}$ from N_E in Eq.(8.69), if possible.
8. Find $N_{11} * (I_{12} + F_{11,21})$ by inverse input augmentation of $(N_{11} * (I_{12} + F_{11,21}))^{AI}$.
9. Derive $I_{12} + F_{11,21}$ from $N_{11} * (I_{12} + F_{11,21})$, if possible.
10. Derive $F_{11,21}$ from $I_{12} + F_{11,21}$, if possible.

Algorithm 8.15

1. Define N_E, N_{11}, N_{21}, I_{12}, I_{22} and $F_{11,21}$.
2. Confirm that N_E satisfies the feedback constraints, if possible.
3. Make N_E equal to $(N_{11} * (I_{12} + F_{11,21}))^{AI}$; $(N_{21} * (F_{21,11} + I_{22}))^{AI}$ in Eq.(8.69).
4. Find $I_{12} + F_{11,21}$ by vertical merging of I_{12} and $F_{11,21}$.
5. Find $N_{11} * (I_{12} + F_{11,21})$ by horizontal merging of N_{11} and $(I_{12} + F_{11,21})$.
6. Find $(N_{11} * (I_{12} + F_{11,21}))^{AI}$ by input augmentation of $N_{11} * (I_{12} + F_{11,21})$.
7. Derive $(N_{21} * (F_{21,11} + I_{22}))^{AI}$ from N_E in Eq.(8.69), if possible.
8. Find $N_{21} * (F_{21,11} + I_{22})$ by inverse input augmentation of $(N_{21} * (F_{21,11} + I_{22}))^{AI}$.
9. Derive $F_{21,11} + I_{22}$ from $N_{21} * (F_{21,11} + I_{22})$, if possible.
10. Derive $F_{21,11}$ from $F_{21,11} + I_{22}$, if possible.

8.6 Summary on Feedback Fuzzy Networks

The examples presented in this chapter illustrate the application of basic operations, their properties and advanced operations in feedback FNs. These examples validate theoretically the linguistic composition approach used in the book. This applies particularly to FNs with multiple local feedback and multiple global feedback which are more complex type of feedback FNs. However, the other two less complex types of feedforward FNs, i.e. FNs with single local feedback and FNs with single global feedback, are also quite useful in that they are usually part of FNs with multiple local feedback and FNs with multiple global feedback.

The different types of feedback FNs represent different type regression mappings, e.g. mono-variable, poly-variable, one-step and many-step. In particular, single feedback shows mono-variable regression whereas multiple feedback shows poly-variable regression. Also, local feedback reflects one-step regression whereas global feedback reflects many-step regression. So, the type of regression in a FN determines the type of FN to be used.

The relationship between different types of FNs and regressions is described in Table 8.1.

Table 8.1 Relationship between types of FNs and regressions

Feedback FN	Mono-variable regression	Poly-variable regression	One-step regression	Many-step regression
Single local	Yes	No	Yes	No
Single global	Yes	No	No	Yes
Multiple local	No	Yes	Yes	No
Multiple global	No	Yes	No	Yes

The next chapter shows further extension and validation of the results from Chapters 4-8 on FNs. In particular, some generalised theoretical examples and applied case studies of FNs are considered.

Chapter 9
Evaluation of Fuzzy Networks

9.1 Preliminaries on Fuzzy Network Evaluation

The application of basic operations, their structural properties and advanced operations is illustrated in Chapters 7-8 for fairly abstract FNs. The examples presented there show the application of the above operations and their properties to the overall structure of these networks. Although the latter may be good testing examples for the underlying FN theory, they reflect only some variations of the main types of network topologies introduced for both feedforward and feedback FNs. Therefore, it is necessary to extend these considerations in a more applied context in terms of generalised examples and case studies.

The current chapter evaluates the theoretical results on FNs from Chapters 4-8 in the context of assessment of structural complexity, composition of HFSs, decomposition of SFSs, indicators of model performance and applications in case studies. One part of the evaluation is based on general network theory as a FN can be viewed as a specific type of network. Another part of the evaluation builds on general systems theory as a FN represents a system of fuzzy systems. The rest of the evaluation is based on comparison with SFSs and HFSs which are widely used as applications of fuzzy logic.

The evaluation of the theoretical results is considered in the context of both analysis and design. However, the analysis part is covered in more detail than the design part in some sections such as the one with the case studies. This is due to the fact that it is usually easier to analyse a FN on the basis of a HFS than to design a FN from a SFS. Moreover, it can be seen from the examples in Chapters 7-8 that network analysis tasks always have a unique solution whereas network design tasks may have multiple solutions or no solution at all.

The evaluation presented in this chapter demonstrates the applicability of FNs for solving some real problems without going into specific details. In this sense, the formal models for FNs used are mainly at network level, i.e. in the form of block schemes and topological expressions. Most formal models at node level such as Boolean matrices and binary relations are embedded implicitly in the block schemes and topological expressions.

All presented case studies are for FNs with a fairly small number of inputs and a single output. However, the extension to case studies of higher dimension is straightforward. The only difference in this extension is the more complex structure of the FN in the case of more inputs and the multiple use of the single equivalent node in the case of multiple outputs.

A. Gegov: Fuzzy Networks for Complex Systems, STUDFUZZ 259, pp. 247–274.
springerlink.com

9.2 Assessment of Structural Complexity

Structural complexity is an attribute of any general type of network. This attribute can be assessed by different measures and it is often linked to important network properties such as robustness. The latter shows the ability of the network to maintain some functionality when some of its links, nodes or even whole clusters are damaged.

In this context, a FN can also be assessed with respect to its structural complexity. Such an assessment may be useful not only for the analysis of existing FNs in terms of robustness but also for the design of new FNs that are robust to different types of damage.

A basic measure of structural complexity for a FN is the number of non-identity nodes. This number includes both feedforward and feedback nodes. Identity nodes are excluded from the considerations as they do not affect the existing complexity. In particular, feedforward identity nodes are used only for manipulating feedforward identity mappings propagating through at least one layer in the underlying grid structure of the network. As far as feedback identity nodes are concerned, they are used only for representing feedback identity mappings which are eventually removed from the underlying grid structure of the network.

Another basic measure of structural complexity for a FN is the number of non-identity connections. This number includes both feedforward and feedback connections. Identity connections are excluded from the considerations as they also do not affect the existing complexity. In this sense, feedforward and feedback identity connections are used only as inputs and outputs for feedforward and feedback identity nodes, respectively.

A more specific measure of structural complexity for a FN is the overall number of cells in the grid structure. This number can be obtained by multiplying the number of horizontal levels by the number of vertical layers. Identity nodes are included in the considerations as they may affect the number of cells, e.g. newly introduced identity nodes may lead to the appearance of new cells in the underlying grid structure of the network. This measure applies to both feedforward and feedback identity nodes.

Another more specific measure of structural complexity for a FN is the number of populated cells in the grid structure. This number can be obtained by enumerating all non-empty cells. Identity nodes are also included in the considerations as they may affect the number of populated cells, e.g. newly introduced identity nodes may be moved to new non-empty cells in the underlying grid structure of the network. This measure also applies to both feedforward and feedback identity nodes.

A more general measure of structural complexity for a FN is the average width, i.e. the average path length from first layer nodes to last layer nodes in the grid structure. This length can be obtained by first summing the number of links between any pair of nodes such that the first node is in the first layer and the last node is in the last layer. The sums for each pair are then summed and divided by the number of pairs. This measure reflects the width of the FN from left to right in a temporal context as the layers in the grid structure represent some kind of temporal hierarchy.

Another more general measure of structural complexity for a FN is the average depth, i.e. the average path length from first level nodes to last level nodes in the grid structure. This length can be obtained by first summing the number of links between any pair of nodes such that the first node is in the first level and the last node is in the last level. The sums for each pair are then summed and divided by the number of pairs. This measure reflects the depth of the FN from top to bottom in a spatial context as the levels in the grid structure represent some kind of spatial hierarchy.

9.3 Composition of Hierarchical Fuzzy Systems

The most common type of fuzzy system with multiple rule bases is the HFS. This fuzzy system is introduced in Chapter 2 and it has two major forms – forward and backward. In the first case, inputs are gradually added to rule bases in subsequent layers in an increasing order, i.e. from the first to the last input. In the second case, inputs are gradually added to rule bases in subsequent layers in a decreasing order, i.e. from the last to the first input.

A HFS can be converted into an initial FN which can then be composed into a final FN. The latter is similar to a SFS with a single rule base which is also introduced in Chapter 2. The two examples below consider the above two-step sequence of conversion and composition for a forward and backward HFS, respectively. Both examples relate to network analysis and they are presented in a general form.

Example 9.1

This example considers an initial HFS in a forward form with a set of 'm' inputs $\{x_1, x_2,\ldots, x_m\}$, a set of 'm-1' nodes $\{N_{11}, N_{12},\ldots, N_{1,m-1}\}$, a set of 'm-2' connections $\{z^1, z^2,\ldots, z^{m-2}\}$ and a single output y. Each connection represents a current iteration of the output and the number of connections is equal to the number of iterations. This initial HFS can be described by the block-scheme in Fig.9.1 and the topological expression in Eq.(9.1) from where it can be seen that each network node has two inputs and one output.

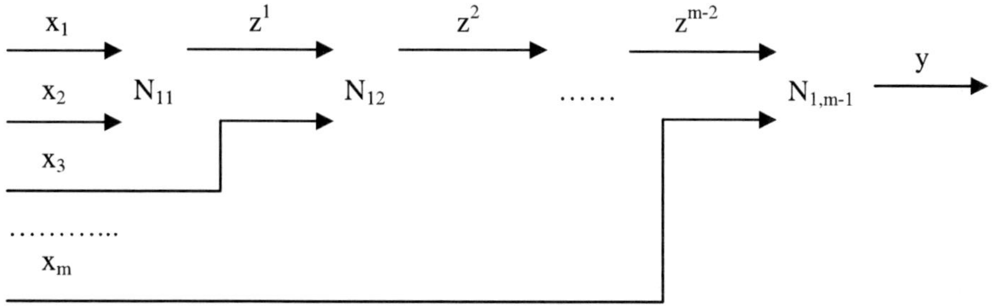

Fig. 9.1 HFS for Example 9.1

$$[N_{11}]\,(x_1, x_2 \,|\, z^1) * [N_{12}]\,(z^1, x_3 \,|\, z^2) * \ldots * [N_{1,m-1}]\,(z^{m-2}, x_m \,|\, y) \tag{9.1}$$

The initial HFS can be converted into an initial FN by representing all identity mappings propagating through any layers in the grid structure with the sets of identity nodes $\{I_{21}\}, \ldots, \{I_{m-1,1}, I_{m-1,2}, \ldots\}$. This initial FN can be described by the block-scheme in Fig.9.2 and the topological expression in Eq.(9.2). It can be seen from there that each network node has two inputs and one output whereas each identity node has one input and one output.

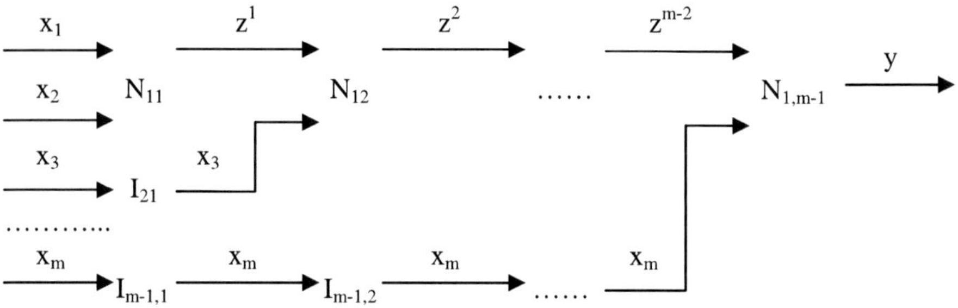

Fig. 9.2 Initial FN for Example 9.1

$$\{[N_{11}]\,(x_1, x_2 \,|\, z^1) + [I_{21}]\,(x_3 \,|\, x_3) + \ldots + [I_{m-1,1}]\,(x_m \,|\, x_m)\} * \tag{9.2}$$

$$\{[N_{12}]\,(z^1, x_3 \,|\, z^2) + \ldots + [I_{m-1,2}]\,(x_m \,|\, x_m)\} *$$

$$\ldots\ldots\ldots\ldots\ldots\ldots *$$

$$[N_{1,m-1}]\,(z^{m-2}, x_m \,|\, y)$$

The initial FN can be composed into a final FN by merging first vertically and then horizontally all network and identity nodes into a single equivalent node $*_{p=1}^{m-1}\,(N_{1p} + +_{q=p+1}^{m-1}\, I_{qp})$. This final FN can be described by the block-scheme in Fig.9.3 and the topological expression in Eq.(9.3) from where it can be seen that the single equivalent node has 'm' inputs and one output.

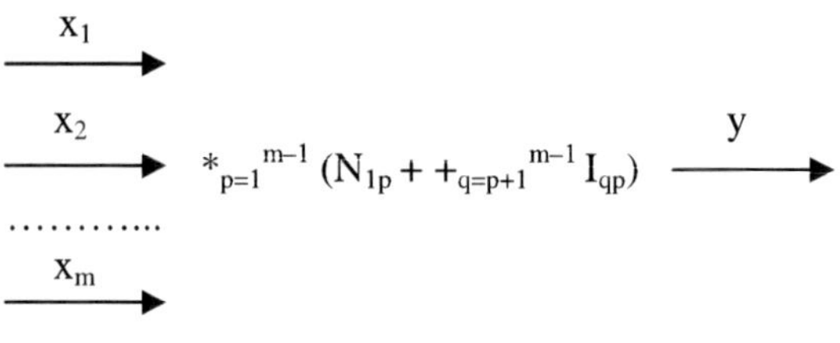

Fig. 9.3 Final FN for Example 9.1

$$[*_{p=1}^{m-1} (N_{1p} + +_{q=p+1}^{m-1} I_{qp})] (x_1, x_2,\ldots, x_m \mid y) \qquad (9.3)$$

If the HFS for Example 9.1 has a set of 'n' outputs $\{y_1, y_2,\ldots, y_n\}$, then it must be presented as a set of 'n' independent HFSs. In this case, the two-step sequence above is repeated for each HFS and its output.

Example 9.2

This example considers a HFS in a backward form with a set of 'm' inputs $\{x_1, x_2,\ldots, x_m\}$, a set of 'm-1' nodes $\{N_{m-1,1}, N_{m-1,2},\ldots, N_{m-1,m-1}\}$, a set of 'm-2' connections $\{z^1, z^2,\ldots, z^{m-2}\}$ and a single output y. Each connection represents a current iteration of the output and the number of connections is equal to the number of iterations. This HFS can be described by the block-scheme in Fig.9.4 and the topological expression in Eq.(9.4) from where it can be seen that each network node has two inputs and one output.

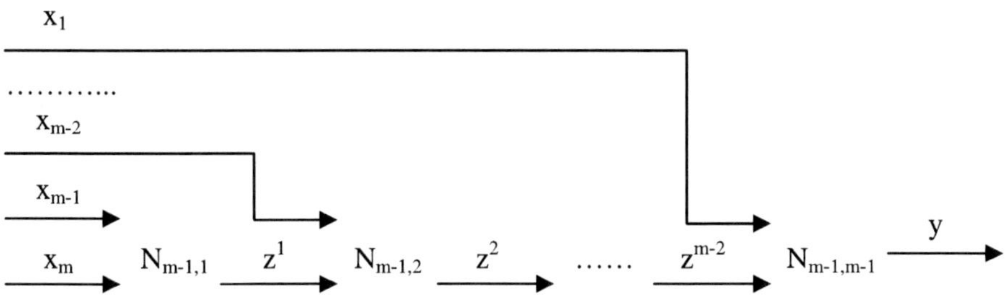

Fig. 9.4 HFS for Example 9.2

$$[N_{m-1,1}] (x_{m-1}, x_m \mid z^1) * [N_{m-1,2}] (x_{m-2}, z^1 \mid z^2) * \ldots * [N_{m-1,m-1}] (x_1 , z^{m-2} \mid y) \qquad (9.4)$$

The HFS can be converted into an initial FN by representing all identity mappings propagating through any layers in the grid structure with the sets of identity nodes $\{I_{11}, I_{12}, \ldots\}, \ldots ,\{I_{m-2,1}\}$. This initial FN can be described by the block-scheme in Fig.9.5 and the topological expression in Eq.(9.5). It can be seen from there that each network node has two inputs and one output whereas each identity node has one input and one output.

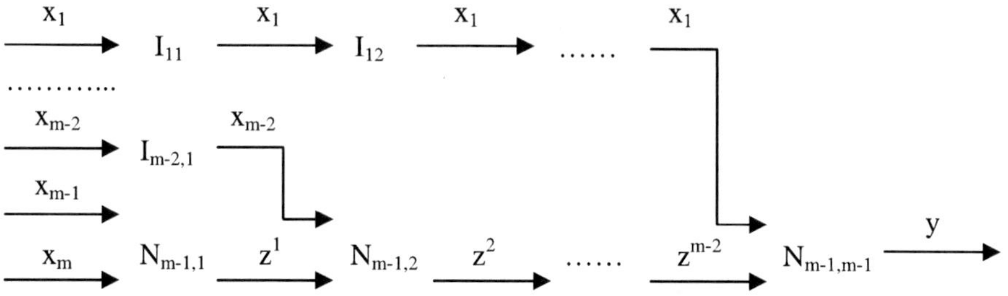

Fig. 9.5 Initial FN for Example 9.2

$$\{[I_{11}] (x_1 \,|\, x_1) + \ldots + [I_{m-2,1}] (x_{m-2} \,|\, x_{m-2}) + [N_{m-1,1}] (x_{m-1}, x_m \,|\, z^1)\} * \tag{9.5}$$

$$\ldots\ldots\ldots\ldots\ldots\ldots\ldots\ldots *$$

$$\{[I_{12}] (x_1 \,|\, x_1) + \ldots + [N_{m-1,2}] (x_{m-2}, z^1 \,|\, z^2)\} *$$

$$[N_{m-1,m-1}] (x_1, z^{m-2} \,|\, y)$$

The initial FN can be composed into a final FN by merging first vertically and then horizontally all network and identity nodes into a single equivalent node $*_{p=1}^{m-1} (+_{q=1}^{m-1-p} I_{qp} + N_{m-1,p})$. This final FN can be described by the block-scheme in Fig.9.6 and the topological expression in Eq.(9.6) from where it can be seen that the single equivalent node has 'm' inputs and one output.

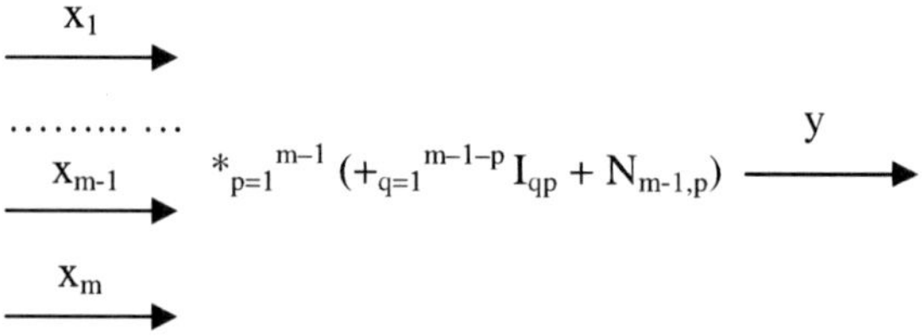

Fig. 9.6 Final FN for Example 9.2

$$[*_{p=1}^{m-1} (+_{q=1}^{m-1-p} I_{qp} + N_{m-1,p}) (x_1, \ldots, x_{m-1}, x_m \,|\, y) \tag{9.6}$$

If the HFS for Example 9.2 has a set of 'n' outputs $\{y_1, y_2, \ldots, y_n\}$, then it must be presented as a set of 'n' independent HFSs. In this case, the two-step sequence above is repeated for each HFS and its output.

9.4 Decomposition of Standard Fuzzy Systems

The most common type of fuzzy system is the SFS. A SFS can be decomposed into an initial FN which can then be converted into a final FN. The latter is similar to a HFS with multiple rule bases. The two algorithms below consider the above two-step sequence of decomposition and conversion into a forward and backward HFS, respectively. Both algorithms relate to network design and they are presented in a general form.

The sequence of decomposition and conversion of a SFS into a HFS using FNs is an inverse image of the sequence of conversion and composition of a HFS into a SFS using FNs. In either case, the initial and the final FNs act as a bridge between the SFS and the HFS.

Algorithm 9.1 follows from Example 9.1. The algorithm can be used with Figs.9.1-9.3 and Eqs.(9.1)-(9.3) whereby the ordering of the figures and the equations is reversed. In this case, $N_{E,k}$ is the single equivalent node for the fuzzy subnetwork with the first 'k' inputs of all 'm' inputs to the SFS. This single equivalent node is given by Eq.(9.7) which is special case of Eq.(9.3).

$$N_{E,k} = *_{p=1}^{k-1} (N_{1p} + +_{q=p+1}^{k-1} I_{qp}) \tag{9.7}$$

Algorithm 9.1

1. Find N_{11} from the first two inputs and the output.
2. If m=2, go to step 9.
3. Set k=3.
4. While k≤m, do steps 5-7.
5. Find $N_{E,k}$ from the first k inputs and the output.
6. Derive $N_{1,k-1}$ from Eq.(9.7) for $N_{E,k}$, if possible.
7. Set k=k+1.
8. Endwhile.
9. End.

Algorithm 9.2 follows from Example 9.2. The algorithm can be used in conjunction with Figs.9.4-9.6 and Eqs.(9.4)-(9.6) whereby the order of the figures and the equations is reversed. In this case, $N_{E,k}$ is the single equivalent node for the fuzzy subnetwork with the last 'k' inputs of all 'm' inputs to the SFS. This single equivalent node is given by Eq.(9.8) which is special case of Eq.(9.6).

$$N_{E,k} = *_{p=1}^{k-1} (+_{q=1}^{k-1-p} I_{qp} + N_{k-1,p}) \tag{9.8}$$

Algorithm 9.2

1. Find $N_{m-1,1}$ from the last two inputs and the output.
2. If m=2, go to step 9.
3. Set k=3.
4. While k≤m, do steps 5-7.
5. Find $N_{E,k}$ from the last k inputs and the output.
6. Derive $N_{m-1,k-1}$ from the formula for $N_{E,k}$, if possible.
7. Set k=k+1.
8. Endwhile.
9. End.

9.5 Indicators of Model Performance

SFSs, HFSs and FNs can be used as models of different processes. These models can be built on the basis of measured data or expert knowledge. The quality of the models can be quantified using different performance indicators. For this purpose, four model performance indicators are discussed further in this section. These are the Feasibility Index (FI), the Accuracy Index (AI), the Efficiency Index (EI) and the Transparency Index (TI). Some of these indicators are similar to indicators already used in fuzzy systems whereas the others are novel and specific to FNs.

The first model performance indicator considered is the FI. This indicator is approximate in that it gives a general indication of the extent to which it is possible to build a model in the first place. The FI is given by Eq.(9.9).

$$FI = (\text{sum}_{i=1}^{n} p_i) / n \tag{9.9}$$

The notations in Eq.(9.9) are as follows: 'n' is number of non-identity nodes, 'p_i' is number of inputs to the i-th non-identity node and 'sum' is a symbol for arithmetic summation. The assumption made here is that it is easier to build a model for a smaller average number of inputs per node in the model. Also, identity nodes are excluded from this indicator as they are virtual nodes for converting a HFS into a FN which do not affect the feasibility. It is obvious from Eq.(9.9) that a lower FI implies better model feasibility.

The second model performance indicator considered is the AI. This indicator is precise in that it gives a specific indication of the mean absolute difference between the model and the data, i.e. the modelling error. The AI is given by Eq.(9.10).

$$AI = \text{sum}_{i=1}^{nl} \text{sum}_{j=1}^{qil} \text{sum}_{k=1}^{vji} (|y_{ji}^{k} - d_{ji}^{k}| / vij) \tag{9.10}$$

The notations in Eq.(9.10) are as follows: 'nl' is the number of nodes in the last layer, 'qil' is the number of outputs from the i-th node in the last layer, 'vji' is the number of discrete values for the j-th output from the i-th node in the last layer, 'y_{ji}^{k}' is the simulated k-th discrete value for the j-th output from the i-th node in the last layer and 'd_{ji}^{k}' is the measured k-th discrete value for the j-th output from the i-th node in the last layer, 'sum' is a symbol for arithmetic summation and '| |' is a symbol for absolute value. In this case, identity nodes are included in this indicator alongside any other nodes in the last layer as their outputs also have to be compared with the data. It is obvious from Eq.(9.10) that a lower AI implies better model accuracy.

The third model performance indicator considered is the EI. This indicator is precise in that it gives a specific indication of the overall number of rules in the model. The EI is given by Eq.(9.11).

$$EI = \text{sum}_{i=1}^{n} (q_i^{FID} . r_i^{FID}) \tag{9.11}$$

The notations in Eq.(9.11) are as follows: 'n' is the number of non-identity network nodes, 'q_i^{FID}' is the number of outputs from the i-th non-identity node with an associated FID sequence, 'r_i' is the number of rules for the i-th non-identity node with an associated FID sequence and 'sum' is a symbol for arithmetic summation. In this case, a model is more efficient if the overall number of rules is smaller as this number is proportional to the overall amount of computations. Also, identity nodes are excluded from this indicator as they are virtual nodes for converting a HFS into a FN which do not affect the efficiency. It is obvious from Eq.(9.11) that a lower EI implies better model efficiency.

The last model performance indicator considered is the TI. This indicator is approximate in that it gives a general indication of the extent to which the model can be inspected from the inside, i.e. as a white box rather than a black box. The TI is given by Eq.(9.12).

$$TI = (p + q) / (n + m) \tag{9.12}$$

The notations in Eq.(9.12) are as follows: 'p' is the overall number of inputs, 'q' is the overall number of outputs, 'n' is the number of non-identity nodes, 'm' is the number of non-identity connections and 'sum' is a symbol for arithmetic summation. The assumption made here is that it is easier to inspect a model from the inside in the presence of fewer inputs and outputs for the overall model as well as in the presence of more submodels and interactions between them. Also, identity nodes are excluded from this indicator as they are virtual nodes for converting a HFS into a FN which do not affect the transparency. It is obvious from Eq.(9.12) that a lower TI implies better model transparency.

9.6 Applications for Case Studies

The applications of some of the theoretical results from Chapters 4-8 and some of the evaluation methods from the preceding sections of this chapter are illustrated for two case studies. The first case study deals with mortgage assessment which is based on expert knowledge from the bank industry [110]. The second case study deals with product pricing which is based on statistical data from the retail industry [55, 122].

Case Study 9.1

This case study is about a decision support system for assessing mortgage applications. The assessment is based on separate evaluations of the applicant and the property. The input factors taken into account for the evaluation of the applicant are their asset and the income. For the evaluation of the property, the input factors taken into account are its price and location. The outputs from these two evaluation stages are the applicant status and the property status. These outputs, together with the interest on the mortgage and the income of the applicant, are fed as input factors for the evaluation of the amount of credit that can be given to the applicant. The output from this third evaluation stage is the credit status.

The decision support system above can be represented by an initial FN. The latter can be described by the block-scheme in Fig.9.7 and the topological expression in Eq.(9.13). The notations used in the figure and the equation are as follows: N_{11} is the rule base for the applicant evaluation, N_{21} is the rule base for the property evaluation, N_{12} is the rule base for the credit evaluation, $x_{11,12}^{1,3}$ is the applicant income, x_{11}^{2} is the applicant asset, x_{12}^{2} is the mortgage interest, x_{21}^{1} is the property location, x_{21}^{2} is the property price, $z_{11,12}^{1,1}$ is the applicant status, $z_{21,12}^{1,4}$ is the property status and y_{12} is the credit status.

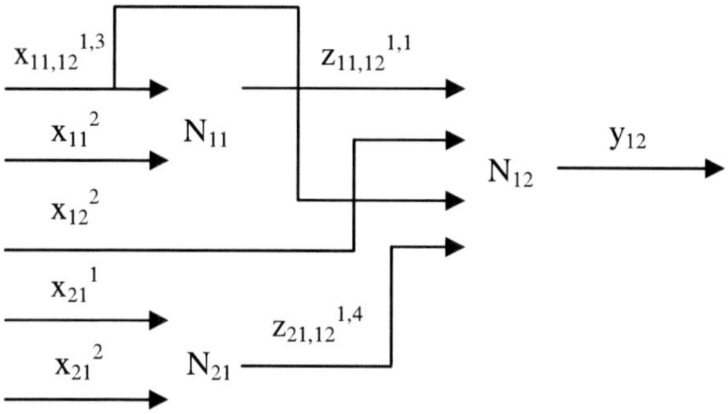

Fig. 9.7 Initial FN for Case Study 9.1

$$\{[N_{11}]\,(x_{11,12}^{1,3}, x_{11}^{2} \mid z_{11,12}^{1,1}) + [N_{21}]\,(x_{21}^{1}, x_{21}^{2} \mid z_{21,12}^{1,4})\} * \qquad (9.13)$$

$$[N_{12}]\,(z_{11,12}^{1,1}, x_{12}^{2}, x_{11,12}^{1,3}, z_{21,12}^{1,4} \mid y_{12})$$

There are two identity mappings propagating through the first layer of the underlying grid structure of the initial FN - $x_{11,12}^{1,3}$ and x_{12}^{2}. These mappings can be presented by the identity nodes I_{01} and $I_{1.5,1}$, respectively. As a result of this presentation, the initial FN can be transformed into a first interim FN. The latter can be described by the block-scheme in Fig.9.8 and the topological expression in Eq.(9.14).

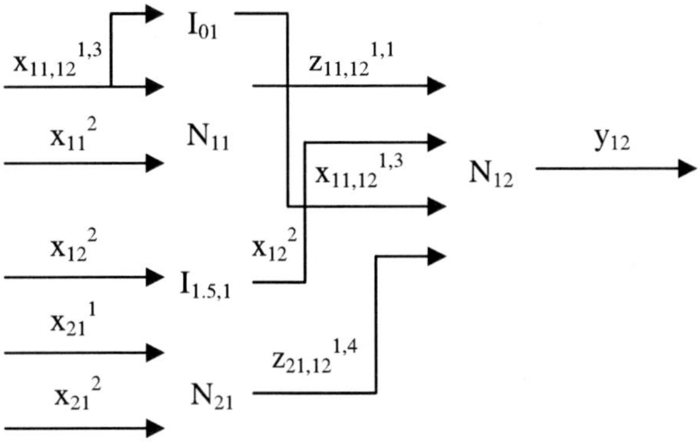

Fig. 9.8 First interim FN for Case Study 9.1

$$\{\{[I_{01}]\ (x_{11,12}^{1,3} \mid x_{11,12}^{1,3})\ ;\ [N_{11}]\ (x_{11,12}^{1,3}, x_{11}^{2} \mid z_{11,12}^{1,1})\} + \quad (9.14)$$

$$[I_{1.5,1}]\ (x_{12}^{12} \mid x_{12}^{12}) +\ [N_{21}]\ (x_{21}^{1}, x_{21}^{2} \mid z_{21,12}^{1,4})\} *$$

$$[N_{12}]\ (z_{11,12}^{1,1},\ x_{12}^{2}, x_{11,12}^{1,3}, z_{21,12}^{1,4} \mid y_{12})$$

The outputs $x_{11,12}^{1,3}$ and $z_{11,12}^{1,1}$ from nodes I_{01} and N_{11} in the first interim FN could be merged if I_{01} is first augmented as I_{01}^{AI} with the input x_{11}^{2}. This augmentation operation transforms the first interim FN into a second interim FN. The latter can be described by the block-scheme in Fig.9.9 and the topological expression in Eq.(9.15).

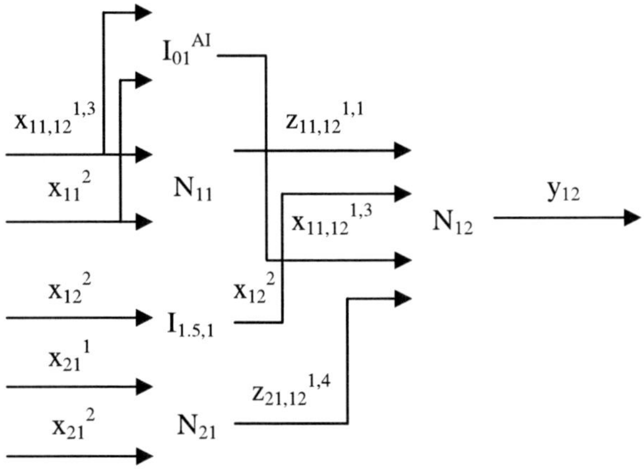

Fig. 9.9 Second interim FN for Case Study 9.1

$$\{\{[I_{01}^{AI}]\,(x_{11,12}^{1,3}, x_{11}^{2} \,|\, x_{11,12}^{1,3})\ ;\ [N_{11}]\,(x_{11,12}^{1,3}, x_{11}^{2} \,|\, z_{11,12}^{1,1})\} + [I_{1.5,1}]\,(x_{12}^{12} \,|\, x_{12}^{12}) + [N_{21}]\,(x_{21}^{1}, x_{21}^{2} \,|\, z_{21,12}^{1,4})\} * [N_{12}]\,(z_{11,12}^{1,1}, x_{12}^{2}, x_{11,12}^{1,3}, z_{21,12}^{1,4} \,|\, y_{12}) \tag{9.15}$$

The outputs $x_{11,12}^{1,3}$ and $z_{11,12}^{1,1}$ from nodes I_{01}^{AI} and N_{11} in the second interim FN can already be merged as both nodes have the same common inputs $x_{11,12}^{1,3}$ and x_{11}^{2}. This merging operation transforms the second interim FN into a third interim FN. The latter can be described by the block-scheme in Fig.9.10 and the topological expression in Eq.(9.16).

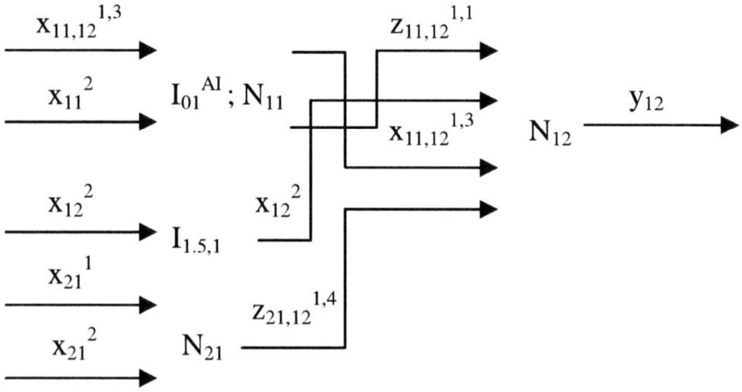

Fig. 9.10 Third interim FN for Case Study 9.1

$$\{[I_{01}^{AI}; N_{11}]\,(x_{11,12}^{1,3}, x_{11}^{2} \,|\, x_{11,12}^{1,3}, z_{11,12}^{1,1}) + [I_{1.5,1}]\,(x_{12}^{12} \,|\, x_{12}^{12}) + [N_{21}]\,(x_{21}^{1}, x_{21}^{2} \,|\, z_{21,12}^{1,4})\} * [N_{12}]\,(z_{11,12}^{1,1}, x_{12}^{2}, x_{11,12}^{1,3}, z_{21,12}^{1,4} \,|\, y_{12}) \tag{9.16}$$

The nodes I_{01}^{AI}; N_{11}, $I_{1.5,1}$ and N_{21} in the third interim FN can be merged vertically. This merging operation transforms the third interim FN into a fourth interim FN. The latter can be described by the block-scheme in Fig.9.11 and the topological expression in Eq.(9.17).

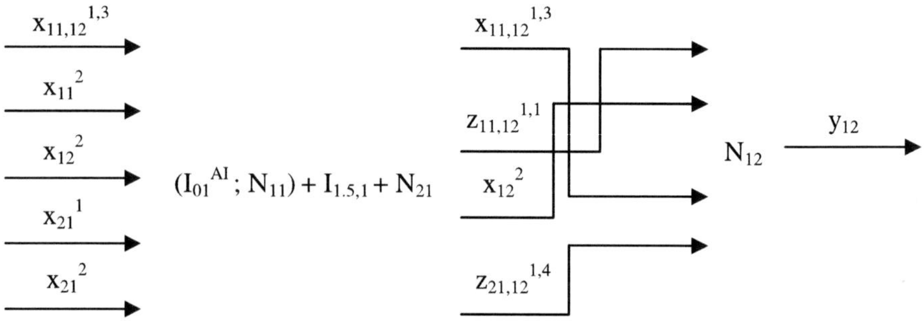

Fig. 9.11 Fourth interim FN for Case Study 9.1

$$[(I_{01}^{AI}; N_{11}) + I_{1.5,1} + N_{21}] \quad (9.17)$$

$$(x_{11,12}^{1,3}, x_{11}^{2}, x_{12}^{12}, x_{21}^{1}, x_{21}^{2} \mid x_{11,12}^{1,3}, z_{11,12}^{1,1}, x_{12}^{12}, z_{21,12}^{1,4}) \; *$$

$$[N_{12}]\; (z_{11,12}^{1,1}, x_{12}^{2}, x_{11,12}^{1,3}, z_{21,12}^{1,4} \mid y_{12})$$

The first three outputs from node $(I_{01}^{AI}; N_{11}) + I_{1.5,1} + N_{21}$ in the fourth interim FN can be permuted such that the first output becomes third, the second output becomes first and the third output becomes second. This permutation operation transforms the fourth interim FN into a fifth interim FN. The latter can be described by the block-scheme in Fig.9.12 and the topological expression in Eq.(9.18).

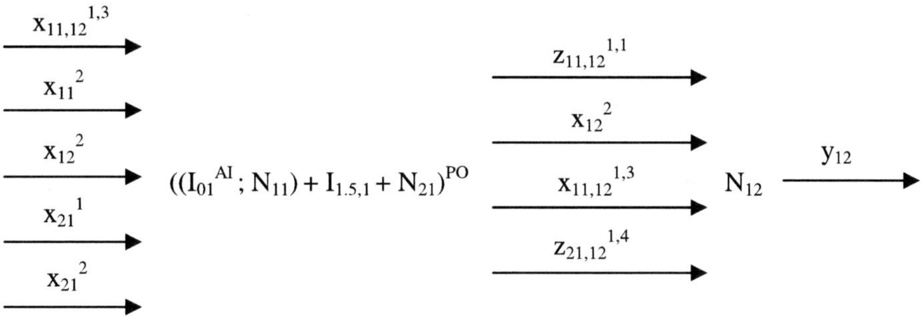

Fig. 9.12 Fifth interim FN for Case Study 9.1

$$[((I_{01}^{AI}; N_{11}) + I_{1.5,1} + N_{21})^{PO}] \quad (9.18)$$

$$(x_{11,12}^{1,3}, x_{11}^{2}, x_{12}^{12}, x_{21}^{1}, x_{21}^{2} \mid z_{11,12}^{1,1}, x_{12}^{12}, x_{11,12}^{1,3}, z_{21,12}^{1,4}) \; *$$

$$[N_{12}]\; (z_{11,12}^{1,1}, x_{12}^{2}, x_{11,12}^{1,3}, z_{21,12}^{1,4} \mid y_{12})$$

The nodes $((I_{01}^{AI}; N_{11}) + I_{1.5,1} + N_{21})^{PO}$ and N_{21} in the fifth interim FN can be merged horizontally. This merging operation transforms the fifth interim FN into a final FN. The latter can be described by the block-scheme in Fig.9.13 and the topological expression in Eq.(9.19).

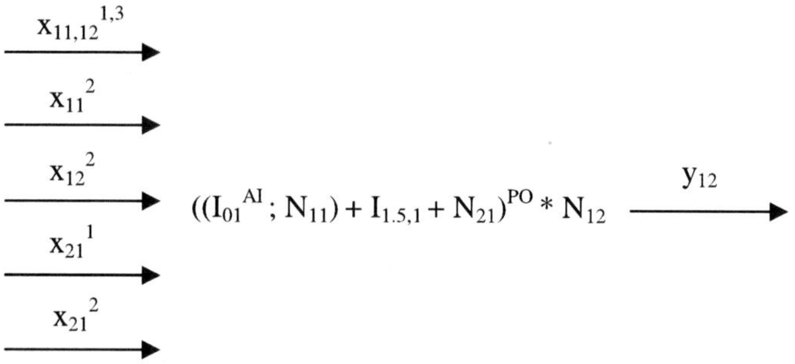

Fig. 9.13 Final FN for Case Study 9.1

$$[((I_{01}^{AI};N_{11})+I_{1.5,1}+N_{21})^{PO}*N_{12}]\ (x_{11,12}^{1,3}, x_{11}^{2}, x_{12}^{12}, x_{21}^{1}, x_{21}^{2} \mid y_{12}) \tag{9.19}$$

The structural complexity of the FN for this case study can be assessed using the measures introduced in Section 9.2. It can be seen from the initial FN in Fig.9.7 that the number of non-identity nodes is 3 and the number of non-identity connections is 2. As far as the other measures of structural complexity of the FN are concerned, they can be found from the first interim FN and are as follows: overall number of cells – 8, number of populated cells – 5, average width – 1, average depth – 0.

Case Study 9.2

This case study is about a decision support system for determining product prices. The determination is based on the maximum cost that a retailer is willing to pay to a manufacturer or a trader for the provision or the delivery of a certain quantity of a product. The input factors taken into account for the determination of the price are the expected selling price of the product, the margin, i.e. the relative difference between the price and the cost of the product, and the expected sell through, i.e. the relative quantity of the product expected to be sold. The output from this process is the max cost, i.e. the maximum cost of the product.

The decision support system above can be represented by a SFS. The latter can be described by the block-scheme in Fig.9.14. The notations used are as follows: N is the rule base for the SFS, x_1 is the expected selling price, x_2 is the margin, x_3 is the expected sell through and y is the max cost.

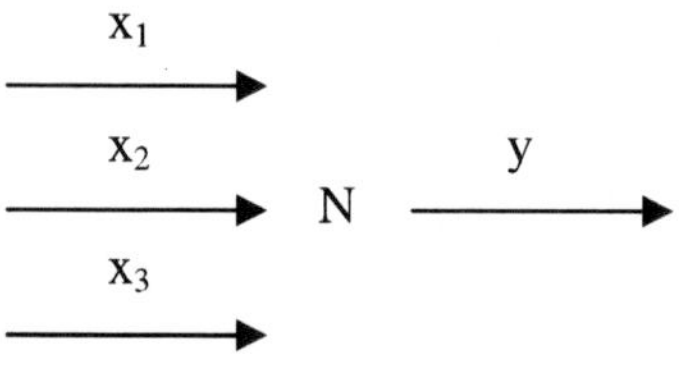

Fig. 9.14 SFS for Case Study 9.2

In addition, the decision support system above can be represented by a HFS. The latter can be described by the block-scheme in Fig.9.15. The notations used are as follows: N_{11} is the first rule base for the HFS, N_{12} is the second rule base for the HFS, x_1, x_2 , x_3 and y have the same meanings as for the SFS whereas z is the provisional max cost, i.e. the provisional maximum cost of the product.

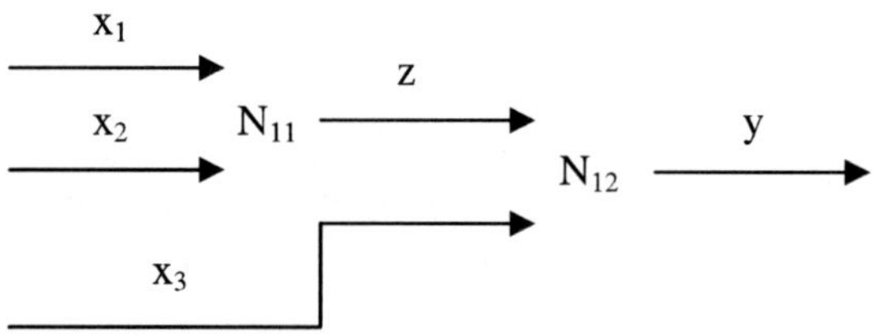

Fig. 9.15 HFS for Case Study 9.2

Finally, the decision support system above can be represented by an initial FN with two levels and two layers. The latter can be described by the block-scheme in Fig.9.16 whereby most notations used are the same as the ones for the HFS. The only new notation here is the identity rule base I_{21} representing the propagation of the identity mapping x_3 through the first layer. In this context, N_{11} and N_{12} are network rules bases.

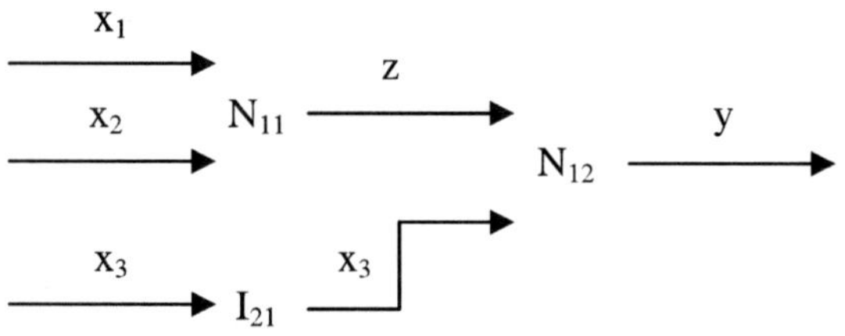

Fig. 9.16 Initial FN for Case Study 9.2

The initial FN can be transformed into a final FN with a single equivalent rule base. The latter can be described by the block-scheme in Fig.9.17 whereby most notations used are the same as the ones for the SFS in Fig.9.14. The only difference here is that the rule base N for the SFS is replaced by the single equivalent rule base $(N_{11} + I_{21}) * N_{12}$ for the FN.

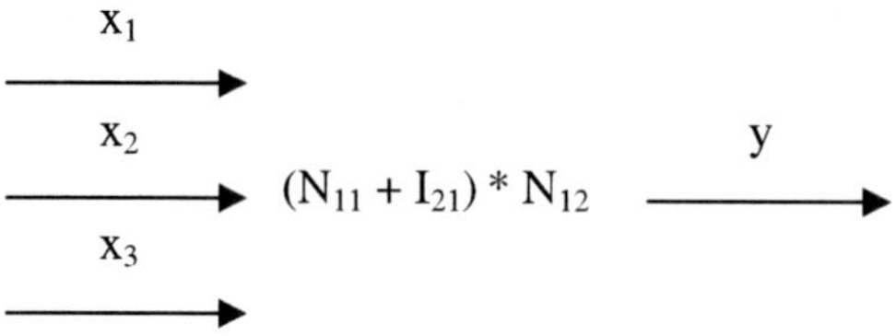

Fig. 9.17 Final FN for Case Study 9.2

The inputs x_1, x_2, x_3 are presented by five linguistic terms each, as shown in Figs.9.18-9.20. These terms represent triangular fuzzy membership functions that cover uniformly the whole variation range for the inputs. For consistency, all variation ranges are normalised between 0 and 100.

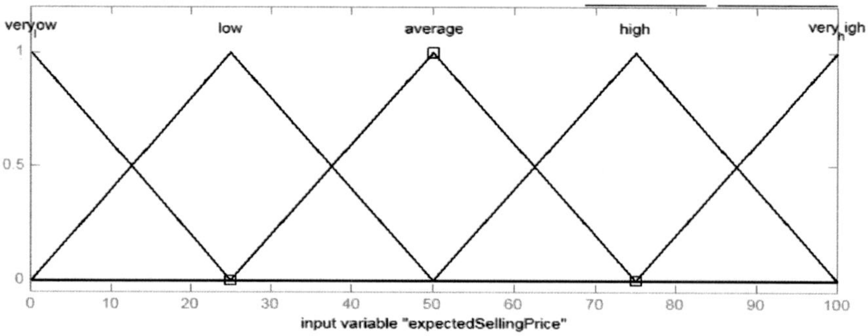

Fig. 9.18 Linguistic terms for first input in Case Study 9.2

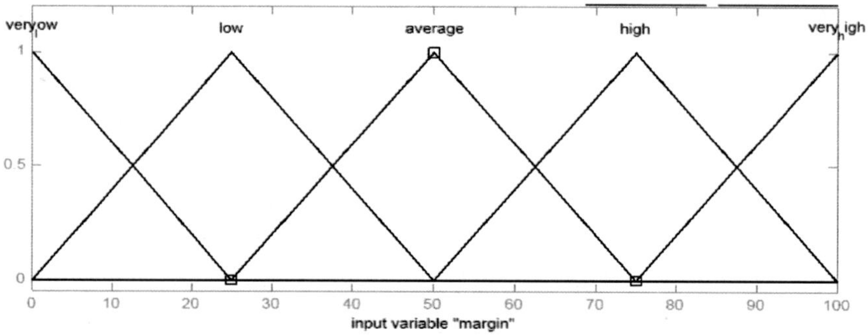

Fig. 9.19 Linguistic terms for second input in Case Study 9.2

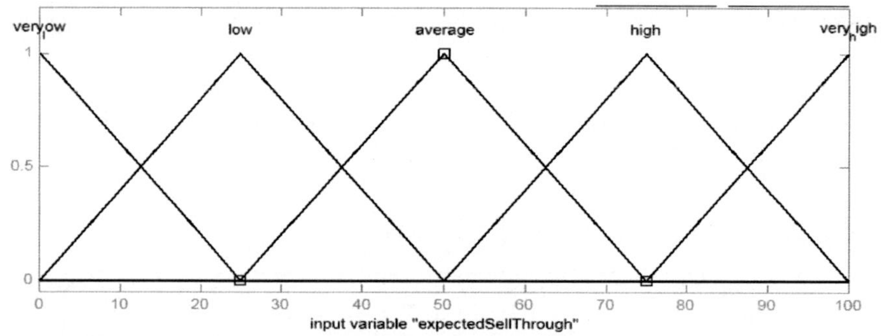

Fig. 9.20 Linguistic terms for third input in Case Study 9.2

The output y and the connection z are presented by eleven linguistic terms each, as shown in Figs.9.21-9.22. These terms also represent triangular fuzzy membership functions that cover uniformly the whole variation range for the output and the connection. Similarly, all variation ranges are normalised between 0 and 100.

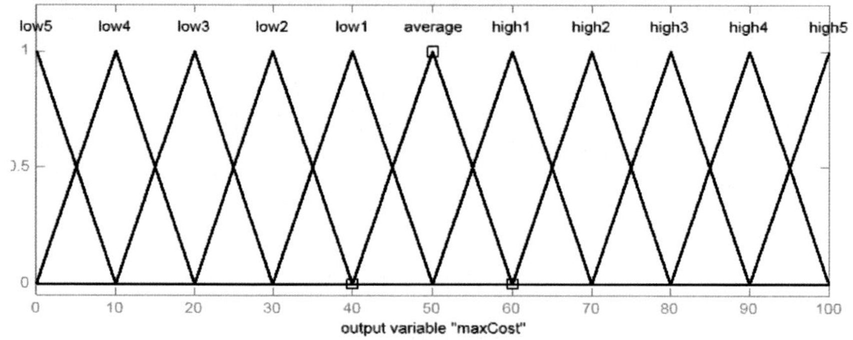

Fig. 9.21 Linguistic terms for output in Case Study 9.2

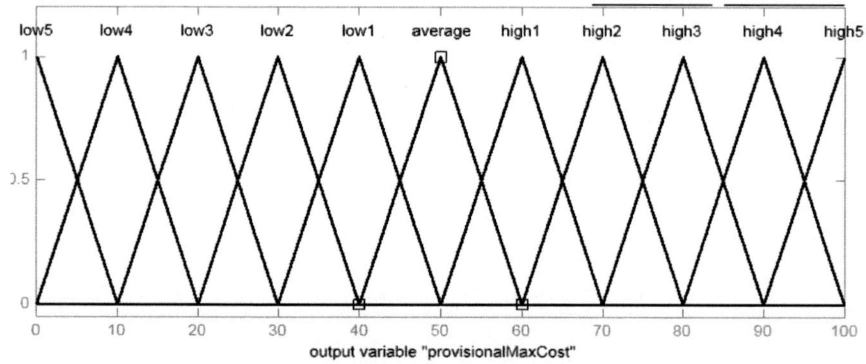

Fig. 9.22 Linguistic terms for connection in Case Study 9.2

The rule base for the SFS is shown as an integer table in two parts in Tables 9.1-9.2. This rule base is derived from statistical data about the product pricing process. The derivation is done using a clustering approach whereby the rules represent an approximation of the input-output data points from the data set for the process.

Table 9.1 First part of rule base for SFS in Case Study 9.2

x_1	x_2	x_3	y	x_1	x_2	x_3	y	x_1	x_2	x_3	y
1	1	1	1	2	1	1	1	3	1	1	1
1	1	2	1	2	1	2	2	3	1	2	4
1	1	3	1	2	1	3	4	3	1	3	6
1	1	4	1	2	1	4	5	3	1	4	9
1	1	5	1	2	1	5	6	3	1	5	11
1	2	1	1	2	2	1	1	3	2	1	1
1	2	2	1	2	2	2	2	3	2	2	3
1	2	3	1	2	2	3	3	3	2	3	5
1	2	4	1	2	2	4	4	3	2	4	7
1	2	5	1	2	2	5	5	3	2	5	9
1	3	1	1	2	3	1	1	3	3	1	1
1	3	2	1	2	3	2	2	3	3	2	2
1	3	3	1	2	3	3	2	3	3	3	4
1	3	4	1	2	3	4	3	3	3	4	5
1	3	5	1	2	3	5	4	3	3	5	6
1	4	1	1	2	4	1	1	3	4	1	1
1	4	2	1	2	4	2	1	3	4	2	2
1	4	3	1	2	4	3	2	3	4	3	2
1	4	4	1	2	4	4	2	3	4	4	3
1	4	5	1	2	4	5	2	3	4	5	4
1	5	1	1	2	5	1	1	3	5	1	1
1	5	2	1	2	5	2	1	3	5	2	1
1	5	3	1	2	5	3	1	3	5	3	1
1	5	4	1	2	5	4	1	3	5	4	1
1	5	5	1	2	5	5	1	3	5	5	1

Table 9.2 Second part of rule base for SFS in Case Study 9.2

x_1	x_2	x_3	y	x_1	x_2	x_3	y
4	1	1	1	5	1	1	1
4	1	2	5	5	1	2	6
4	1	3	9	5	1	3	11
4	1	4	11	5	1	4	11
4	1	5	11	5	1	5	11
4	2	1	1	5	2	1	1
4	2	2	4	5	2	2	5
4	2	3	7	5	2	3	9
4	2	4	9	5	2	4	11
4	2	5	11	5	2	5	11
4	3	1	1	5	3	1	1
4	3	2	3	5	3	2	4
4	3	3	5	5	3	3	6
4	3	4	7	5	3	4	9
4	3	5	9	5	3	5	11
4	4	1	1	5	4	1	1
4	4	2	2	5	4	2	2
4	4	3	3	5	4	3	4
4	4	4	4	5	4	4	5
4	4	5	5	5	4	5	6
4	5	1	1	5	5	1	1
4	5	2	1	5	5	2	1
4	5	3	1	5	5	3	1
4	5	4	1	5	5	4	1
4	5	5	1	5	5	5	1

The two rule bases for the HFS are shown as integer tables in Tables 9.3-9.4. These rule bases are derived from statistical data about the two subprocesses within the product pricing process. The derivation is also done using a clustering approach whereby the rules represent an approximation of the input-output data points from the data sets for the subprocesses.

Table 9.3 First rule base for HFS in Case Study 9.2

x_1	x_2	z	x_1	x_2	z	x_1	x_2	z
1	1	1	3	1	6	5	1	11
1	2	1	3	2	5	5	2	9
1	3	1	3	3	4	5	3	6
1	4	1	3	4	2	5	4	4
1	5	1	3	5	1	5	5	1
2	1	4	4	1	9	-	-	-
2	2	3	4	2	7	-	-	-
2	3	2	4	3	5	-	-	-
2	4	2	4	4	3	-	-	-
2	5	1	4	5	1	-	-	-

Table 9.4 Second rule base for HFS in Case Study 9.2

z	x_3	y	z	x_3	y	z	x_3	y
1	1	1	5	1	1	9	1	1
1	2	1	5	2	3	9	2	5
1	3	1	5	3	5	9	3	9
1	4	1	5	4	7	9	4	11
1	5	1	5	5	9	9	5	11
2	1	1	6	1	1	10	1	1
2	2	2	6	2	4	10	2	6
2	3	2	6	3	6	10	3	10
2	4	3	6	4	9	10	4	11
2	5	3	6	5	11	10	5	11
3	1	1	7	1	1	11	1	1
3	2	2	7	2	4	11	2	6
3	3	3	7	3	7	11	3	11
3	4	4	7	4	10	11	4	11
3	5	5	7	5	11	11	5	11
4	1	1	8	1	1	-	-	-
4	2	3	8	2	5	-	-	-
4	3	4	8	3	8	-	-	-
4	4	6	8	4	11	-	-	-
4	5	7	8	5	11	-	-	-

The rule base for the FN is shown as an integer table in two parts in Tables 9.5-9.6. This rule base is derived using merging operations on the two nodes from the HFS and the associated identity node.

Table 9.5 First part of rule base for FN in Case Study 9.2

x_1	x_2	x_3	y	x_1	x_2	x_3	y	x_1	x_2	x_3	y
1	1	1	1	2	1	1	1	3	1	1	1
1	1	2	1	2	1	2	3	3	1	2	4
1	1	3	1	2	1	3	4	3	1	3	6
1	1	4	1	2	1	4	6	3	1	4	9
1	1	5	1	2	1	5	7	3	1	5	11
1	2	1	1	2	2	1	1	3	2	1	1
1	2	2	1	2	2	2	2	3	2	2	3
1	2	3	1	2	2	3	3	3	2	3	5
1	2	4	1	2	2	4	4	3	2	4	7
1	2	5	1	2	2	5	5	3	2	5	9
1	3	1	1	2	3	1	1	3	3	1	1
1	3	2	1	2	3	2	2	3	3	2	3
1	3	3	1	2	3	3	2	3	3	3	4
1	3	4	1	2	3	4	3	3	3	4	6
1	3	5	1	2	3	5	3	3	3	5	7
1	4	1	1	2	4	1	1	3	4	1	1
1	4	2	1	2	4	2	2	3	4	2	2
1	4	3	1	2	4	3	2	3	4	3	2
1	4	4	1	2	4	4	3	3	4	4	3
1	4	5	1	2	4	5	3	3	4	5	3
1	5	1	1	2	5	1	1	3	5	1	1
1	5	2	1	2	5	2	1	3	5	2	1
1	5	3	1	2	5	3	1	3	5	3	1
1	5	4	1	2	5	4	1	3	5	4	1
1	5	5	1	2	5	5	1	3	5	5	1

Table 9.6 Second part of rule base for FN in Case Study 9.2

x_1	x_2	x_3	y	x_1	x_2	x_3	y
4	1	1	1	5	1	1	1
4	1	2	5	5	1	2	6
4	1	3	9	5	1	3	11
4	1	4	11	5	1	4	11
4	1	5	11	5	1	5	11
4	2	1	1	5	2	1	1
4	2	2	4	5	2	2	5
4	2	3	7	5	2	3	9
4	2	4	10	5	2	4	11
4	2	5	11	5	2	5	11
4	3	1	1	5	3	1	1
4	3	2	3	5	3	2	4
4	3	3	5	5	3	3	6
4	3	4	7	5	3	4	9
4	3	5	9	5	3	5	11
4	4	1	1	5	4	1	1
4	4	2	2	5	4	2	3
4	4	3	3	5	4	3	4
4	4	4	4	5	4	4	6
4	4	5	5	5	4	5	7
4	5	1	1	5	5	1	1
4	5	2	1	5	5	2	1
4	5	3	1	5	5	3	1
4	5	4	1	5	5	4	1
4	5	5	1	5	5	5	1

The output surfaces for the SFS, the HFS and the FN are shown in Figs.9.23-9.26 in three dimensions. In this case, a separate surface represents each of the two rule bases of the HFS. The rule bases for the SFS and the FN are represented by one surface each whereby the third input is fixed to the value in the middle of its variation range, i.e. 50.

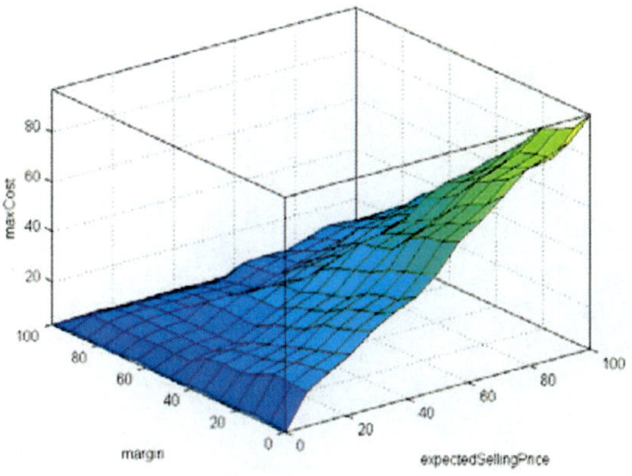

Fig. 9.23 Output surface for SFS in Case Study 9.2

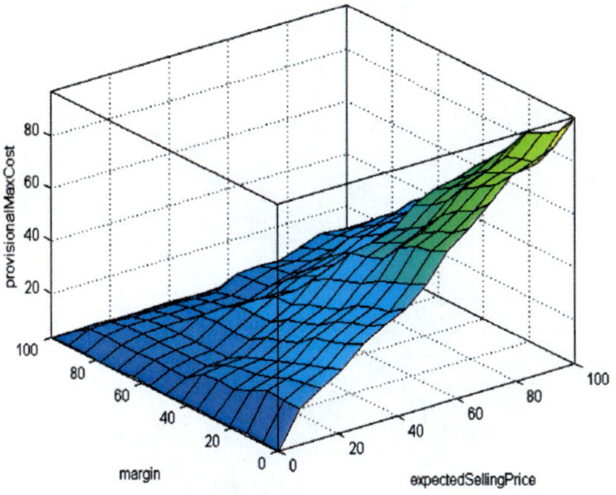

Fig. 9.24 Output surface for first rule base of HFS in Case Study 9.2

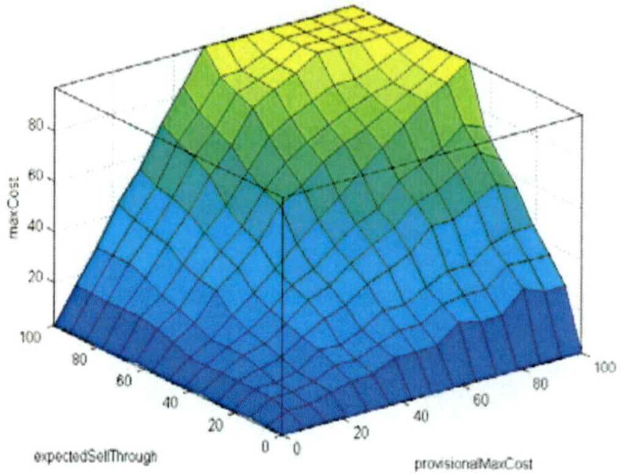

Fig. 9.25 Output surface for second rule base of HFS in Case Study 9.2

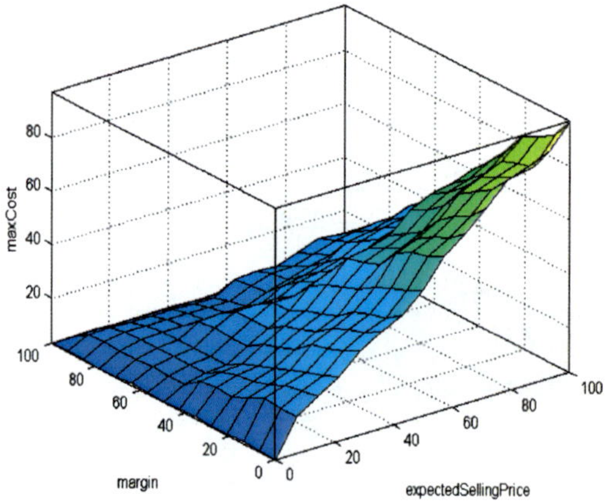

Fig. 9.26 Output surface for FN in Case Study 9.2

The simulation results for the SFS, the HFS and the FN are shown in Figs.9.27-9.29 where the data and the model output are presented in blue and green, respectively. In this case, each of the three models is simulated for all 125 possible permutations of the discrete crisp values of the inputs 0, 25, 50 75, 100.

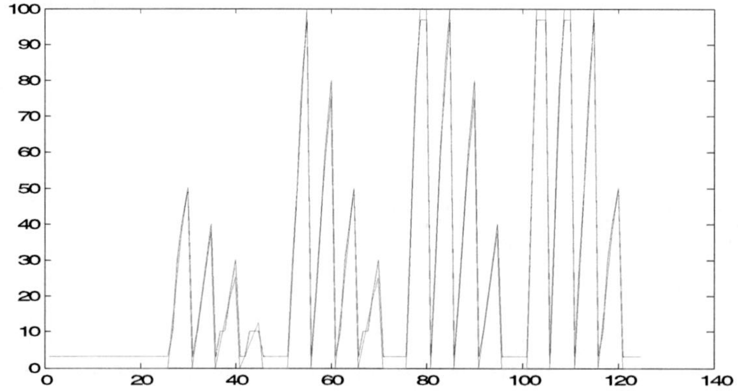

Fig. 9.27 Simulation results for SFS in Case Study 9.2

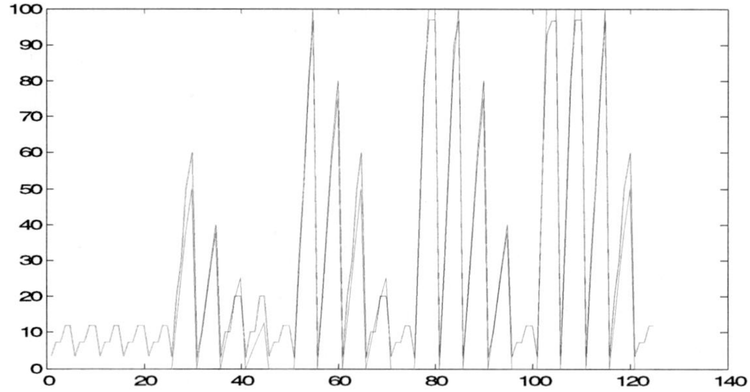

Fig. 9.28 Simulation results for HFS in Case Study 9.2

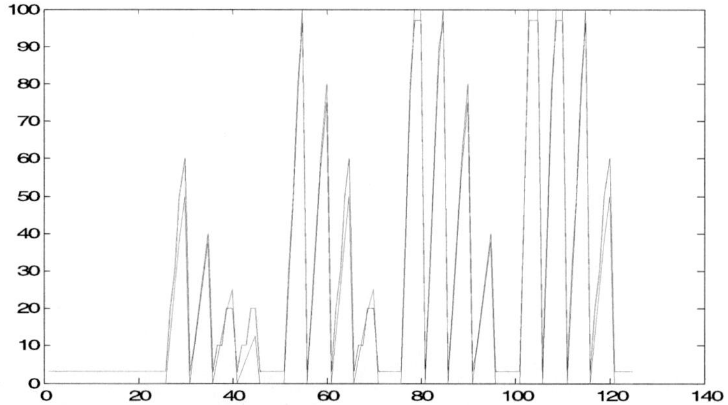

Fig. 9.29 Simulation results for FN in Case Study 9.2

The comparative evaluation of the SFS, the HFS and the FN is shown in Table 9.7. This evaluation uses the performance indicators introduced by Eqs. (9.9)-(9.12).

Table 9.7 Comparative evaluation of SFS, HFS and FN for Case Study 9.2

Performance	SFS	HFS	FN
Feasibility	3	2	2
Accuracy	2.86	5.57	3.64
Efficiency	125	80	125
Transparency	4	1.33	1.33

Table 9.7 shows that in terms of feasibility the FN is superior to the SFS and equivalent to the HFS. With regard to accuracy, the FN is inferior to the SFS but superior the HFS. As far as efficiency is concerned, the FN is equivalent to the SFS but inferior to the HFS. And finally, in terms of transparency, the FN is superior to the SFS and equivalent to the HFS.

The accuracy of the FN can be further improved due to the associated horizontal merging operation applied to some of its nodes. During this operation, the number of linguistic terms for the connection can be varied while preserving the overall number of rules in the single equivalent node for the FN. This variation may lead to the reduction of the approximation error from the linguistic composition of the associated nodes.

Table 9.30 shows how the accuracy of the FN changes while varying the linguistic terms for the connection from 0 to 50. The table presents the modelling error in a logarithmic scale whereby the best accuracy for the FN of 2.86 is first achieved for 17 linguistic terms of the connection. This improved accuracy of the FN is equal to the accuracy of the SFS. Further on during the above variation process, the accuracy of the FN remains almost steady with very small changes above the optimal value of 2.86.

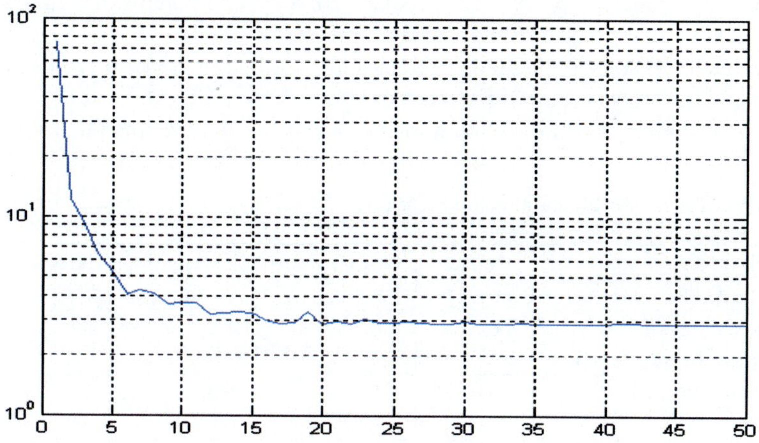

Fig. 9.30 Change of accuracy for FN in Case Study 9.2

The results in Table 9.30 can be interpreted in the context of composing a HFS into a SFS. During this composition process, the accuracy is maximised at a fixed loss of efficiency due to the increased number of rules in the final SFS. At the same time, there is a full preservation of feasibility and transparency as the FN is associated directly with the initial HFS.

Similarly, the results in Table 9.30 can be extended in the context of decomposing a SFS into a HFS. During this decomposition process, the efficiency can be improved at a fixed rate while minimising the loss of accuracy. This can be achieved by utilising the multiple solutions for the node identification operation in horizontal merging. In this case, the number of rules in the final HFS is reduced for the solution that guarantees the smallest increase of the error for the initial SFS. At the same time, there is no change of feasibility and transparency as the FN is associated directly with the initial SFS.

9.7 Summary on Fuzzy Network Evaluation

The results presented in this chapter illustrate several evaluation methods for FNs and some applications of these methods. All methods are presented only in a metric based context, i.e. in a quantifiable manner without any comparative references. These method include the ones based on assessment of structural complexity, composition of HFSs, decomposition of SFSs and indicators of model performance. However, the application of these methods for the case studies is presented also in a comparison based context, i.e. with comparative references to SFSs and HFSs.

Also, some of the evaluation methods presented and their applications are extensions of existing methods and applications while others are quite novel. For example, the methods based on assessment of structural complexity, indicators of model performance and the applications for the case studies are extensions of existing methods and applications whereas the methods based on composition of HFSs and decomposition of SFSs are novel.

The different types of evaluation methods for FNs are described in Table 9.8.

Table 9.8 Types of evaluation methods for FNs

Evaluation methods	Metric based	Comparison based	Extension	Novel
Assessment of structural complexity	Yes	No	Yes	No
Composition of HFSs	Yes	No	No	Yes
Decomposition of SFSs	Yes	No	No	Yes
Indicators of model performance	Yes	No	Yes	No
Applications for case studies	Yes	Yes	Yes	No

The next chapter gives a general conclusion to the results presented in all preceding chapters. In particular, it highlights these results in the context of theoretical significance, methodological impact and application areas.

Chapter 10
Conclusion

10.1 Theoretical Significance of Fuzzy Networks

The first objective of this book is the introduction of a theoretical framework for FNs as a novel type of fuzzy systems. In this context, Chapters 3-6 provide a solid basis for such a framework by means of formal models, basic operations, structural properties and advanced operations for FNs.

The theoretical framework described in the book represents a novel application of discrete mathematics and systems theory. In particular, the framework uses the concepts of Boolean matrix and binary relation for formal modelling of FNs. These concepts are widely used in discrete mathematics and some of its applications such as graph theory and network theory. At the same time, the framework uses some concepts from systems theory such as sequential and parallel subsystems in series and in parallel. Subsystems are widely used in systems theory and some of its applications such as cybernetics and connectionism.

The theoretical framework is illustrated by numerous examples. The examples facilitate the understanding of this framework and demonstrate its versatility. They also provide a good basis for extending the framework to FNs with more complex topologies.

10.2 Methodological Impact of Fuzzy Networks

The second objective of this book is the presentation of an applied methodology for using the theoretical framework for FNs. In this context, Chapters 7-8 provide a solid basis for such a methodology by means of feedforward and feedback FNs.

The applied methodology described in the book represents an extension of SFSs and HFSs. In particular, the methodology considers a FN as a compact way of representing a HFS whereby structure as a complexity attribute is dealt with during the composition process to improve accuracy as a system property. At the same time, the methodology considers a FN as a detailed way of representing a SFS whereby dimensionality as a complexity attribute is dealt with during the decomposition process to improve efficiency as a system property.

In the context of the consideration above, a FN considers nonlinearity and uncertainty as complexity attributes in the same way as a SFS or a HFS. In this case, feasibility and transparency as system properties are not affected during the above composition and decomposition processes. In particular, after composing a HFS

A. Gegov: Fuzzy Networks for Complex Systems, STUDFUZZ 259, pp. 275–277.
springerlink.com

into a SFS by means of a FN, the feasibility and transparency of the HFS are preserved for the SFS. Likewise, after decomposing a SFS into a HFS by means of a FN, the feasibility and transparency of the SFS are not changed for the HFS.

Apart from being an extension, the methodology also acts as a bridge between SFSs and HFSs. This is done by using a FN to compose a HFS into a SFS or to decompose a SFS into a HFS whereby either accuracy or efficiency as performance indicators can be improved significantly at the expense of a reasonable deterioration of the other performance indicator. This bridging capability improves the flexibility of fuzzy systems as models depending on the preferences and requirements to these models.

The applied methodology is illustrated by numerous examples. The examples facilitate the understanding of this methodology and demonstrate its versatility. They also provide a good basis for extending the methodology to FNs with more complex connections.

10.3 Application Areas of Fuzzy Networks

The third objective of this book is the consideration of subject areas for utilising the applied methodology for FNs. In this context, Chapter 9 provides a solid basis for such subject areas by means of two case studies.

The subject areas described in the book are from the bank and the retail industries. However, the results can be used in many other application areas where the knowledge or data about the process to be modelled can be provided in a modular fashion, i.e. for each interacting subprocess by means of an individual rule base. Such modular processes are quite common in many areas such as decision making, manufacturing, communications, transport and finance [3, 8, 14, 40, 84, 120, 125, 131, 132, 133, 147, 150, 158, 159, 165, 168]. In this case, the interacting modules can be decision making units, manufacturing cells, communication nodes, traffic junctions or financial institutions.

Although the results in this book are presented mainly for fuzzy rule based systems, most of them can be used in other application areas, e.g. deterministic and probabilistic rule based systems. In this sense, the presented approach can be easily extended to any types of rule based systems and rule based networks [47, 99, 114].

10.4 Philosophical Aspects of Book Contents

The focus of this book is on the scientific concept of a FN. As such, this concept has also some philosophical aspects.

One aspect is that a FN is a generalisation of a fuzzy system, i.e. it is a system of systems. Therefore, a FN is characterised by a higher level of abstraction than a fuzzy system. A system of systems is like a set of sets whereby each element in the upper level set is not an individual object but a set from the lower level sets.

Just as a set of sets facilitates the creation of more complex data structures, a system of systems facilitates the creation of more complex information structures. In this context, these information structures are actually knowledge structures used for artificial intelligence and knowledge engineering [10, 17, 31, 89, 103, 104].

Another aspect is that a FN is an extension of a fuzzy system. Just as a binary set is a special case of a fuzzy set all of whose elements are with membership degrees 0 and 1, a fuzzy system is a special case of a FN that has a single node and no connections. Similarly, a FN is a general case of a fuzzy system whose subsystems are accounted for explicitly, just as a fuzzy set is a general case of a binary set whose elements are assigned membership degrees.

Finally, a FN can also be used for modelling what is usually perceived as the most complex natural system in the world - the universe. In this context, a FN provides a way of looking at the universe as a collection of galaxies rather than taking it as a single entity. But while this type of approach has been around for quite a long time in cosmology and has contributed significantly to the understanding of the universe, it is still in its infancy in fuzzy logic. So, the author hopes that this book will open new horizons that will enhance substantially the level of understanding of complex processes in the context of fuzzy modelling.

References

[1] Aja-Fernandez, S., Alberola-Lopez, C.: Fast inference in fuzzy systems using transition matrices. IEEE Transactions on Fuzzy Systems 12(2), 170–182 (2004)

[2] Aja-Fernandez, S., Alberola-Lopez, C.: Matrix modelling of hierarchical fuzzy systems. IEEE Transactions on Fuzzy Systems 16(3), 585–599 (2008)

[3] Alex, R.: Fuzzy point estimation and its application on fuzzy supply chain analysis. Fuzzy sets and systems 158(14), 1571–1587 (2007)

[4] Attia, A.F.: Hierarchical fuzzy controllers for an astronomical telescope tracking. Applied Soft Computing 9(1), 135–141 (2009)

[5] Aznarte, J.L.M., Benitez, J.M., Castro, J.L.: Smooth transition autoregressive models and fuzzy rule-based systems: functional equivalence and conse-quences. Fuzzy sets and systems 158(24), 2734–2745 (2007)

[6] Baczynski, M., Jayaram, B.: Fuzzy implications. Springer, Berlin (2008)

[7] Behounek, L., Cintula, P.: From fuzzy logic to fuzzy mathematics: a methodological manifesto. Fuzzy Sets and Systems 157(5), 647–669 (2006)

[8] Bojadziev, G., Bojadziev, M.: Fuzzy logic for business, finance and management. World Scientific, Singapore (2007)

[9] Bortolan, G., Pedrycz, W.: Linguistic neurocomputing: the design of neural networks in the framework of fuzzy sets. Fuzzy Sets and Systems 128(3), 389–412 (2002)

[10] Brachman, R., Levesque, H.: Knowledge representation and reasoning. Morgan Kauffman, New York (2004)

[11] Buckley, J.J.: Simulating fuzzy systems. Springer, Berlin (2005)

[12] Bucolo, M., Fortuna, L., La Rosa, M.: Complex dynamics through fuzzy chains. IEEE Transactions on Fuzzy Systems 12(3), 289–295 (2004)

[13] Campello, R.J.G.B., do Amaral, W.C.: Hierarchical fuzzy relational models: linguistic interpretation and universal approximation. IEEE Transactions on Fuzzy Systems 14(3), 446–453 (2006)

[14] Cao, Y., Ying, M.: Observability and decentralized control of fuzzy discrete-event systems. IEEE Transactions on Fuzzy Systems 14(2), 202–216 (2006)

[15] Carvalho, J.P., Tome, J.: Qualitative optimization of fuzzy causal rule bases using fuzzy Boolean nets. Fuzzy sets and systems 158(17), 1931–1946 (2007)

[16] Cassilas, J., Cordon, O., del Jesus, M.J., Herrera, F.: Genetic tuning of fuzzy rule deep structures preserving interpretability and its interaction with fuzzy rule set reduction. IEEE Transactions on Fuzzy Systems 13(1), 13–29 (2005)

[17] Chaudhury, S., Singh, T., Goswami, P.S.: Distributed fuzzy case based reasoning. Applied Soft Computing 4(4), 323–343 (2004)

[18] Chen, B., Liu, X.: Fuzzy approximate disturbance decoupling of MIMO nonlinear systems by backstepping and application to chemical processes. IEEE Transactions on Fuzzy Systems 13(6), 832–847 (2005)
[19] Chen, B., Liu, X., Tong, S.: Adaptive fuzzy output tracking control of MIMO nonlinear uncertain systems. IEEE Transactions on Fuzzy Systems 15(2), 287–300 (2007)
[20] Chen, B., Tong, S., Liu, X.: Fuzzy approximate disturbance decoupling of MIMO nonlinear systems by back stepping approach. Fuzzy sets and systems 158(10), 1097–1125 (2007)
[21] Chen, M.Y., Linkens, D.A.: Rule-base self-generation and simplification for data-driven fuzzy models. Fuzzy Sets and Systems 142(2), 243–265 (2004)
[22] Chen, S.S., Chang, Y.C., Su, S.F., Chung, S.L., Lee, T.T.: Robust static out-put-feedback stabilization for nonlinear discrete-time systems with time delay via fuzzy control approach. IEEE Transactions on Fuzzy Systems 13(2), 263–272 (2005)
[23] Chen, T.: A fuzzy back propagation network for output time prediction in a wafer fab. Applied Soft Computing 2(3), 211–222 (2003)
[24] Chen, Y., Yang, B., Abraham, A., Peng, L.: Automatic design of hierarchical Takagi-Sugeno type fuzzy systems using evolutionary algorithms. IEEE Transactions on Fuzzy Systems 15(3), 385–397 (2007)
[25] Cheong, F., Lai, R.: Designing a hierarchical fuzzy logic controller using the differential evolution approach. Applied Soft Computing 7(2), 481–491 (2007)
[26] Chiu, C.S.: Mixed feedforward/feedback based adaptive fuzzy control for a class of MIMO nonlinear systems. IEEE Transactions on Fuzzy Systems 14(6), 716–727 (2006)
[27] Choi, H.H.: Output feedback stabilization of uncertain fuzzy systems using variable structure system approach. Fuzzy sets and systems 160(19), 2812–2823 (2009)
[28] Choong, Y.W., Laurent, A., Laurent, D.: Mining multiple-level fuzzy blocks from multidimensional data. Fuzzy sets and systems 159(12), 1535–1553 (2008)
[29] Tseng, C.-S., Chen, B.-S.: H-infinity decentralized fuzzy model reference tracking control design for nonlinear interconnected systems. IEEE Transactions on Fuzzy Systems 9(6), 795–809 (2001)
[30] Cordon, O., Herrera, F., Zwir, I.: Linguistic modelling by hierarchal systems of linguistic rules. IEEE Transactions on Fuzzy Systems 10(1), 2–20 (2002)
[31] Cordon, O., Herrera, F., Zwir, I.: A hierarchical knowledge based envi-ronment for linguistic modelling: models and iterative methodology. Fuzzy Sets and Systems 138(2), 307–341 (2003)
[32] De Barros, J.C., Dexter, A.L.: On-line identification of computationally undemanding evolving fuzzy models. Fuzzy sets and systems 158(17), 1997–2012 (2007)
[33] De Oliveira, J.V., Gomide, F.: Formal methods for fuzzy modelling and control. Fuzzy Sets and Systems 121(1), 1–2 (2001)
[34] De Souza, F.J., Vellasco, M.M.R., Pacheco, M.A.C.: Hierarchical neuro-fuzzy quadtree models. Fuzzy Sets and Systems 130(2), 189–205 (2002)
[35] Destercke, S., Guillaume, S., Charnomordic, B.: Building an interpretable fuzzy rule base from data using orthogonal least squares – application to a depol-lution problem. Fuzzy sets and systems 158(18), 2078–2094 (2007)
[36] Devillez, A., Billaudel, P., Lecolier, G.V.: A fuzzy hybrid hierarchical clustering method with a new criterion able to find the optimal partition. Fuzzy Sets and Systems 128(3), 323–338 (2002)

[37] Di Nola, A., Lettieri, A., Perfilieva, I., Novak, V.: Algebraic analysis of fuzzy systems. Fuzzy sets and systems 158(1), 1–22 (2007)
[38] Dong, J., Yang, G.H.: Static output feedback H-infinity control of a class of nonlinear discrete-time systems. Fuzzy sets and systems 160(19), 2844–2859 (2009)
[39] Dong, Y., Zhuang, Y., Chen, K., Tai, X.: A hierarchical clustering algorithm based on fuzzy graph connectedness. Fuzzy Sets and Systems 157(13), 1760–1774 (2006)
[40] Du, X., Ying, H., Lin, F.: Theory of extended fuzzy discrete-event sys-tems for handling ranges of knowledge uncertainties and subjectivity. IEEE Transactions on Fuzzy Systems 17(2), 316–328 (2009)
[41] Dvorak, A., Novak, V.: Formal theories and linguistic descriptions. Fuzzy Sets and Systems 143(1), 169–188 (2004)
[42] Evsukoff, A., Branco, A.C.S., Galichet, S.: Structure identification and parameter optimization for non-linear fuzzy modelling. Fuzzy Sets and Systems 132(2), 173–188 (2002)
[43] Evsukoff, A.G., Galichet, S., De Lima, B.S.L.P., Ebecken, N.F.F.: Design of interpretable fuzzy rule-based classifiers using spectral analysis with structure and parameter optimization. Fuzzy Sets and Systems 160(7), 857–881 (2009)
[44] Fernandez-Caballero, A.: Contribution of fuzziness and uncertainty to modern artificial intelligence. Fuzzy Sets and Systems 160(2), 129–129 (2009)
[45] Fiordaliso, A.: Autostructuration of fuzzy systems by rules sensitivity analysis. Fuzzy Sets and Systems 118(2), 281–296 (2001)
[46] Fiordaliso, A.: A constrained Takagi-Sugeno fuzzy system that allows for better interpretation and analysis. Fuzzy Sets and Systems 118(2), 307–318 (2001)
[47] Gan, T.: Advanced modelling of complex systems by rule based networks. MSc Thesis, University of Portsmouth (2009)
[48] Gao, Y., Er, M.J.: NARMAX time series model prediction: feedforward and recurrent fuzzy neural network approaches. Fuzzy Sets and Systems 150(2), 331–350 (2005)
[49] Garibaldi, J.M., Ozen, T.: Uncertain fuzzy reasoning: a case study in modelling expert decision making. IEEE Transactions on Fuzzy Systems 15(1), 16–30 (2007)
[50] Gegov, A.: Complexity management in fuzzy systems: a rule base compression approach. Springer, Berlin (2007)
[51] Gegov, A., Gobalakrishnan, N.: Advanced inference in fuzzy systems by rule base compression. Mathware and Soft Computing 14(3), 201–216 (2007)
[52] Gegov, A., Maketas, D.: Formal presentation of fuzzy systems with multiple sensor inputs. Sensors and Transducers, 366–373 (2005)
[53] Gegov, A., Virk, G., Azzi, D., Haynes, B., Alkadhimi, K.: Soft-computing based predictive modelling of building management systems. Knowledge Based Intelligent Engineering Systems 5(1), 41–51 (2001)
[54] Gegov, A., Parashkevova, D., Ljubenov, K.: Analysis of systems under probabilistic and fuzzy uncertainty using multivalued logic. General Systems 33(2-3), 153–162 (2004)
[55] Gegov, E.: Intelligent system for fuzzy modelling of retail pricing. BSc Thesis, University College London (2008)
[56] Gera, Z.: Computationally efficient reasoning using approximated fuzzy intervals. Fuzzy sets and systems 158(7), 689–703 (2007)
[57] Giiven, M.K., Passino, K.M.: Avoiding exponential parameter growth in fuzzy systems. IEEE Transactions on Fuzzy Systems 9(1), 194–199 (2001)

[58] Gil-Aluja: Fuzzy sets in the management of uncertainty. Springer, Berlin (2004)
[59] Gobi, A.F., Pedrycz, W.: Logic minimization as an efficient means of fuzzy structure discovery. IEEE Transactions on Fuzzy Systems 16(3), 553–566 (2008)
[60] Gonzalez-Olvera, M.A., Tang, Y.: A new recurrent neurofuzzy network for identification of dynamic systems. Fuzzy sets and systems 158(10), 1023–1035 (2007)
[61] Gottwald, S.: Mathematical aspects of fuzzy sets and fuzzy logic: some reflections after 40 years. Fuzzy Sets and Systems 156(3), 357–364 (2005)
[62] Guillaume, S.: Designing fuzzy inference systems from data: an inter-pretability-oriented review. IEEE Transactions on Fuzzy Systems 9(3), 426–443 (2001)
[63] Guillaume, S., Charnomordic, B.: Generating an interpretable family of fuzzy partitions from data. IEEE Transactions on Fuzzy Systems 12(3), 324–335 (2004)
[64] Hagras, H.A.: A hierarchical type-2 fuzzy logic control architecture for autonomous mobile robots. IEEE Transactions on Fuzzy Systems 12(4), 524–539 (2004)
[65] Hajek, P.: Computational complexity of t-norm based propositional fuzzy logics with rational truth constants. Fuzzy Sets and Systems 157(5), 677–682 (2006)
[66] Hall, L.O.: Rule chaining in fuzzy expert systems. IEEE Transactions on Fuzzy Systems 9(6), 822–828 (2001)
[67] Herrera, L.J., Pomares, H., Rojas, I., Valenzuela, O., Prieto, A.: TaSe - a Taylor series-based fuzzy system model that combines interpretability with accu-racy. Fuzzy Sets and Systems 153(3), 403–427 (2005)
[68] Hong, T.P., Kuo, C.S., Wang, S.L.: A fuzzy AprioriTid mining algorithm with reduced computational time. Applied Soft Computing 5(1), 1–10 (2004)
[69] Hoppner, F.: Speeding up fuzzy c-means: using a hierarchical data or-ganisation to control the precision of membership means. Fuzzy Sets and Systems 128(3), 365–376 (2002)
[70] Horng, Y.J., Chen, S.M., Chang, Y.C., Lee, C.H.: A new method for fuzzy information retrieval based on fuzzy hierarchical clustering and fuzzy inference techniques. IEEE Transactions on Fuzzy Systems 13(2), 216–228 (2005)
[71] Hsiao, F.H., Hwang, J.D., Cheng, C.W., Tsai, Z.R.: Robust stabilization of nonlinear multiple time-delay large-scale systems via decentralized fuzzy control. IEEE Transactions on Fuzzy Systems 13(1), 152–163 (2005)
[72] Huang, S.H., Xing, H.: Extract intelligible and concise fuzzy rules from neural networks. Fuzzy Sets and Systems 132(2), 233–243 (2002)
[73] Huang, T.C.K.: Developing an efficient knowledge discovering model for mining fuzzy multi-level sequential patterns in sequence databases. Fuzzy sets and systems 160(23), 3359–3381 (2009)
[74] Igel, C., Temme, K.H.: The chaining syllogism in fuzzy logic. IEEE Transactions on Fuzzy Systems 12(6), 849–853 (2004)
[75] Jelleli, T.M., Alimi, A.M.: On the applicability of the minimal config-ured hierarchical fuzzy control and its relevance to function approximation. Applied Soft Computing 9(4), 1273–1284 (2009)
[76] Duan, J.-C., Chung, F.-L.: Cascaded fuzzy network model based on syllogistic fuzzy reasoning. IEEE Transactions on Fuzzy Systems 9(2), 293–306 (2001)
[77] Jin, Y.: Advanced fuzzy systems design and applications. Springer, Berlin (2003)
[78] Joo, M.G., Lee, J.S.: Universal approximation by hierarchical fuzzy system with constraints on the fuzzy rule. Fuzzy Sets and Systems 130(2), 175–188 (2002)
[79] Joo, M.G., Lee, J.S.: A class of hierarchical fuzzy systems with constraints on the fuzzy rules. IEEE Transactions on Fuzzy Systems 13(2), 194–203 (2005)

[80] Joo, M.G., Sudkamp, T.: A method of converting a fuzzy system to a two-layered hierarchical fuzzy system and its run-time efficiency. IEEE Transactions on Fuzzy Systems 17(1), 93–103 (2009)

[81] Juang, C.F.: Temporal problems solved by dynamic fuzzy network based on genetic algorithm with variable-length chromosomes. Fuzzy Sets and Systems 142(2), 199–219 (2004)

[82] Juang, C.F., Chung, I.F., Hsu, C.H.: Automatic construction of feedforward/recurrent fuzzy systems by clustering-aided simplex particle swarm optimization. Fuzzy sets and systems 158(17), 1979–1996 (2007)

[83] Kao, C., Chyu, C.L.: A fuzzy linear regression model with better explanatory power. Fuzzy Sets and Systems 126(3), 401–409 (2002)

[84] Karaboga, D., Bagis, A., Haktanir, T.: Controlling spillway gates of dams by using fuzzy logic controller with optimum number of rules. Applied Soft Computing 8(1), 232–238 (2008)

[85] Kasabov, N.: Adaptation and interaction in dynamical systems: model-ling and rule discovery through evolving connectionist systems. Applied Soft Computing 6(3), 307–322 (2006)

[86] Kim, E.: Output feedback tracking control of robot manipulators with model uncertainty via adaptive fuzzy logic. IEEE Transactions on Fuzzy Systems 12(3), 368–378 (2004)

[87] Kim, K.J., Cho, S.B.: Personalized mining of web documents link struc-tures and fuzzy concept networks. Applied Soft Computing 7(1), 398–410 (2007)

[88] Kluska, J.: Analytical methods in fuzzy modelling and control. Springer, Berlin (2009)

[89] Kolman, E., Margaliot, M.: Knowledge-based neurocomputing: a fuzzy logic approach. Springer, Berlin (2009)

[90] Kouikoglou, V.S., Phillis, Y.A.: On the monotonicity of hierarchical sum-product based fuzzy systems. Fuzzy sets and systems 160(24), 3530–3538 (2009)

[91] Kumar, M., Stoll, R., Stoll, N.: A robust design criterion for interpretable fuzzy models with uncertain data. IEEE Transactions on Fuzzy Systems 14(2), 314–328 (2006)

[92] Kuo, R.J., Chen, C.H., Hwang, Y.C.: An intelligent stock trading decision support system through integration of genetic algorithm based fuzzy neural net-work and artificial neural network. Fuzzy Sets and Systems 118(1), 21–45 (2001)

[93] Landago, M., Rio, M.J., Perez, A.: A note on smooth approximation ca-pabilities of fuzzy systems. IEEE Transactions on Fuzzy Systems 9(2), 229–237 (2001)

[94] Lee, M.L., Chung, H.Y., Yu, F.M.: Modelling of hierarchical fuzzy systems. Fuzzy Sets and Systems 135(2), 343–361 (2003)

[95] Lendek, Z., Babuska, R., De Schutter, B.: Stability of cascaded fuzzy sys-tems and observers. IEEE Transactions on Fuzzy Systems 17(3), 641–653 (2009)

[96] Li, T.H.S., Lin, K.J.: Composite fuzzy control of nonlinear singularly perturbed systems. IEEE Transactions on Fuzzy Systems 15(2), 176–187 (2007)

[97] Li, Y.M., Shi, Z.K., Li, Z.H.: Approximation theory of fuzzy systems based upon genuine many-valued implications – MIMO cases. Fuzzy Sets and Systems 130(2), 159–174 (2002)

[98] Liang, X., Pedrycz, W.: Logic-based fuzzy networks: a case study in system modelling with triangular norms and uninorms. Fuzzy sets and systems 160(24), 3475–3502 (2009)

[99] Ligeza, A.: Logical foundations for rule-based systems. Springer, Berlin (2006)

[100] Liu, P., Li, H.: Approximation analysis of feedforward regular fuzzy neural network with two hidden layers. Fuzzy Sets and Systems 150(2), 373–396 (2005)

[101] Liu, Y.J., Tong, S.C., Wang, W.: Adaptive fuzzy output tracking control for a class of uncertain nonlinear systems. Fuzzy sets and systems 160(18), 2727–2754 (2009)

[102] Lodwick, W.A., Santos, J.: Constructing consistent fuzzy surfaces from fuzzy data. Fuzzy Sets and Systems 135(2), 259–277 (2003)

[103] Luger, G.: Artificial intelligence: structures and strategies for complex problem solving. Addison Wesley, New York (2002)

[104] Ma, J., Chen, S., Xu, Y.: Fuzzy logic from the viewpoint of machine intelligence. Fuzzy Sets and Systems 157(5), 628–634 (2006)

[105] Mar, J., Lin, H.T.: A car-following collision prevention control device based on the cascaded fuzzy inference system. Fuzzy Sets and Systems 150(3), 457–473 (2005)

[106] Mitrakis, N.E., Theokaris, J.B., Petridis, V.: A multilayered neuro-fuzzy classifier with self-organizing properties. Fuzzy sets and systems 158(23), 3132–3159 (2008)

[107] Moreno-Garcia, J., Castro-Schez, J., Jimenez, L.: A fuzzy inductive al-gorithm for modelling dynamical systems in a comprehensible way. IEEE Transactions on Fuzzy Systems 15(4), 652–672 (2007)

[108] Moreno-Velo, F.J., Baturone, I., Barriga, A., Sanchez-Solano, S.: Automatic tuning of complex fuzzy systems with X-fuzzy. Fuzzy sets and systems 158(18), 2026–2038 (2007)

[109] Mucientes, M., Casillas, J.: Quick design of fuzzy controllers with good interpretability in mobile robotics. IEEE Transactions on Fuzzy Systems 15(4), 636–651 (2007)

[110] Negnevitsky, M.: Artificial intelligence: a guide to intelligent systems. Pearson, Harlow (2002)

[111] Novak, V., Perfilieva, I., Behounek, L., Cintula, P.: Formal methods for fuzzy mathematics, approximation and reasoning – part 1. Fuzzy sets and systems 159(14), 1727–1728 (2008)

[112] Novak, V., Perfilieva, I., Behounek, L., Cintula, P.: Formal methods for fuzzy mathematics, approximation and reasoning – part 2. Fuzzy sets and systems 160(8), 1003–1004 (2009)

[113] Noval, V., Lehmke, S.: Logical structure of fuzzy if-then rules. Fuzzy Sets and Systems 157(15), 2003–2029 (2006)

[114] Nowell, D.: Complex systems modelling by rule based networks. BSc Thesis, University of Portsmouth (2009)

[115] Oh, S.K., Pedrycz, W.: Self-organising polynomial neural networks based on polynomial and fuzzy polynomial neurons: analysis and design. Fuzzy Sets and Systems 142(2), 163–198 (2004)

[116] Paiva, R.P., Dourado, A.: Interpretability and learning in neuro-fuzzy systems. Fuzzy Sets and Systems 147(1), 17–38 (2004)

[117] Pal, N.R., Eluri, V.K., Mandal, G.K.: Fuzzy logic approaches to struc-ture preserving dimensionality reduction. IEEE Transactions on Fuzzy Systems 10(3), 277–286 (2002)

[118] Park, J.H., Park, G.T., Kim, S.H., Moon, C.J.: Output-feedback control of uncertain nonlinear systems using a self-structuring adaptive fuzzy observer. Fuzzy Sets and Systems 151(1), 21–42 (2005)

[119] Pedrycz, W.: Logic-based fuzzy neurocomputing with unineurons. IEEE Transactions on Fuzzy Systems 14(6), 860–873 (2006)
[120] Peidro, D., Mula, J., Poler, R., Verdegay, J.L.: Fuzzy optimization for supply chain planning under supply, demand and process uncertainties. Fuzzy sets and systems 160(18), 2640–2657 (2009)
[121] Perfilieva, I.: Logical foundations of rule-based systems. Fuzzy Sets and Systems 157(5), 615–621 (2006)
[122] Petrov, N.: Mathematical modelling of complex systems using fuzzy networks theory. BSc Thesis, Technical University of Sofia (2008)
[123] Piegat, A.: Fuzzy modelling and control. Springer, Berlin (2001)
[124] Pomares, H., Rojas, I., Gonzalez, J., Prieto, A.: Structure identification in complete rule-based fuzzy systems. IEEE Transactions on Fuzzy Systems 10(3), 349–359 (2002)
[125] Ravi, V., Kurniawan, H., Thai, P.N.K., Kumar, R.: Soft computing system for bank performance prediction. Applied Soft Computing 8(1), 305–315 (2008)
[126] Rong, H.J., Sundararajan, N., Huang, G.B., Saratchandran, P.: Sequential adaptive fuzzy inference system for nonlinear system identification and prediction. Fuzzy Sets and Systems 157(9), 1260–1275 (2006)
[127] Ross, T.: Fuzzy logic with engineering applications. Wiley, Chichester (2004)
[128] Roubos, H., Setnes, M.: Compact and transparent fuzzy models and classifiers through iterative complexity reduction. IEEE Transactions on Fuzzy Systems 9(4), 516–524 (2001)
[129] Zhang, R., Phillis, Y.A.: Admission control and scheduling in simple series parallel networks using fuzzy logic. IEEE Transactions on Fuzzy Systems 9(2), 307–314 (2001)
[130] Sala, A., Guerra, T.M., Babuska, R.: Perspectives of fuzzy systems and control. Fuzzy Sets and Systems 156(3), 432–444 (2005)
[131] Sanchez, J.D.A.: Calculating insurance claim reserves with fuzzy regression. Fuzzy Sets and Systems 157(23), 3091–3108 (2006)
[132] Serguieva, A., Hunter, J.: Fuzzy interval methods in investment risk appraisal. Fuzzy Sets and Systems 142(3), 443–466 (2004)
[133] Shakouri, H., Nadimi, G.R.: Bankruptcy prediction modeling with hybrid case-based reasoning and genetic algorithms approach. Applied Soft Computing 9(2), 590–598 (2009)
[134] Shaocheng, T., Bin, C., Yongfu, W.: Fuzzy adaptive output feedback control for MIMO nonlinear systems. Fuzzy Sets and Systems 156(2), 285–299 (2005)
[135] Sindelar, R., Babuska, R.: Input selection for nonlinear regression models. IEEE Transactions on Fuzzy Systems 12(5), 688–698 (2004)
[136] Stach, W., Kurgan, L., Pedrycz, W., Reformat, M.: Genetic learning of fuzzy cognitive maps. Fuzzy Sets and Systems 153(3), 371–401 (2005)
[137] Sun, Q., Li, R., Zhang, P.: Stable and optimal adaptive fuzzy control of complex systems using fuzzy dynamic model. Fuzzy Sets and Systems 133(1), 1–17 (2003)
[138] Taniguchi, T., Tanaka, K., Ohtake, H., Wang, H.O.: Model construction, rule reduction and robust compensation for generalized form of Takagi-Sugeno fuzzy systems. IEEE Transactions on Fuzzy Systems 9(4), 525–538 (2001)
[139] Tao, C.W., Taur, J.S., Wang, C.M., Chen, U.S.: Fuzzy hierarchical swing-up and sliding position controller for the inverted pendulum-cart system. Fuzzy sets and systems 159(20), 2763–2784 (2008)

[140] Taur, J.S., Tao, C.W.: Nested design of fuzzy controllers with partial fuzzy rule base. Fuzzy Sets and Systems 120(1), 1–15 (2001)
[141] Treesatayapun, C.: Fuzzy rules emulated network and its application on nonlinear control systems. Applied Soft Computing 8(2), 996–1004 (2008)
[142] Trillas, E., Alsina, C.: On the law [(p and q) implies r] = [(p implies r) or (q implies r)] in fuzzy logic. IEEE Transactions on Fuzzy Systems 10(1), 84–88 (2002)
[143] Tsekouras, G., Sarimveis, H., Kavakli, E., Bafas, G.: A hierarchical fuzzy-clustering approach to fuzzy modelling. Fuzzy Sets and Systems 150(2), 245–266 (2005)
[144] Tseng, C.S.: Model reference output feedback fuzzy tracking control design for nonlinear discrete-time systems with time-delay. IEEE Transactions on Fuzzy Systems 14(1), 58–70 (2006)
[145] Tseng, C.S.: A novel approach to H-infinity decentralized fuzzy-observer-based fuzzy control design for nonlinear interconnected systems. IEEE Transactions on Fuzzy Systems 16(5), 1337–1350 (2008)
[146] Tseng, C.S., Chen, B.S., Li, Y.F.: Robust fuzzy observer-based fuzzy control design for nonlinear systems with persistent bounded disturbances: a novel decoupled approach. Fuzzy sets and systems 160(19), 2824–2843 (2009)
[147] Tseng, F.M., Tseng, G.H., Yu, H.C., Yuan, B.J.C.: Fuzzy ARIMA model for forecasting the foreign exchange market. Fuzzy Sets and Systems 118(1), 9–19 (2001)
[148] Wan, F., Shang, H., Wang, L.X., Sun, Y.X.: How to determine the minimum number of fuzzy rules to achieve given accuracy: a computational geometric approach to SISO case. Fuzzy Sets and Systems 150(2), 199–209 (2005)
[149] Wang, H., Kwong, S., Jin, Y., Wei, W., Man, K.F.: Multi-objective hierarchical genetic algorithm for interpretable fuzzy rule-based knowledge extraction. Fuzzy Sets and Systems 149(1), 149–186 (2005)
[150] Wang, J., Shu, Y.F.: Fuzzy decision modeling for supply chain management. Fuzzy Sets and Systems 150(1), 107–127 (2005)
[151] Wang, J.W., Cheng, C.H., Huang, K.C.: Fuzzy hierarchical TOPSIS for supplier selection. Applied Soft Computing 9(1), 377–386 (2009)
[152] Wang, M., Chen, B., Dai, S.L.: Direct adaptive fuzzy tracking control for a class of perturbed strict-feedback nonlinear systems. Fuzzy sets and systems 158(24), 2655–2670 (2007)
[153] Wang, M., Chen, B., Liu, X., Shi, P.: Adaptive fuzzy tracking control for a class of perturbed strict-feedback nonlinear time-delay systems. Fuzzy sets and systems 159(8), 949–967 (2008)
[154] Wang, R.J.: Nonlinear decentralized state feedback controller for uncertain fuzzy time-delay interconnected systems. Fuzzy Sets and Systems 151(1), 191–204 (2005)
[155] Wang, S., Chung, F.L., Bin, S.H., Dewen, H.: Cascaded centralized TSK fuzzy system: universal approximator and high interpretation. Applied Soft Computing 5(2), 131–257 (2005)
[156] Wang, W.J., Lin, W.W.: Decentralized PDC for large-scale fuzzy systems. IEEE Transactions on Fuzzy Systems 13(6), 779–786 (2005)
[157] Wang, W.J., Luoh, L.: Stability and stabilization of fuzzy large-scale systems. IEEE Transactions on Fuzzy Systems 12(3), 309–315 (2004)
[158] Wang, Y., Wang, S., Lai, K.K.: A new fuzzy support vector machine to evaluate credit risk. IEEE Transactions on Fuzzy Systems 13(6), 820–831 (2005)

[159] Wu, B., Tseng, N.F.: A new approach to fuzzy regression models with application to business cycle analysis. Fuzzy Sets and Systems 130(1), 33–42 (2002)
[160] Wu, Y., Dexter, A.: A computationally efficient method for identifying generic fuzzy models. Fuzzy sets and systems 160(17), 2567–2578 (2009)
[161] Xiong, N., Litz, L.: Reduction of fuzzy control rules by means of premise learning – method and case study. Fuzzy Sets and Systems 132(2), 217–231 (2002)
[162] Xu, C., Shin, Y.C.: Design of a multilevel fuzzy controller for nonlinear systems and stability analysis. IEEE Transactions on Fuzzy Systems 13(6), 761–778 (2005)
[163] Xu, C., Shin, Y.C.: Interaction analysis of MIMO nonlinear systems based on fuzzy basis function network model. Fuzzy sets and systems 158(18), 2013–2015 (2007)
[164] Xu, S., Lam, J.: Robust H-infinity control for uncertain discrete time-delay fuzzy systems via output feedback controllers. IEEE Transactions on Fuzzy Systems 13(1), 82–93 (2005)
[165] Yee, S.K., Milanovic, J.V.: Fuzzy logic controller for decentralized stabilization of multimachine power systems. IEEE Transactions on Fuzzy Systems 16(4), 971–981 (2008)
[166] Yeh, Z.M., Li, K.H.: A systematic approach for designing multistage fuzzy control systems. Fuzzy Sets and Systems 143(2), 251–273 (2004)
[167] Yoneyama, J.: H-infinity output feedback control for fuzzy systems with immeasurable premise variables: discrete-time case. Applied Soft Computing 8(2), 949–958 (2008)
[168] Yu, L., Wang, S., Lai, K.K.: A neural-network-based nonlinear meta-modeling approach to financial time series forecasting. Applied Soft Computing 9(2), 563–574 (2009)
[169] Yu, W., Rodriguez, F.O., Moreno-Armendariz, M.A.: Hierarchical fuzzy CMAC for nonlinear systems modeling. IEEE Transactions on Fuzzy Systems 16(5), 1302–1314 (2008)
[170] Zeng, X.J., Keane, J.A.: Approximation capabilities of hierarchical fuzzy systems. IEEE Transactions on Fuzzy Systems 13(5), 659–672 (2005)
[171] Zeng, X.J., Goulermas, J.Y., Liatsis, P., Wang, D., Keane, J.A.: Hierarchical fuzzy systems for function approximation on discrete input spaces with appli-cations. IEEE Transactions on Fuzzy Systems 16(5), 1197–1215 (2008)
[172] Zhang, H., Dang, C., Li, C.: Decentralized H-infinity filter design for discrete-time interconnected fuzzy systems. IEEE Transactions on Fuzzy Systems 17(6), 1428–1440 (2009)
[173] Zhang, J.Y., Liu, Z.Q., Zhou, S.: Dynamic domination in fuzzy causal networks. IEEE Transactions on Fuzzy Systems 14(1), 42–57 (2006)
[174] Zhang, T., Feng, G., Liu, H., Lu, J.: Piecewise fuzzy anti-windup dynamic output feedback control of nonlinear processes with amplitude and rate actuator saturations. IEEE Transactions on Fuzzy Systems 17(2), 253–264 (2009)
[175] Zhang, T.P.: Stable adaptive fuzzy sliding mode control of interconnected systems. Fuzzy Sets and Systems 122(1), 5–19 (2001)
[176] Zhang, Z., Yue, W., Tao, R., Zhou, S.: An improved self-organizing CPN-based fuzzy system with adaptive back-propagation algorithm. Fuzzy Sets and Systems 130(2), 227–236 (2002)
[177] Zhou, S., Liu, Z.Q., Zhang, J.Y.: Fuzzy causal networks: general model, inference and convergence. IEEE Transactions on Fuzzy Systems 14(3), 412–420 (2006)
[178] Zhou, S.M., Gan, J.Q.: Low-level interpretability and high-level inter-pretability: a unified view of data-driven interpretable fuzzy system modelling. Fuzzy sets and systems 159(23), 3091–3131 (2008)

Index